Lecture Notes in Computer Science

Edited by G. Goos and J. Hartmanis

712

Lecture Notes in Computer Science
Edited by G. Goos and J. Hartmanis

Advisory Board: W. Brauer D. Gries J. Stoer

P. Venkat Rangan (Ed.)

Network and Operating System Support for Digital Audio and Video

Third International Workshop
La Jolla, California, USA, November 12-13, 1992
Proceedings

Springer-Verlag
Berlin Heidelberg New York
London Paris Tokyo
Hong Kong Barcelona
Budapest

Series Editors

Gerhard Goos
Universität Karlsruhe
Postfach 69 80
Vincenz-Priessnitz-Straße 1
D-76131 Karlsruhe, Germany

Juris Hartmanis
Cornell University
Department of Computer Science
4130 Upson Hall
Ithaca, NY 14853, USA

Volume Editor

P. Venkat Rangan
Depart. of Computer Science & Engineering, University of California at San Diego
9500 Gilman Drive, La Jolla, California 92093-0114, USA

CR Subject Classification (1991): H.5.1, C.2, D.4, H.4.3

ISBN 3-540-57183-3 Springer-Verlag Berlin Heidelberg New York
ISBN 0-387-57183-3 Springer-Verlag New York Berlin Heidelberg

Typesetting: Camera-ready by authors
Printing and binding: Druckhaus Beltz, Hemsbach/Bergstr.
45/3140-543210 - Printed on acid-free paper

Preface

Technological advances are revolutionizing computers and networks to support digital video and audio, leading to new design spaces in computer systems and applications. Under the surface of exciting multimedia technologies lies a mine of research problems. The goal of this workshop, which was sponsored by IEEE Computer and Communication Societies, in cooperation with ACM SIGCOMM, SIGOIS and SIGOPS, was to bring together the leading researchers in all aspects of multimedia computing, communication, storage, and applications.

The field of multimedia has witnessed an explosive growth in the last few years. This workshop, the third in the series (the first was in November 1990 at Berkeley, California, and the second in November 1991 in Heidelberg, Germany) attracted a record number of 128 submissions from four continents (America, Asia, Australia and Europe). Each submission was reviewed by three program committee members. The selection of papers was extremely competitive, with only 26 full papers and 14 short papers being accepted. There were seven full paper sessions and five short paper sessions. In addition, there was an illuminating invited keynote banquet address by Professor Edward A. Fox of Virginia Polytechnic and State University on the topic "Progressing Towards a Hypermedia National Library". In order to keep the workshop environment conducive to in-depth interactions among active researchers, attendance at the workshop was by invitation only. The number of attendees at the workshop was about 95 from 15 countries.

This book constitutes the formal proceedings of the workshop. Authors of the best papers of the workshop have been encouraged to submit full papers for early publication in the ACM/Springer-Verlag journal *Multimedia Systems* (the first journal to focus exclusively on issues relating to multimedia), and also to ACM Multimedia 93, the first ACM international conference on multimedia, August 2-6, 1993, Anaheim, California.

I would like to thank the program committee members for their timely and in-depth reviews, which contributed greatly to constituting what was a very solid workshop program. My thanks also go to my students, Harrick Vin and Srinivas Ramanathan, and my secretary, Ida O'Neil, for their help and involvement in making arrangements for this workshop.

In closing, I am pleased to present this book reporting on the state-of-the-art research in multimedia systems.

June 1993

P. Venkat Rangan
Program Chair

Program Committee

Table of Contents

Session 1: Network and Operating System Support for Multimedia

Chair: Domenico Ferrari, University of California at Berkeley

The first session of the workshop consisted of four presentations that covered a wide variety of topics: from a new type of local-area network for multimedia applications to LAN support for video conferencing, from an implementation of an internetwork real-time protocol to priority inversion countermeasures in the context of multimedia applications.

Kevin Jeffay of the University of North Carolina described one of the many current projects intended to explore what can be done with present-day networks and protocols in the area of multimedia applications. Unlike other efforts, this one relied on a real-time operating system, YARTOS, developed by Jeffay and his coauthors. As long as their results are not interpreted in a much more general and long-term prospective than warranted by their here-and-now, ad-hoc scope (for example, if they are not used to "demonstrate" that admission control is not needed because "the system works even without it"), these studies are useful, as they often enhance our understanding of the requirements of continuous-media applications and our experience with them. The case described in the presentation was the one of desktop conferencing between two users across an Ethernet or another LAN. Particular attention in the talk was paid to the novel transport protocol designed to manage adaptively the scarce and time-variant available bandwidth, deliver audio packets reliably, reduce end-to-end delay and jitter, and deal satisfactorily with packet losses. The protocol runs on top of IP, and uses some more bandwidth than UDP, but is obviously better than UDP for conferencing applications.

The second presentation was given by Gerard Smit of the University of Twente in the Netherlands. It provided an overview of, and some of the details about, a new type of local-area network called Rattlesnake, which has been designed considering the needs of real-time applications as well as those of the more traditional ones. Like DEC's AN1, it is characterized by an aggregate bandwidth much greater than that of each link; however, Rattlesnake is based on a fixed topology, that of a Kautz graph, and is not self-reconfiguring; on the other hand, AN1 does not make any distinction between real-time and non-real-time traffic, whereas Rattlesnake relies on hybrid TDM to provide circuit switching to the former and packet switching to the latter. Also, cut-through is used for real-time packets, while the non-real-time ones are stored in the switches and then forwarded. The choice of hybrid TDM was, in the opinion of some of the attendees, not sufficiently motivated in the presentation, as the main argument seemed to ignore the existence of a number of schemes that can provide hard real-time guarantees even with ATM or, more generally, packet switching. The properties of Kautz topologies were considered interesting; in particular, the fixed degree, the fault tolerance, and the self-routing capabilities even in presence of link failures.

Ralf Herrtwich of the IBM European Networking Center in Heidelberg, Germany, talked about the implementation of the ST-II protocol that was done at ENC. He summarized the reasons for choosing ST-II as the internetwork protocol in the Heidelberg Transport System (HeiTS), which has been built to support a distributed multimedia platform. He also discussed the additions and enhancements made by him and his coauthor to ST-II to complete it and make it better suited to the needs of multimedia traffic: in particular, resource management mechanisms, characterized by optimistic (non-worst-case) reservation and graceful degradation, and a feedback scheme to allow rates to be controlled without having to keep all clocks in synchrony. This implementation of ST-II is probably the most complete and interesting to date. It is to be hoped that its authors will soon report on their experiments and experience with it.

The fourth and last paper of the session was presented by Akira Nakamura of Cambridge University in England. The connection between the two parts of the presentation was not easily grasped by the audience. In its first part, much longer than the second, the paper describes a new algorithm to deal with the priority inversion problem; the algorithm is less conservative than the priority ceiling protocol, which may cause waiting on a shared lock even when it is not necessary. In the second part, the question is raised whether the scheduling problems that may be caused by shared locks are really encountered in multimedia applications. The authors of the paper argue, unfortunately without providing any experimental evidence, that these applications may encounter lock conflicts, if any, between processing stages for the same stream rather than for access to data structures shared by several independent streams. Thus, in a sense, operating systems should not be extended, but rather reduced, and in any case, rethought and redesigned, for multimedia applications. The arguments presented in the second part of the talk sparked a heated debated between those who agreed and those who disagreed with the main thesis of the authors; the discussion was unfortunately cut short by the expiration of the session's deadline.

Adaptive, Best-Effort Delivery of Digital Audio and Video Across Packet-Switched Networks*

K. Jeffay, D.L. Stone, T. Talley, F.D. Smith

University of North Carolina at Chapel Hill
Department of Computer Science
Chapel Hill, NC 27599-3175 USA
{jeffay,stone,talley,smithfd}@cs.unc.edu

Abstract: We present an overview of a "best-effort" transport protocol that supports conferencing with digital audio and video across interconnected packet switched networks. The protocol delivers the highest quality conference service possible given the current load in the network. Quality is defined in terms of synchronization between audio and video, the number of frames played out of order, and the end-to-end latency in the conference. High quality conferences are realized through four transport and display mechanisms and a real-time implementation of these mechanisms that integrates operating system services (*e.g.*, scheduling and resource allocation, and device management) with network communication services (*e.g.*, transport protocols). In concert these mechanisms dynamically adapt the conference frame rate to the bandwidth available in the network, minimize the latency in the displayed streams while avoiding discontinuities, and provide quasi-reliable delivery of audio frames.

1. Introduction

The focus of this work is on the real-time transmission of live digital audio and video across interconnected packet-switched networks. Our goal is to support high-fidelity audio/video conferences, *e.g.*, conferences with high quality audio and full-motion color video. The approach one adopts for the real-time communication of audio and video will depend on numerous factors including the architecture of the audio/video subsystem, the encoding and compression technologies employed in the audio/video subsystem, the available network bandwidth, and the degree of physical layer support in the network for real-time communication. As a starting point we consider existing local area networks (*i.e.*, ethernets, token rings, and FDDI rings) interconnected by bridges or routers, and an audio/video system that acquires and compresses individual frames of video at NTSC rates.

One technical challenge in this environment is to manage and accommodate jitter in the audio/video streams seen by the processes responsible for displaying the streams. In order to sustain a high-fidelity conference, frames of audio/video data must arrive at a displaying workstation so as to be played at a precise rate (*e.g.*, one frame every 33 ms for NTSC video). Although these frames can be generated at the desired rate by the originating audio/video hardware, the exact rate at which frames arrive at a receiver can be grossly distorted by poor operating system scheduling and resource allocation on the transmitting and receiving workstations, and varying load in the network. Jitter is problematic because it can cause discontinuities in the playing of audio/video streams and can increase the latency of the conference. Jitter therefore directly impacts conference quality.

---

* This work supported by the National Science Foundation (grant numbers CCR-9110938 and ICI-9015443), and by the Digital Equipment Corporation and the IBM Corporation.

It is instructive to view a video conferencing system as a distributed pipeline. Frames of audio/video data are acquired, digitized, and compressed at one machine, transmitted over a network to a second machine where they are decompressed and displayed. Ideally each stage of this pipeline should operate in real-time, that is, each stage should process a frame before the previous stage outputs the following frame. Through the use of a real-time operating system that provides computation and communication services specifically tailored to the needs of applications that process continuous-time media streams such as digital audio and video [4], we can closely control the processing of audio/video frames on a transmitting workstation to ensure that frames progress from acquisition and compression hardware to the network interface with minimal jitter. Correspondingly, on the receiving machine, we ensure the arriving frames are delivered to the audio/video subsystems for display with jitter no worse than in the arriving stream. However, in our environment we cannot control other uses of the network and hence we cannot provide the same level of control over the transmission of individual frames.

We have developed a set of techniques for the transmission and display of audio/video frames that yield the highest quality conference possible given the conditions present in the network. First, in the transport layer, audio is transmitted redundantly (*i.e.*, units of audio data are transmitted multiple times). This allows a conference receiver to function well in the face of packet loss. Second, real-time scheduling and queue management techniques are used in the transport and network interface layers to manage conference latency and automatically adapt the data rate of the conference to varying network bandwidth. At a conference receiver the display of audio and video data follows a protocol to (further) manage conference latency and ameliorate deviations in inter-arrival times of audio/video data. The third mechanism dynamically varies the synchronization between displayed audio and video. This allows a receiver to trade off synchronization between audio and video for low conference latency and the ability to accommodate large deviations in interarrival times of audio/video data. Lastly, a data aging mechanism is used to reduce the latency in a conference after periods of high network load. Together, these mechanisms form a transport and display protocol for conferencing. It is argued that these mechanisms, in concert, provide the best possible conference service given the available network bandwidth at any point in time.

These techniques have been implemented in a multimedia transport protocol (MTP) that has been used to transmit live digital audio and video across packet-switched networks using Internet (IP) protocols [7]. We have investigated the effects of these transport and display techniques on the quality of the audio and video streams displayed at the receiver under a variety of network configurations and loading conditions. In this paper we survey two of the techniques we have developed to ameliorate the effects of jitter. These are the dynamic variation of audio/video synchronization and the adaptation of the conference frame rate to that currently sustainable in the network. The following section presents our requirements for audio/video transmission and describes the technologies on which our work is based, and describes the salient communications problems this work addresses. We follow in Section 3 with a description of the transport and display mechanisms that address jitter. We conclude in Section 4 with a review of our results.

2. Requirements and Methodology

2.1 Transport and Display Requirements

Users judge conferences by three metrics: fidelity of, and latency in, the displayed audio and video streams, and the synchronization between displayed audio and video. We measure fidelity of a stream in terms of the number of discontinuities that occur when the stream is displayed. A stream of audio/video is a totally ordered sequence of frames. A *frame* is the logical data unit by which audio/video is acquired and displayed. A *discontinuity* in a displayed stream occurs when frame n is not followed by frame $n + 1$. Since we cannot control how the network is used by others, we cannot bound the number of discontinuities that will occur over any interval in time. The goal of our transport and display mechanisms is simply to minimize the number of discontinuities in the audio and video streams.

Related to fidelity is the synchronization between audio and video frames (*i.e.*, "lip sync"). Our mechanisms bound the amount of synchronization loss between displayed audio and video streams. The third metric used to judge a conference is end-to-end stream latency. Factors such as the lengths of the hardware audio and video pipelines, the latency in the operating system and network influence latency in our environment. Our protocol endeavors to ensure an end-to-end (acquisition to display) latency of no more 250 ms [2]. Our approach to latency management is described in [7].

Jitter control is required to minimize the number of discontinuities in the displayed audio and video streams. The actual requirements for allowable discontinuities differ dramatically between audio and video. As most communication in a conference is spoken conversations, users demand much more of the audio used in a conference than they demand of the video. Moreover, because there is less temporal coherence in the audio stream than in the corresponding video stream, users are quite sensitive to discontinuities in the audio caused by late or lost packets. In contrast, users can tolerate significant discontinuities in the video stream. For example, if video frame n is not present at the receiver when it is time for it to be played, the preceding frame, frame $n - 1$, can be played in its place without adverse effect. By contrast, if a corresponding amount of audio data (*e.g.*, 33 ms worth) does not arrive on time, a replay of the previous 33 ms of audio data (or the playing of silence) is easily noticed.

While audio frames could have a duration significantly less than 33 ms, it is paramount to note that the requirement for reliable transmission of audio is independent of the size of an audio frame. *The important measure is not the duration of an audio frame but rather the total duration of all audio frames that are sent in a single network packet as this defines the amount of audio data that can potentially be lost or delayed in the network.* As the load in a network increases, the critical resource becomes access to the network. If the time to access the network increases beyond the audio frame generation time, a sender must transmit multiple audio frames in a network packet. If a packet is lost or delayed, a discontinuity will occur with a duration equal to the cumulative lengths of the audio frames in the lost packet. In our experiments we have routinely observed network access times exceeding 33 ms. Under these conditions, discontinuities perceivable to the user occur. For this reason we require tighter control of the jitter in the audio stream.

2.2 Experimental Configuration

We have constructed a tested for experimenting with the transmission of live digital audio and video across IP networks. For audio and video acquisition and display we use IBM-Intel ActionMedia 750 hardware and IBM PS/2 workstations. The PS/2s run a real-time operating system we have developed (YARTOS [5]) and support UDP/IP. The network configuration used for experiments consists of two 16 megabit token rings interconnected by a series of 10 megabit ethernets. Packets are initially transmitted on one token ring, routed to an ethernet, bridged to a second ethernet, routed to a second token ring and then displayed.

The ActionMedia system uses a two stage pipeline to acquire and display frames of video. The acquisition pipeline consists of digitization and compression stages; the display pipeline consists of decompression and display stages. The acquisition and display stages are 33 ms long. The length of the compression and decompression stages is a function of scene complexity and the algorithm employed. In our use of the system these stages typically are in the range 20-25 ms. A compressed, color video frame (at 240x256 resolution) is between 6000-8000 bytes. Under program control, each video frame is digitized, compressed, transmitted over the network, received, decompressed, and displayed as shown in Figure 1. Audio frames are digitized and compressed in real-time (*i.e.*, in a single stage). The length of this stage is programmable. To reduce processing overhead, we typically use values of 16.5 or 33 ms. A 33 ms audio frame is a little over 500 bytes. The data rate for a full-fidelity audio stream is approximately 128 kilobits per second.

There is a significant disparity between the lengths of the audio and video pipelines. The acquisition and compression stages run sequentially for each video frame. Therefore, the earliest a video frame can be transmitted is approximately 60 ms after it is introduced to the system (assuming no operating system or application overhead). In contrast, the audio data corresponding to a video frame can be transmitted 33 ms after entering the system. A key facet of our transport and display mechanisms will be an exploitation of this disparity.

3 Transport and Display Mechanisms

We have implemented an unreliable connection-oriented stream transport protocol, called MTP (Multimedia Transport Protocol), on top of UDP/IP. The novel aspects of the protocol are in the way data is managed within the transport layer. We adopt a real-time systems philosophy in the design and implementation of the protocol. By this we mean that we provide the transport layer with information on the timing characteristics and semantics of the data to be transmitted, so that MTP can manage the resources available to it (*e.g.*, CPU cycles, buffers, network transmission slots) to meet the real-time requirements of applications. This implies that the transport layer is more tightly integrated with higher layers (*e.g.*, the application layer) and lower layers (*e.g.*, the network interface layer) than in traditional protocols. For example, when an MTP connection is established, the application specifies (1) audio and video frame times, (2) number of buffers MTP should use for queueing audio and video frames, and (3) the number of times audio frames should be transmitted. The frame times are used by the transport layer to compute deadlines for transmitting packets, and to intelligently combine audio and video data into network packets. Packets are scheduled at the network interface using a deadline-based scheduling algorithm [6].

The application controlling the audio and video generation processes invokes an asynchronous transmit operation whenever an audio or video frame has been generated. Outgoing frames are queued in the transport system. The protocol manages separate queues for audio and video frames. Priority is given to audio frames and, to the extent possible, audio frames are never fragmented.

The essence of this approach is to view the audio/video acquisition processes, the network subsystem, and the display processes as a distributed pipeline as shown in Figure 2. At the sender and receiver, the stages of the ActionMedia pipeline can be tightly synchronized (given sufficient real-time support from the operating system) and data can be made to flow between stages as it is generated (i.e., with 0 or 1 buffers). The network transmission and reception processes cannot be so controlled and hence queues must be maintained in the network subsystem to interface the network with the audio/video processes. The jitter control mechanisms we develop are a set of queue manipulation policies that address the issues of what data should be enqueued/dequeued and when it should be enqueued/dequeued. Given the time-critical nature of audio and video data, queueing decisions should be made as late as possible. For example, the ideal time to decide which audio and video should be transmitted next is to wait until access to the network is actually obtained (e.g., when a free token arrives at a station on a token ring). Similarly, the ideal time to decide which frame of video should be displayed is during the vertical blanking interval of the display. Delaying decisions as long as possible maximizes the amount of data that is available to make a decision. The implication of this is that the transport protocol must be integrated with the network device driver.

The two jitter control mechanisms we present address the issue of *what data* to enqueue and dequeue. There are two queues to manage: a transport queue at the sender that contains compressed audio and video frames that are ready to be transmitted, and a display queue at a receiver that contains complete audio and video frames that are ready to enter the display pipeline. We distinguish between two types of jitter: jitter caused by short-term increases in network load (e.g., bursts), and jitter caused by longer-term increases in network load. In the former case we ameliorate the effects of jitter by buffering audio frames at the display and allowing the audio stream to be played out of synchronization with the video stream (e.g., play the audio ahead of its corresponding video) when frames arrive late. In the latter case, the bandwidth available to the conference may decrease below that required to support full-fidelity audio and video streams. In this case the conference frame rate must be adapted to that currently sustainable in the network.

For purposes of illustrating the effect of each mechanism on jitter, we use an audio frame time of 33 ms and a video frame time of 66 ms. For video this means that 33 ms of video (i.e., an NTSC frame) is acquired during each 66 ms interval. At the receiver each video frame will be displayed for 66 ms. This results in a video frame rate of 15 frames per second.

3.1 Display Queue Management

The goal in display queue management is to (1) minimize the number of discontinuities in the displayed audio/video streams, (2) display audio/video data with minimal latency, and (3) bound the synchronization differential between displayed audio and video. These goals typically cannot be achieved simultaneously and hence there are trade-offs between latency, discontinuities, and lip synchronization that must be managed. Here we concentrate on one queue management technique: dynamically varying audio/video synchronization.

Consider the problem of bursty network traffic. Network bursts increase latency and cause discontinuities by introducing jitter. Figure 3 shows the effect of a network burst on video latency. Figure 3 plots latency versus conference duration (measured in number of frames displayed). Instantaneous increases in latency indicate discontinuities. Latency is broken down into:

- the time spent in the ActionMedia pipeline at the originating machine,
- the time spent in the transport protocol layer (including time spent in the transport queue),
- the time spent at the network interface waiting to access the network,
- the physical network transmission time,
- the time spent queued at the receiver waiting to enter the display pipeline, and
- the time spent in the display pipeline itself.

The time spent at the network interface is combined with the physical transmission time into a measure of network congestion called *per frame network access time*. This is the total time required to transmit an audio or video frame as measured from the time the transport system invokes the network-adapter device-driver to transmit the first packet containing data for the frame until the last packet containing data for the frame has been transmitted. On the receiver, the time spent waiting to enter the display pipeline is further broken down into two components:

- the time spent synchronizing with the display pipeline, and
- the time spent queueing at the entrance to the display pipeline.

We make a distinction between these two components of latency because the former is beyond the control of the display software while the latter is directly controllable. A frame of video is displayed every 33 ms, synchronized with the vertical blanking interval of the display. When a frame arrives at a receiver it may have to wait up to 33 ms before it can enter the display pipeline. Therefore, even if video frames were transmitted instantaneously from sender to receiver, audio and video latency may differ between two otherwise identical conferences by as much as 33 ms. The time spent waiting to synchronize with the display pipeline is called the *display pipeline synchronization time* (DPST). In order to compare the effects of transport and display policies across different executions of the system, the DPST must be factored out. Therefore we explicitly represent this term. The time a frame spends queueing at the entrance to the display pipeline is defined as the time spent queued on the receiver minus the DPST. By definition this time will be a multiple of a frame time.

In Figure 3 video is initially being displayed with latency of 260 ms. After approximately 80 video frames have been generated, load is introduced in the network. The effect of this load is to increase the network access time at the sender and delay the arrival of frames at the receiver. The initial increase in network access time is compensated for by a decrease in the DPST at the receiver and thus the latency in the stream remains constant. As frames arrive later and later, the DPST eventually becomes 0 indicating that frames are entering the display pipeline immediately upon arrival. At this point the next frame that arrives late (*i.e.*, more than 66 ms after the previous frame) will cause a discontinuity in the video stream and an increase in latency. As more traffic is introduced into the network, the network access time continues to increase. For frames 80-100, the access time has increased from approximately 30 to 70 ms per frame (recall that multiple network accesses are required to transmit a video frame). This increase results in a queue of unsent frames at the sender indicating that the sender is no longer able to transmit data in real-time. If no action is taken and the load in the network does not subside, frames will be queued at the sender for longer and longer durations. This is illustrated by an increase in the time spent in the transport queue starting around frame 70. Eventually a second discontinuity and corresponding increase in latency occurs. By the time the 100th frame is displayed, frames are arriving at the receiver approximately 150 ms after being generated and are displayed with 133 ms more latency than the initial frames.

As the network burst subsides the sender is able to transmit faster than real-time and empties its queue. However, since the receiver can only display frames at a constant rate, a queue forms at the display. Moreover, because the receiver displays frames at a constant rate, the latency in the conference does not change. A temporary increase in the network access time of between 10-15 ms (per packet) results in a sustained increase in latency of 133 ms. However, the conference is now resilient to future network bursts of equal magnitude. After frame 125, the frames queued at the receiver insulate the display process from additional bursts. In general, the increase in network access time required to induce a discontinuity in a displayed stream will be a function of the current display queue length and the amount of time required to synchronized with the display pipeline. If there is no display queue, i.e., frames are displayed with minimal latency, a stream displayed with a DPST of t ms will have a discontinuity if a frame arrives more than t ms late. If one frame is queued at the receiver then the receiver can tolerate the late arrival of a single frame by at most one frame time plus the current DPST.

The network burst in Figure 3 increased video latency and created discontinuities. If audio is synchronized with video then the audio stream suffers equally. The effects of bursts on the audio stream can be avoided by exploiting the disparity in the lengths of the audio and video pipelines and allowing the receiver to play audio frames before their corresponding video frames are displayed. Figure 4 shows the effect of the network burst on audio latency. Note that audio frames are queued at the display as each frame must wait for its corresponding video frame to be decompressed. Because of this, there are a sufficient number of audio frames in the display queue to play audio without any discontinuities during the burst. Note that after a network burst, audio is being played with 133 ms less latency than the video, i.e., the audio is 2 video-frame times ahead of its corresponding video.

By dynamically varying the synchronization between the audio and video streams we are able to minimize discontinuities in the audio stream due to network bursts. This technique has two limiting effects. First, assuming a situation with stable network load and audio/video displayed with minimum possible latency, audio can be played ahead of video by no more the difference in the lengths of the combined acquisition/display pipelines for audio and video plus the difference in the network access times for audio and video frames. Audio can be played ahead of video by at most 66 ms when the video stream is displayed with minimum latency. Therefore, if audio and video are initially synchronized, the audio stream can tolerate temporary increases in network access times of at least 66 ms per frame (plus the current DPST) without creating an audio discontinuity and corresponding increase in latency. If there is additional latency in the video stream due to queueing or high per frame network access (as in Figure 3) then audio can be played ahead of video by more than 66 ms.

The second factor limiting the deviation in audio video synchronization is the limit of human perception and tolerance of synchronization between audio and video. Audio should be played ahead of video only to the extent that audio discontinuities are perceived as more annoying the than loss of audio/video synchronization. We conjecture that users will tolerate audio being played by 100 ms ahead video. Informal user studies indicate that this is the level at which the loss of synchronization is first noticeable in our system (given a particular video compression algorithm and its resulting image quality).

3.2 Transport Queue Management

The goal in transport queue management is to (1) adapt the outgoing frame rates to the bandwidth currently available to the sender, (2) deliver audio and video data with minimal latency, and (3) ensure reliable delivery of audio frames. Again, there are trade-offs between these goals. For example, one can choose to maximize throughput or minimize latency. Here we concentrate on one transport queue issue: the transmission of audio frames multiple times to ameliorate the effect of packet loss in the network.

Ideally audio and video frames should be transmitted as they are generated. When the network access time to send a frame becomes larger than the frame time, queues form in the transport system. The lengths of these queues and the policies for managing them can dramatically effect the throughput and latency of the conference. For example, if a frame cannot be transmitted when it is generated, enqueueing the frame only serves to increase its latency. However, if frames that cannot be transmitted immediately are discarded, the throughput of the stream may unnecessarily decrease (with a corresponding increase in the number of discontinuities). When the network access time exceeds the frame time, transport queues of length greater than 1 cannot increase the throughput of frames (unless multiple frames can be transmitted in a single network packet) and only serves to increase the latency in the displayed stream. A transport queue of length greater than one is useful only if the additional latency it introduces can be tolerated by users of the system. If this is the case and if increases in network access times are relatively short lived, then longer queues are of some utility as they reduce the likelihood that the sender will drop a frame. More frames will be sent to the receiver who is in a better position to decide whether or not frames should be displayed or skipped (to reduce latency). Our experiments indicate that a transport queue of length 1 is useful in all cases and that there are (rapidly) diminishing returns to lengthening the queue. The conference quickly reacts to changes in available bandwidth without increasing latency.

The audio stream is not as susceptible to the effect of queue length since, for the networks we consider, multiple audio frames can be transmitted in a single packet. Should the available bandwidth decrease below that required for the full audio frame rate, more aggressive application level mechanisms, such as changing the coding or compression scheme, will have to be employed to sustain the conference.

3.3 Performance

We have used the transport and display techniques outlined above to transmit live audio and video across interconnected packet-switched networks. Here we describe the results of one experiment: a conference spanning two token ring and two ethernet networks. We compare the performance of MTP to a straightforward use of UDP to transmit audio/video frames for the three measures of conference quality. The experiment uses an audio frame time of 16.5 ms, a video frame time of 33 ms, and a packet size of 5000 bytes. This gives a full 30 (video) frame per second conference with a uni-directional data rate of 2 megabits per second. In the MTP case, an MTP connection is established with a video transport queue length of 1 frame and an audio queue of 3 frames. Each audio frame is transmitted twice. At the receiver, audio and video is initially display in exact synchronization. Audio is never allowed to get ahead of video by more than 100 ms. In the UDP case, the conferencing application combines audio and video frames and presents them together to the transport system. This is done to minimize the number of packets required to transmit audio/video frames and thereby increase the resiliency of the transmission process to network bursts. This has the effect, however, of effectively requiring the receiver to display audio and video in exact synchronization. For the UDP conference the network device driver at the sender uses a queue of length 3.

Figure 5 shows the performance of UDP and MTP in the face of congestion on the sender's token ring. Figures 5a and 5b show latency and discontinuities. Vertical lines are plotted at 10 second intervals indicating the minimum and maximum latency during the interval. A line connects the mean in each 10 second interval. Figure 5c shows the synchronization loss in the displayed streams. It shows the amount (in 16.5 ms frame times) that audio was played ahead of video. Vertical lines again indicate the minimum maximum synchronization loss in a 10 second interval. A line connects the mean loss. The congestion in this experiment is due to file transfers and the simulation of additional conferences. UDP experiences large fluctuations in latency and hence has a high number (thousands) of audio/video discontinuities. Latency averages 300ms. MTP experiences much smaller fluctuations in video latency and has an average latency of approximately 250 ms. However, significantly more video discontinuities are experienced. This is because the overhead of transmitting audio redundantly occasionally increases the network access time per video frame beyond 33 ms and frames are dropped at the sender. However, this is not noticeable. (The average video frame rate is 28 frames per second.) MTP audio fidelity is quite good with only a small number discontinuities and an average latency of 225 ms. Audio is played an average of 16 - 33 ms ahead of video and is occasionally 100 ms ahead. The UDP conference is aesthetically quite poor; primarily because of the number of audio discontinuities. UDP drops 1600 audio frames at the sender while MTP drops only 1. This is because MTP treats audio and video separately and adapts its packetization scheme to available bandwidth. In spite of the fact that MTP has more video discontinuities than UDP, the MTP conference is comparable to a conference on an unloaded network.

4. Summary and Conclusions

Our goal is to support real-time communication of audio/video frames across interconnected packet switched networks. Our approach has been to understand the cost, and feasibility of, supporting multimedia conferencing without admission control, resource reservation, or real-time support from the physical layer. To this end, we have developed a "best effort" transport protocol that provides the highest quality conference service possible given the bandwidth available to the conference.

Quality is defined in terms of synchronization between audio and video, the number of frames played out of order, and the end-to-end latency in the conference.

We described two transport and display mechanisms that provide a measure of jitter control. The mechanisms are: a facility for varying the synchronization between displayed audio and video to achieve high-fidelity audio, and a network congestion monitoring mechanism that is used to control audio/video latency and dynamically adapt the conference frame rate to the bandwidth available in the network.

Our work demonstrates empirically that it is possible to establish and maintain high-fidelity conferences across congested packet-switched networks without requiring special services from the network such as admission control, jitter control [3], resource reservation [1], or isochronous transmission. While such services are clearly useful and important, they equally clearly come at an additional cost; be it the installation of new wiring or other network hardware, the modification of routing software at all network interconnection points, or a reduction in the total bandwidth available conferences. It is therefore useful to understand the cost of supporting high-fidelity conferencing (as measured in terms of the required workstation communications and application software) in the absence of such services. While it is not possible to precisely state this cost, we have provided evidence that, even in the absence of direct network support, high-fidelity conferences are achievable at modest cost.

5. References

[1] Anderson, D.P., Herrtwich, R.G., Schaefer, C., 1990. *SRP: A Resource Reservation Protocol for Guaranteed Performance Communication in the Internet*, University of California Berkeley, Dept. of Electrical Eng. and Computer Science Technical Report, TR-90-006, February 1990.

[2] Ferrari, D., 1990. *Client Requirements for Real-Time Communication Services*, IEEE Communications, (November), pp. 65-72.

[3] Ferrari, D., 1991. *Design and Application of a Jitter Control Scheme for Packet-Switch Internetworks*, 2nd Intl. Workshop on Network and Operating System Support for Digital Audio and Video, Heidelberg, Germany, pp. 81-84.

[4] Jeffay, K., Stone, D., Poirier, D., 1992. *YARTOS: Kernel support for efficient, predictable real-time systems*, in "Real-Time Programming," W. Halang and K. Ramamritham, eds., Pergamon Press, Oxford, UK.

[5] Jeffay, K., Stone, D.L., and Smith, F.D., 1992. *Kernel Support for Live Digital Audio and Video. Computer Communications*, 16, 6 (July), pp. 388-395.

[6] Jeffay, K., 1992. *Scheduling Sporadic Tasks with Shared Resources in Hard-Real-Time Systems*, Proc. 13th IEEE Real-Time Systems Symp., Phoenix, AZ, December 1992 (to appear).

[7] Jeffay, K., Stone, D.L., and Smith, F.D., 1992. *Transport and Display Mechanisms For Multimedia Conferencing Across Packet-Switched Networks*. In submission.

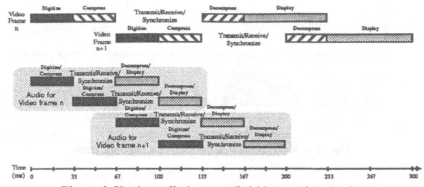

Figure 1 Hardware limits on audio/video synchronization (66ms video frame times, 33 ms audio frame times).

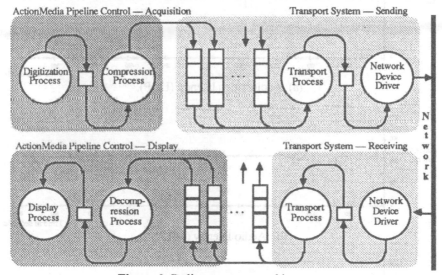

Figure 2 Delivery system architecture.

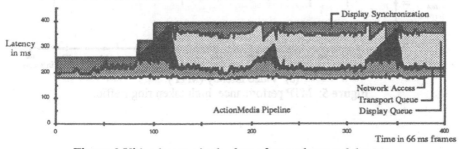

Figure 3 Video latency in the face of several network bursts.

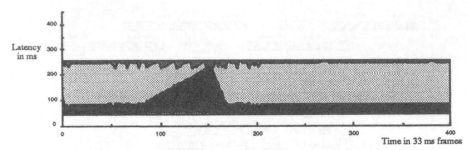

Figure 4 Audio latency in the face of a network burst.

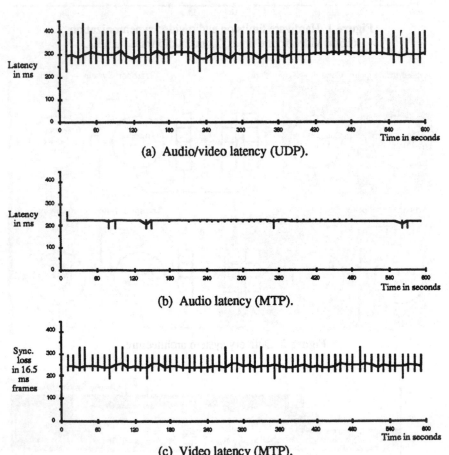

(a) Audio/video latency (UDP).

(b) Audio latency (MTP).

(c) Video latency (MTP).
Figure 5: MTP performance, high token ring traffic.

The Architecture of Rattlesnake:

a Real-Time Multimedia Network

Gerard J.M. Smit, Paul J.M. Havinga
University of Twente, dept. Computer Science
P.O. Box 217
7500 AE Enschede, the Netherlands

e-mail: smit@cs.utwente.nl

Keywords: ATM, multimedia communications, low latency networks, real-time, Kautz graphs.

Abstract

This paper presents the architecture and implementation of the Rattlesnake network. It will be used as a platform for ATM communication, and will provide communication facilities for demanding distributed real-time multimedia applications.

Workstations are connected via point-to-point (TAXI) links to a switching fabric. The fabric consists of switching elements, interconnected in a Kautz topology. Kautz networks have desirable features such as: small diameter, fixed degree, fault tolerant and have a self routing capability even if some links fail.

The transfer mode is based on hybrid TDM, a combination of STM (circuit switching) and ATM (packet switching). For hard real-time traffic (e.g. voice and video) STM like end-to-end logical connections can be set up to guarantee a bounded latency. For non real-time traffic (e.g. file transfer), that has a more bursty nature, we use ATM with store-and-forward routing to achieve a high network utilisation. The nosy worms protocol is used to avoid deadlock.

The network uses real-time virtual channels. These channels can be claimed to reserve bandwidth and to guarantee a bounded latency. For the implementation we use off-the-shelf programmable components (FPGAs).

1. Introduction

This paper describes the design and implementation of a local area network called Rattlesnake for interconnecting high performance workstations and dedicated servers (e.g. file servers, communication servers and computation servers). All servers, workstations and devices communicate via ATM cells only. Rattlesnake is a subproject of the Huygens research project [Mullender 91]. The Huygens project started in 1991[1]. It addresses multimedia systems and applications and embodies all levels between Operating Systems support and the hardware.

Integration of different services on the same network require LANs to transmit not only data but also voice and video traffic. We expect the following two major services to be supported:

- *Real-time low-latency services.*
  Low-latency services are necessary for voice and video transfer, process control,

---

1. The Huygens project is funded by ESPRIT BRA via the Pegasus and BROADCAST projects and the ESPRIT Research Network in Distributed Computing Systems Architectures.

remote sensing etc. The data for these services is usually worthless if it does not arrive in time. A video sequence, for instance, must be retrieved at a high and constant rate; frames retrieved too late are no longer useful and can be ignored. In general the loss of a few frames will be tolerated. Other critical applications, like process control, require reliable and hard real-time connections where certain deadlines must be met. After all, criticallity usually means that the application must do certain things in time, despite failures.

Non real-time transactions and bulk data transfer.
Transactions in distributed operating systems require low latency, and the amount of data to be transmitted is usually small or moderate. Examples include database queries and remote procedure calls (RPC).
Bulk data transfer is a service that carries a large amount of data with relatively low latency constraints.

The bandwidth of many existing networks is by far not enough for distributed multimedia applications. Their throughput and latency is becoming a bottleneck in demanding real-time applications.
The fundamental advantage of our network is that we claim that our network will have a *high aggregate bandwidth* with a moderate link bandwidth. Where in networks with a ring or bus topology the aggregate network bandwidth is limited to the link bandwidth, in our project the aggregate bandwidth will be many times the link bandwidth and grows with the number of workstations attached to the network.

A drawback of most state-of-the-art networks is that a bounded packet delay cannot be guaranteed. In our project the *real-time behaviour* is an essential design issue. We can guarantee a bounded latency, therefore our network can be used in distributed real-time applications such as distributed process control and distributed multimedia applications.

2. Kautz graphs

This section describes Kautz graphs, the network used inside the switching fabric.

2.1 Definition of Kautz graphs [Kautz 68]

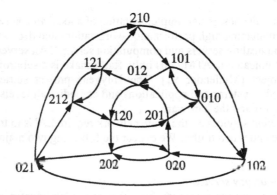

Fig. 1: Example of a Kautz graph (K(2,3)).

A Kautz digraph $K(d,k)$ with in-degree and out-degree d and diameter k is the digraph whose vertices are labelled with words $(x_1,...,x_k)$ of length k from an alphabet of $d+1$ letters by removing those words in which there are two consecutive identical letters (x_i <> x_{i+1}, for $1 <= i <= k-1$). There is an arc from a vertex x to a vertex y if and only if the last $k-1$ letters of x are the same as the first $k-1$ letters of y. An undirected Kautz graph can be found by removing the orientation of the digraph.

2.2 Properties of Kautz digraphs

Kautz graphs have a number of interesting properties [Bermond 89]. In this paper we only mention the properties that are of interest for the design of the Rattlesnake architecture.

- The number of vertices $N = d^k + d^{k-1}$. This implies that Kautz graphs $K(d,k)$ interconnect considerably more processors than other topologies with degree d and diameter k.

	k=4	k=8	number of vertices
cube	81	6561	$N = d^k$
torus	48	192	$N = k^d$
Kautz	108	8748	$N = d^k + d^{k-1}$

Table 1: Number of vertices of several topologies for d=3.

- The degree of a Kautz graph is *fixed* and independent of N. By increasing k, networks of arbitrarily large size can be built with a fixed (=2d) number of connections per node. Because the degree is fixed, nodes can be implemented using (VLSI) components. Where Kautz networks have a fixed degree, other networks, such as the binary hypercube, require the number of connections per node to *increase* with (the logarithm of) the number of nodes. The number of wires in a Kautz graph equals d x N, whereas a binary cube needs (log N x N)/2 wires.

- The *diameter* of the network is k ($< ^d$log N).

- A Kautz graph can emulate standard computation graphs such as a linear array, ring and tree. A tree can be used to implement *multicast* efficiently in the architecture.

- A Kautz digraph is *fault tolerant*. The connectivity of $K(d,k)$ equals d. The diameter of the network in case of faulty nodes has also been studied by Imase et al. [Imase 86]. They showed the existence of d vertex disjoint paths between any pair of vertices in $K(d,k)$, one of a length of at most k, $d-3$ of a length of at most $k+1$ and two of a length of at most $k+2$.

diameter (k)	number of nodes (N)	mean shortest route length	mean route length
3	36	2.2	3.4
5	324	3.8	5.4
7	2,916	5.7	7.4
9	26,244	7.7	9.3

Table 2: Route length vs. diameter for d=3

- Another interesting property of the network is the fact that it admits *self routing* of messages, both when the network is fault free as well as when some nodes or links are faulty. A straightforward generic route of length k can be found by simple concatenation of source and destination word. However, in general there are routes with length $< k$. An algorithm for generating all node disjoint routes is straightforward [Smit 91].

Example 1 (see figure 1):
In the graph we find the route Rg = < 120201 > from (120) to (201) via node (202) and (020). This route has length 3 (= k). However, there is a shorter route: Rs = <1201> of length 1.

A drawback of Kautz graphs is that in general these graphs (just like hypercubes) are not planar and have a coarse modularity (i.e. when k increases by one, the number of nodes N multiplies by d). Furthermore there is no known straightforward one-to-one mapping of the class of algorithms for binary hypercubes onto Kautz graphs because Kautz graphs are not node-symmetric.

3. Implementation of the Rattlesnake switching fabric

3.1 Rattlesnake global overview

From a global perspective our network has a star topology; i.e. there is a central switching fabric (also called *hub* or switch box) and all servers and workstations are connected with point-to-point links to this switch box (see figure 2). It can connect approximate 100 workstations or servers, typically located within a floor of a building. A similar approach can be found in the Autonet project [Schroeder 90].
A star topology has a number of advantages. Firstly, within the switching fabric low diameter, high connectivity networks can be used. This means that a number of connections can be handled simultaneously and hence we get a high aggregate bandwidth. Secondly, as most of the complexities are contained within the fabric, the interfaces in the workstations can be simple and low cost. Because the distances within the switchbox are relatively small we can use parallel and synchronous links. Thirdly, we are able to use low cost, high speed point-to-point serial links with state-of-the-art components such as TAXI chips.

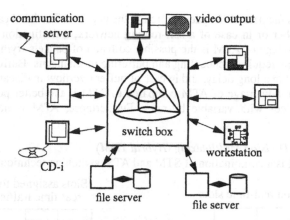

Figure 2: Global architecture of a distributed
multimedia system.

The Rattlesnake switching fabric is composed of a large number of identical basic switching building blocks. Each basic switch maps one-to-one to a node of the internal Kautz network of the switching fabric. Consequently, inside the switching fabric we can use the parallelism and (fault-tolerant) routing properties of Kautz graphs.

A message generated by a source station travels through the switching elements of the switching fabric and reaches the destination station. Switching elements contain logic to forward messages from an input link to an output link. A cross-bar (or bus) inside each switch connects the links.

There are a number of important issues in the design of a switching fabric such as: the switching mode (circuit switching, packet switching or hybrid switching), the routing strategy (wormhole routing, store-forward routing or virtual cut through) and the scheduling policy (FIFO, priority).

3.2 Switching modes

Well known switching modes are:

- *Synchronous Transfer Mode (STM)*

  The STM or circuit switching mode uses frames. Time is divided into equal parts, in which a frame is transmitted. A frame is divided into slots. Each source is allowed to use one or more slots of a frame. STM has the disadvantage that the bandwidth is divided into fixed slots and frames. When an application has claimed a slot but does not use it this bandwidth is wasted. A major advantage of STM is that it guarantees bandwidth with a fixed, bounded delay.

- *Asynchronous Transfer Mode (ATM)*

  The ATM[1], or packet switch mode, approach abandons the concept of frame references. ATM achieves a more flexible bandwidth sharing by allowing the terminals to seize bandwidth when a packet is ready for transmission. Each packet has a label

---

1. The CCITT has recommended ATM as the target transfer mode for broadband ISDN networks.
It is expected that most devices and applications will generate ATM packets in the future.

that identifies the path of the packet through the network. The label can be a virtual channel number or in case of a self-routing network, a destination address. The main disadvantage of ATM is the possible collision of packets trying to seize the same slot. This requires scheduling and buffering mechanisms. Buffering of packets may result in a long delay, and in case of buffer overflow in discarding of packets. The main advantage of ATM compared to STM is the better performance in case of services with variable bit-rates. Furthermore, ATM is not technology dependent.

Hybrid Time Division Multiplexing (Hybrid TDM)
Hybrid TDM is a combination of STM and ATM switching techniques [Hui 90].

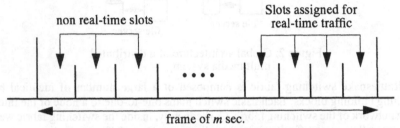

Figure 3: Hybrid Time Division Multiplexing

It combines the flexibility of ATM with the capability of assigning time slots of STM. Each frame has a fixed maximum number of slots. Part of these slots are assigned to hard real-time services (STM with guaranteed bandwidth), and the rest to non hard real-time services (ATM packets). An advantage of hybrid TDM is that if a real-time slot is not needed by its source, it can be used by non real-time services.
The size of a slot is independent of the ATM cell size. This means that one ATM cell is fragmented into several slots, spread out over several frames.

3.3 Routing strategy

The Rattlesnake network uses Hybrid TDM with a combination of wormhole routing (for real-time services) and store-forward routing (for non real-time services).
Depending on the network it is applied to, routing may be subject to deadlock and fairness problems. Deadlocks occur in such situations where messages in a given set have not arrived to their destinations but cannot progress any longer because of circular dependency between the requests to the resources they need. Fairness can occur when message flows are not serviced fairly, making messages to be stuck in one part of the network, although they are not deadlocked.
The fairness problem can be tackled at the design stage of the routing mechanism, whereas deadlock is more directly related to the structure of the network.
There exists a number of deadlock free routing algorithms. Most algorithms are based on preventing cycles in the dependency graph. This can be done by introducing virtual channels. However, this approach *restricts* the routing of packets: it reduces the number of possible paths that a packet may take. In the Rattlesnake network we use the nosy worms protocol [Whobrey 88] to avoid deadlock and which does not have routing restrictions.
All traffic between workstations and servers will use ATM cells only. In the sequel the terms *cell* as well as *message* are used to indicate an ATM cell. First the cell is stored

the external memory associated with the Rattlesnake switch of the source (see fig. 4), then the cell is transferred to the destination, where it is stored in the external memory of the destination switch.

3.3.1 Routing for real-time services

For real-time services we use wormhole routing. It operates by advancing the head of packets directly from incoming to outgoing channels. Only a few control digits (called *flits*) are buffered at each switch.

The performance of wormhole routing can be improved by introducing virtual channels [Smit 92]. *Virtual channels*, implementing a number of virtual channels on one physical link, was first introduced as a technique to avoid deadlocks in networks. Dally [Dally 90] showed that virtual networks increase the connectivity of networks and have performance advantages. In the iWARP [Borkar 90] processor virtual links are used to implement real-time channels. Some virtual channels can be pre-defined and fixed during a communication session. So there is a guaranteed bandwidth between nodes, that can be used for real-time communications.

The general architecture of the Rattlesnake switch is depicted in figure 4. The switching

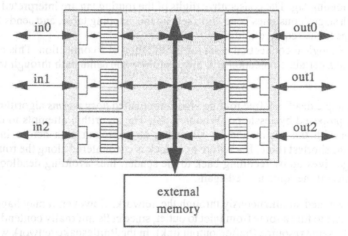

Fig. 4: Structure of the Rattlesnake Switching Element.

elements are connected to each other via bi-directional links. Via these links the switches can send flits to other switches.

A switching element contains k buffers at each link. These buffers are used to implement the virtual channels on a link. Because wormhole routing is used only a few flits are buffered in each node. When establishing a connection from source to destination, each switch assigns an input buffer and an output buffer for this circuit. The routing information is stored in mapping tables (see figure 5). They are used to perform the virtual channel translation and to select the outlink. Each entry in the mapping table has only local significance and identifies the local virtual channel translation.

To find a path from inlet to outlet of the switching fabric we use the self-routing property of Kautz networks. The route to be followed is contained in a single string of digits,

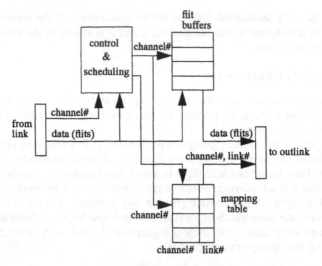

Figure 5: Structure of the input part of a link.

called the routing tag. The consecutive digits of the routing tag are interpreted stage-by-stage. Each stage "consumes" one digit, sets up the mapping table, and sends the rest of the message to the required outlink.

The path of a logical connection is set for the duration of a connection. This means that messages of a certain connection will always follow the same path through the switching fabric.

We use a simple deadlock free routing algorithm (called *nosy worms* algorithm) similar to the one proposed by Jesshope [Whobrey 88]. The algorithm attempts to establish a connection with the destination by allocating an unbroken path across intermittent nodes via the shortest route. If a failure or a block is encountered along the route the set-up message gives up by recoiling back to the sender; thus avoiding deadlock. After a random interval the route is tried again.

Cells are switched simultaneously through the network. However, it may happen that it is not possible to find a route from inlet to outlet, since cells internally contend with each other for the same resource (buffer, output link). In the Rattlesnake network we use several ways to reduce the internal contention rate. First of all, its internal network topology (Kautz graph) offers multiple (node disjoint) paths between inlets and outlets. Another path will be selected when, during the connection phase a setup message is recoiled to the sender.

Furthermore, the physical links between the switching elements have multiple virtual channels. This implies that the chance that a chain of unused virtual channels from source to destination can be found increases significantly [Smit 92]. This effect is somewhat similar to having more node disjoint paths from source to destination.

3.3.2 Routing for non real-time services

To support non real-time services, the network uses a store-and-forward routing method. An entire message, typically one ATM cell, is buffered in all intermediate nodes.

The buffers for the store-and-forward routing are located on external memory near the switch. A path from source to destination is established by reserving buffers in each intermediate node of the path. The size and number of buffers is application dependent. The required buffer space in the intermediate nodes is reserved first. The reservation of buffers also uses the modified nosy-worms protocol (see 3.3.1) to avoid deadlocks. This means that when there is no buffer space available on the shortest route from source to destination, it will try another node disjoint path.

3.3.3 Scheduling policy

In the Rattlesnake switch we use a fair scheduling mechanism, in which the state of all flit-buffers of a link is tested periodically. So, inside the switch there is no difference between real-time and non real-time traffic. When a flit-buffer is full, it will be emptied in bounded time.
The message buffers in the external memory are scheduled with a priority FIFO queuing mechanism. Real-time messages coming from or going to the attached workstation have priority over non real-time messages.

3.4 Current status

In the current prototype we use Xilinx Field Programmable Gate Arrays. Xilinx's FPGA architecture is similar to other gate arrays, with an interior matrix of configurable logic blocks and a surrounding ring of I/O interface blocks. The functions of the FPGA (logic blocks, I/O blocks and their interconnection resources) are controlled by a configuration program stored in an on-chip memory. The configuration program is loaded from an external memory.

The reason for using FPGAs is *flexibility*. A limitation of the current FPGAs is that they have only a small on chip memory (6 Kbit for a XC4005). Wormhole routing, where only a few flits need to be stored in the switch, can be implemented efficiently in FPGAs due to the limited memory requirement. If the switch is used in store-and-forward fashion, off-chip memory is needed. The available FPGA memory is sufficient to implement wormhole routing with a limited number of virtual channels.
We have implemented a Rattlesnake switching element with 4 virtual channels and 4 uni-directional links in a single XC4005 FPGA. We used VHDL as a design tool and a VHDL synthesizer from VIEWlogic to generate the function code for the Xilinx chips. The design uses 96% of the available CLBs[1] and has 102 pins.
Currently we are working on a design with 16 virtual channels per link and 6 uni-direction links, using a XC4010 FPGA. The mean latency of a real-time flit is expected to be less than 5μs per switch, the maximum will be 10μs per switch.

4. Conclusion

In this paper we have presented the architecture of a high-performance, low-latency network suitable for hard real-time multimedia applications. The Rattlesnake network uses Hybrid TDM with a combination of wormhole routing (for real-time services) and store-and-forward routing (for non real-time services). The nosy worms protocol is used for real-time as well as non real-time services to avoid dead-lock.

---

1. A CLB (Configurable Logic Block) is the basic building block of a Xilinx FPGA. It contains programmable combinatorial logic and two storage registers

The Rattlesnake switch uses real-time virtual channels to guarantee a bounded latency.

The network of the switching fabric is based on a Kautz topology. Kautz graphs have interesting properties such as: small diameter, the degree is independent of the network size, the network is fault-tolerant and has a simple routing algorithm.

The fundamental advantage of our network is that we claim a high aggregate bandwidth with a moderate link bandwidth.

A prototype of a switching element has been implemented with a standard FPGA. The design of the switching fabric with FPGAs, allows us to experiment with switching mode, routing strategy and scheduling policy in a multimedia environment.

References

[Bermond 89] Bermond J.C., Homobono N., Peyrat C.:"Large Fault-Tolerant Inter-connection Networks", Graphs and Combinatorics, 1989.

[Borkar 90] Borkar S. et al.:"Supporting Systolic and Memory Communication in iWarp", Proc. 17th ACM/IEEE Symposium on Computer Architecture, 1990, pp 70-81.

[Dally 90] Dally W.J.:"Virtual-channel Flow Control", Proc. 17th ACM/IEEE Symposium on Computer Architecture, 1990, pp 60-67.

[Hui 90] Hui J.Y.:" Switching and traffic theory for integrated broadband networks.", Dordrecht, The Netherlands: Kluwer Academic Publishers, 1990.

[Imase 86] Imase M., Soneoka T., Okada K.: "A fault-tolerant processor inter-connection network" (original in Japanese); translated in Systems and Computers in Japan, vol 17, no 8 pp 21-30, 1986.

[Kautz 68] Kautz W.H.:"Bounds on directed (d,k) graphs. Theory of cellular logic networks and machines", AFCRL-68-0668 Final report, pp 20-28, 1968.

[Mullender 91] Mullender S.J.: "The Huygens Project", internal memo University of Twente dept. Computer Science, 1991.

[Schroeder 90] Schroeder M.D., Birrell A.D. et al.;"Autonet: a High-speed, Self-configuring Local Area Network Using Point-to-point Links", Digital Systems Research Center, Palo Alto, CA, April 1990.

[Smit 91] Smit G.J.M., Havinga P.J.M., Jansen P.G.: "An algorithm for generating node disjoint routes in Kautz digraphs", Proceedings Fifth International Parallel Processing Symposium, Anaheim, CA, 1991.

[Smit 92] Smit G.J.M., Havinga P.J.M.:"Performance analysis if routing algorithms for the Rattlesnake network", to appear in: proceedings of MASCOTS'93, International Workshop on Modelling, Analysis and Simulation of Computer and Telecommunication Systems.

[Whobrey 88] Whobrey D.: "A communications chip for multiprocessors", Proc. CONPAR 88 pp 464-473, 1988.

Beyond ST-II:
Fulfilling the Requirements of Multimedia Communication

Ralf Guido Herrtwich
Luca Delgrossi

IBM European Networking Center
Tiergartenstr. 8
D-6900 Heidelberg 1

rgh@dhdibm1.bitnet
luca@dhdibm1.bitnet

Abstract: The Heidelberg Transport System (HeiTS) uses the Internet ST-II protocol as its communication core. ST-II provides adequate functions for multimedia communication, most notably fast packet forwarding, multiple target addressing, and bandwidth management. Some desirable additions to the protocol include graceful service degradation and queueing feedback for rate adjustment in the nodes. A transport protocol on top of ST-II should provide different reliability classes and take care of segmentation and reassembly.

1 Introduction

As part of its work towards a distributed multimedia programming platform, the IBM European Networking Center in Heidelberg has developed the Heidelberg Transport System (HeiTS) for multimedia communication over the last year [1, 2]. An AIX version of HeiTS on IBM's RS/6000 machines is available since summer 1992; it was ported recently to PS/2s running OS/2 Version 2.0.

The core of HeiTS is the Experimental Internet Stream Protocol, Version 2 – ST-II for short [3]. There exist several implementations of the protocol worldwide. The most prominent code (because it is available in the public domain) comes from BBN and runs on SUN workstations. We did not base our implementation [4] on this or any other existing code because we wanted to have a user-level implementation of the protocol (to make it easier to port the protocol to non-UNIX systems and to integrate it with microkernel operating systems – still keeping the option to migrate the protocol code into the kernel). For the port to OS/2 this decision was crucial. We found the performance impact of this decision perfectly acceptable [4].

Our choice for ST-II as the main protocol in HeiTS was based on the following criteria:

- We wanted to have a protocol tuned to the needs of both transporting audiovisual data and supporting multimedia applications (which are, in fact, two completely different requirements). ST-II with its functions for sending data streams over multicast routing trees with underlying guaranteed-bandwidth channels meets some of the main requirements.

- Most audiovisual applications we had in mind send large chunks of data over some period of time. Because of this large amount of data, the protocol to be used should allow for efficient data transport. Handling of data PDUs in ST-II can indeed be very fast (although this is somewhat at the expense of quick and easy connection establishment).

- With Internet protocols being the most prominent protocol family worldwide, any new protocol for multimedia communication should fit easily into this protocol family (e.g., it should use the familiar IP addressing schemes). ST-II fulfills this requirement.

- We were looking for a non-proprietary protocol that would eventually have other supporters than just IBM. ST-II is already being provided and supported by others.

- Finally, we simply found it a bad idea to start from scratch. A lot of effort has gone into the specification of ST-II (with some more effort needed to correct mistakes and inconsistencies in the existing specification, cf. [5]), so we saw no reason to replicate this effort without having gained practical knowledge about which features are really needed, which can be omitted and which are missing.

We are still in the process of examining the feasibility of HeiTS for sample multimedia applications – and there is no end in sight as new applications are continuously added. This paper summarizes our findings so far. Section 2 describes the functions provided by ST-II and briefly evaluates how essential they are for multimedia communication. Section 3 determines some useful additions to the ST-II functions within the network layer. Section 4 examines what kind of transport functions are required on top of ST-II.

2 Functions Provided by ST-II

ST-II was designed for packetized audio and video communication across an internet as it is known today. It does neither require a particular broadband technology to be in place nor is it aimed at a hardware implementation.[1] This also makes the protocol useful in a transition phase from today's to tomorrow's communication networks – already allowing multimedia communication on the existing infrastructure.

2.1 Fast Data Communication

ST-II tries to accomplish a fast forwarding of data packets. For this reason, an ST-II data packet contains minimal header information. The 8-byte header identifies the packet as an ST-II packet (distinguishing it from IP packets and defining the ST-II version) and contains a priority, the packet size, a hop identification (HID) and a header checksum. Depending on a flag set, it may also contain a time stamp. This time stamp can provide information for scheduling messages in the routers. It may also be useful as synchronization information if it is made available to higher layers.

The HID represents addressing information for the packet. It uniquely identifies a stream for each group of adjacent ST-II agents. During connection setup, adjacent agents negotiate this 2-byte identification in the following way: The sending agent proposes an HID to all receiving agents within the connect message. If they already use this HID they can initiate the choice of another HID by submitting the set of free HIDs. Through iteration of this negotiation, a final HID is selected.[2] Choosing one common HID for all next hops is also a prerequisite for efficient multicast.

---

1. It does, however, require some bandwidth management technology that needs to be added to the existing networks. One core function of ST-II, bandwidth reservation, could otherwise not be achieved (cf. Section 2.3).
2. Considering a stream flowing to 10 next hops each having 1000 streams established, the probability of choosing a valid random 2-byte HID is 85.9% on the first try, 98.1% on the second and 99.8% on the third [3].

2.2 Communication with Multiple Targets

Typical distributed multimedia applications require communication paths where data is distributed from one source to several sinks. Classic examples are television broadcasts or video conferences. ST-II takes this requirement into account by establishing routing trees from one origin to multiple targets for every data stream. The routing tree exactly describes the path a message will take across various subnetworks. If underlying subnetworks provide multicast this function can be efficiently utilized to reach multiple targets (or intermediate hops) on the same subnetwork.

ST-II does not depend on any particular routing model, but its bandwidth reservation model (see next section) is best suited for a static routing technique that forwards each packet. If a node on the route fails ST-II agents have the chance to overcome failures through automatic rerouting provided that the new route has enough bandwidth available.

Targets can be added and removed by the origin while data is flowing. (Mechanisms to ensure that a new target can correctly interpret the data it receives at the time it is connected are outside of the scope of ST-II.) Targets can remove themselves from the stream, but cannot join the stream by themselves. Anonymous listening to streams is, thus, not supported.

2.3 Bandwidth Reservation

The temporal properties of audiovisual data are crucial for their correct processing. In general, any multimedia system has to ensure that data is available at its intended presentation time. The straightforward way to do this is to set aside some bandwidth for each multimedia stream and to guarantee that this bandwidth is never used by other traffic. This is the approach underlying ST-II. However, it is important to note that ST-II in itself does neither perform resource reservation nor does it protect any reserved bandwidth. This has to be accomplished by modules surrounding ST-II. When implementing the protocol, we found it at least as complicated to put the resource management routines in place as to implement the actual ST-II protocol code [6].[3]

ST-II provides the negotiation framework to obtain a certain quality of service (QOS) for connections. QOS is expressed as a so-called flow specification. Its main components are parameters to define the minimum bandwidth required to run the connection, the accumulated (end-to-end) mean delay of the connection and the accumulated (end-to-end) delay variance. Users can also specify desired PDU sizes and rates as well as limits on these two values and on cost and delay. The limiting values are used to decide whether a connection establishment is successful, i.e., if the values cannot be guaranteed the connection is not established.

Point-to-point links of a given total throughput are well-suited for reservations as assumed by ST-II. Reservation is more complicated for existing local-area networks that provide multiple access points. Neither traditional Ethernet nor vanilla Token Ring provide bandwidth reservation functions. For Token Ring, we have developed and implemented a bandwidth management scheme suited for ST-II [8]. Synchronous mode FDDI which is now being incorporated in a second generation of FDDI adapters is much better suited for reservation. So are DQDB and B-ISDN.

---

3. It is worth mentioning that to our knowledge neither BBN's original implementation nor the code from the Swedish Institute for Computer Science contain resource management routines. As part of the DARTnet project, BBN has now implemented the Virtual Clock algorithm [7] under ST-II.

3 Network Layer Additions

We identified some functions as missing from the network layer that would make ST-II even better suited for multimedia applications. We are working on integrating these mechanisms in the next release of our software.

3.1 Graceful Service Degradation

ST-II does not imply any particular bandwidth reservation scheme. Yet, the underlying assumption is that always "sufficient" bandwidth is allocated for an application so that messages are never lost or delayed. This leads to a pessimistic reservation model [9] that potentially underutilizes resources and unnecessarily refuses connections. The requirements of an application are often better met by gracefully degrading service quality as the load in the network increases [10]. Rather than aborting a multimedia transmission, the quality of the transmission is altered.

As a first step towards a more flexible behaviour of the transport system, we introduce optimistic reservation in ST-II [9]. For an optimistically established connection, not a maximum, but a much smaller reservation is made so that the resource is intentionally overbooked. Packets are forwarded on a best-effort rather than on a guaranteed basis. Still they are preferred over discrete-media communication. It is up to the higher layers to deal with potential errors resulting from this transmission mode (see Section 4.1).

Automatic degradation is also possible if the network has information about the way messages are encoded. If an ST-II agent was aware of a particular encoding scheme it could automatically adjust the data quality to the available bandwidth (and also to the abilities of a certain receiver). Hierarchical encoding as in MPEG would be ideal for such an operation if only it was more robust in regard to transmission errors. With more flexibility in regard to guaranteed performance, it is no longer necessary to provide the same QOS to all targets of a connection.

3.2 Queueing Feedback

ST-II suggests to use rate control for sending messages on the network. Any rate-controlled system has to cope with the problem of clock drift: If the receiver has a tendency to work slower there is a chance that its queues fill up. If the sender is slower the queues run dry. One solution to the problem is to provide fine clock synchronization throughout the system [11].

As clock synchronization may lead to autonomy problems in the nodes, we favor an approach where explicit messages are exchanged between adjacent ST-II agents to adjust the speed of their operation (this approach was suggested by [12]).[4] If a receiving agent's queue is filled over a high water mark, a message is sent to the sending agent to slow down (or to stop). If a low water mark is reached a message to speed up (or to restart) is sent. This method is particularly useful if it extends beyond the network to any program handling a multimedia data stream. Queues at sinks such as a decompression board are usually larger than those in the network.

One option of queueing feedback is to use it as a means for pausing recorded multimedia streams when fetching them from a remote file server. As the viewer hits the pause button, the

---

4. This adjustment could also be performed in one of the higher layers. In this case, however, it would just function end-to-end and would not include the router nodes where ST-II agents reside. The ST-II solution, however, requires the introduction of a new control message from the targets to the origin which currently does not exist.

picture immediately freezes. As data accumulates in the queues, the pause message is propagated back to the source. Once the viewer depauses, again the system reaction is immediate as data from the queues is played back. Restart is completed when the message reaches the source.

4 Transport Layer Additions

The original ST-II protocol description allows a variety of protocols on top of ST-II. Apart from two special protocols designed to run on top of the first version of ST, NVP (network voice protocol [13]) and PVP (packetized video protocol [14]), it also considers the use of TCP and UDP on top of ST-II. Neither choice is useful for audiovisual data (although it may be appropriate for bulk data transfer): TCP's error correction mechanisms are not flexible enough to be used for all kinds of video. UDP datagrams do not benefit from the underlying connection-oriented service. As for NVP and PVP, we have several reasons to object to them: We find it not useful to have different protocols for audio and video. Also, the protocols are somewhat out of date as they refer to an earlier version of ST-II. They make certain assumptions about the encoding of audio and video not applicable to modern compression schemes. Most importantly, they contain functions for conference management that we envision to be implement above the transport layer.

We have developed the Heidelberg Transport Protocol, HeiTP [15, 16], as a multimedia transport protocol on top of ST-II. It provides an end-to-end abstraction from the network layer mechanisms and nodes. As a lot of the function required for multimedia transport is already provided by the network layer, HeiTP is light-weight in comparison. It uses an out-band signalling approach that maps HeiTP control PDUs onto ST-II control PDUs, not data PDUs. For this purpose, we have introduced a new ST-II control PDU originating from the targets that was not available in ST-II. For compatibility reasons, this PDU can also be sent across UDP and/or IP (which seems to have been the original intention by the ST-II inventors).

4.1 Reliability

The reliability a multimedia stream needs for transmission depends on the kind of media, the encoding of the media (and perhaps the hardware compressing and decompressing it), and the user requirements on media quality. On a superficial level, it looks as if audiovisual data does not require error handling mechanisms at all: A corrupt audio sample or a wrong video frame portion may not even be noticed because of the high data presentation rate.

For uncompressed digital video, this statement is true. With digital video compression, errors in frame displays usually become more persistent: As new frame content is mostly encoded as a delta to previous content, an error remains visible as long as a new self-contained (key) frame is transmitted. In encoding methods such as MPEG a typical key frame rate is 4 per second. It would have to be higher to be invisible to viewers, but then the desired compression factor could not be achieved. As far as audio is concerned, the human ear notices even brief disturbances. Listeners become easily annoyed at inadequate audio quality.

HeiTP provides four modes of error handling: It can ignore, discard, indicate, and correct corrupt data. To detect corruption, checksumming is used. Among the reliability modes, error correction for time-critical multimedia data is the most difficult one to implement. The typical end-to-end error correction mechanism – retransmission – may cause the corrected message to arrive too late at the destination to be useful. This would cause another error handling mechanism of HeiTP, the detection of late messages, to come into play. Late messages can be indicated or discarded.

We are studying the use of forward error correction (FEC) techniques as an alternative to message retransmission [17]. Apart from timeliness, an advantage of FEC is that no unpredictable load is put on the network. This suits the bandwidth reservation model of ST-II well. However, FEC is based on the idea to introduce data redundancy; it is, of course, useless if the redundancy exceeds the compression factor unless compression was merely done for storage purposes. FEC may, thus, be more useful for audio than for video.

Error correction by retransmission is difficult for communication that involves multiple targets. To satisfy individual correction requests, selective retransmission per target is desirable. However, if several targets request correction of the same message (e.g., because the message was destroyed on a common route portion), this would be inefficient. We will experiment with different retransmission modes. As a first step, corrected messages are sent to all targets and those targets that have not requested retransmission discard it.

4.2 Segmentation and Reassembly

Another main function of HeiTP is the adjustment of data unit sizes between application and network. In general, the application provides a logical data unit (e.g., a video frame) to the transport system. This LDU is then segmented to make it fit into the network packets. Depending on a flag set by the receiver, the packets are reassembled at the receiving side or not. Not reassembling packets is useful for continuous streams with high temporal resolution: For an audio stream, e.g., the LDU concept is less important than for video.

From a resource management standpoint, segmentation also involves the adjustment of application QOS parameters to network QOS parameters [18]. Packet rates have to be calculated from LDU rates. In particular for variable bit rate streams this is not a straightforward process: A constant LDU rate at the transport interface yields an irregular packet delivery at the network interface. This irregularity causes a problem for resource reservation. How we deal with this problem is subject of an upcoming paper.

5 Closing Remark

ST-II (possibly extended) and HeiTP provide a good basis for multimedia data exchange. Yet, distributed multimedia applications require communication support beyond mere data transport. To facilitate the control of different multimedia data streams passing through a network, we are developing HeiMAT, the Heidelberg Multimedia Application Toolkit. HeiMAT takes care of the joint setup of connections for multimedia applications such as video conferences.

We would like to acknowledge the contributions of other members of the HeiProjects. Special thanks are due to Christian Halstrick as the principal designer and implementor of HeiTP and to Frank Hoffmann who is our chief ST-II programmer.. Carsten Vogt guaranteed to perform all work on resource management well beyond a best-effort basis.

References

[1] D. Hehmann, R.G. Herrtwich, R. Steinmetz: Creating HeiTS: Objectives of the Heidelberg High-Speed Transport System, GI-Jahrestagung, Darmstadt, 1991.

31

[2]   D. Hehmann, R.G. Herrtwich, W. Schulz, T. Schütt, R. Steinmetz: Implementing HeiTS: Architecture and Implementation Strategy of the Heidelberg High-Speed Transport System, Second International Workshop on Network and Operating System Support for Digital Audio and Video, Lecture Notes in Computer Science 614, Springer-Verlag, Heidelberg, 1992.

[3]   Experimental Internet Stream Protocol, Version 2 (ST-II), RFC 1190, Internet Network Working Group, 1990.

[4]   L. Delgrossi, R.G. Herrtwich, F.O. Hoffmann: An Implementation of ST-II for the Heidelberg Transport System, in preparation.

[5]   C. Partridge, S. Pink: An Implementation of the Revised Internet Stream Protocol (ST-II), Internetworking: Research and Experience, Vol. 3, No. 1, 1992.

[6]   C. Vogt, R.G. Herrtwich, R. Nagarajan: HeiRAT – The Heidelberg Resource Administration Technique – Design Philosophy and Goals, Technical Report, IBM European Networking Center, 1992.

[7]   L. Zhang: Virtual Clock – A New Traffic Clontrol Algorithm for Packet Switching Networks, SIGCOMM '90 Symposium, Philadelphia, 1992

[8]   R. Nagarajan, C. Vogt: Guaranteed-Performance Transport of Multimedia Traffic Over the Token Ring, Technical Report, IBM European Networking Center, Heidelberg, 1992.

[9]   R.G. Herrtwich: The Role of Performance, Scheduling, and Resource Reservation in Multimedia Systems, International Workshop on Operating Systems of the 90s and Beyond, Lecture Notes in Compute Science 563, Springer-Verlag, Heidelberg, 1991.

[10] J.G. Hanko, E.M. Kuerner, J.D. Northcutt, G.A. Wall: Workstation Support for Time-Critical Applications, Second International Workshop on Network and Operating System Support for Digital Audio and Video, Lecture Notes in Computer Science 614, Springer-Verlag, Heidelberg, 1992.

[11] D. Ferrari: Distributed Delay Jitter Control in Packet-Switching Internetworks, Technical Report, International Computer Science Institute, Berkeley, 1991.

[12] S. Drake: Personal Communication, IBM Almaden Research Center, 1991.

[13] D. Cohen: A Network Voice Protocol NVP-II, University of Southern California, Los Angeles, 1981.

[14] R. Cole: PVP – A Packet Video Protocol, University of Southern California, Los Angeles, 1981.

[15] C. Halstrick: Entwurf und Implementierung eines Multimedia-Transportprotokolls, Diplomarbeit, Universität Erlangen-Nürnberg, 1992.

[16] L. Delgrossi, C. Halstrick, R.G. Herrtwich, H. Stüttgen: HeiTP – A Transport Protocol for ST-II, Globecom 92, Orlando, to appear.

[17] E. Biersack: A Performance Study of Forward Error Correction in ATM Networks, Second International Workshop on Network and Operating System Support for Digital Audio and Video, Lecture Notes in Computer Science 614, Springer-Verlag, Heidelberg, 1992.

[18] D. Ferrari: Client Requirements for Real-Time Communication Services, Technical Report, International Computer Science Institute, Berkeley, 1990.

An Approach to Real-time Scheduling - but is it Really a Problem for Multimedia?

Roger Needham and Akira Nakamura

Computer Laboratory, University of Cambridge
New Museum Site, Pembroke Street, Cambridge, England, CB2 3QG, U.K.

Abstract. This paper details work done on how to avoid bad behaviour caused by priority inversion and lock blocking in real-time processing in multimedia environments. We will show how they can be cured by more complex algorithms. (1) Priority inheritance protocol, (2) the priority ceiling protocol, (3) the semaphore dependency protocol deal with increasing complex situations. The question is why should there be complex situations? Our suggestion is that there should not be complex situations in multimedia systems, therefore and that simple priority inheritance ought to be good enough.

1 Introduction

Workstations or other computers which handle continuous media traffic have to do so subject to real-time constraints of various sorts. It is typical to require to present frames of video at regular intervals, or to provide a smooth flow of bits to drive a loudspeaker. If it is desired to have a number of streams of data going at one time, it is important to know whether the needed work can be done in a timely manner, and to ensure that it is done that way.

Over the years there has been a considerable amount of research in this area under the heading of real-time systems. (The latest operating system, SUN-OS5.0 [Khanna92] provides real-time synchronisation.) Complications arise from interference between processes because of shared locks, so that straightforward synchronisation may not work. This work considers improvements to the current known algorithms for synchronisation, and then, by contrast, considers whether the implicit complication is worthwhile.

Problems arise from shared locks when particular locks are shared between processes of different priority. The classical case is when a process of high priority cannot proceed because a lock it needs is held by a process of low priority; the latter cannot run because a process of intermediate priority is using the processor. The high priority process is thus arbitrarily delayed. The earliest attempts to solve this problem are twenty years old, depending on what is now know as priority inheritance, in which the holder of a lock runs with the priority of the most senior process waiting. Inadequacies in this were discussed by [Sha90]. They proposed a new protocol called the "priority ceiling protocol" which avoids the problem. The name comes from the association with each lock of a highest or ceiling priority - that of the most senior process which ever holds it.

The priority ceiling protocol is conservative in that it will cause waiting when it may not be necessary. We propose a less conservative protocol, observing that static analysis of the use of locks is already required for the priority ceiling work. The static analysis of programs can also lead to a set of lock dependency list, and using the dependency lists, the new scheduling rule is proposed.

In section 2, we give an overview of the real-time synchronisations, give the detail of the proposed protocol called the dependency protocol, and compares the ceiling protocol with the semaphore dependency protocol. In section 3, we consider how both the semaphore dependency protocol and the ceiling protocol work in a realistic view of a multimedia computation.

It is traditional in this literature to refer to *'semaphores'* rather than *'locks'*, and to ease comparison with earlier work we do so here although it is a longer word and strictly refers to something more particular.

2 Real-Time Synchronisation

Priorities are denoted by integers; higher integers represent higher priorities. S_i are semaphores and T_j are tasks with priority P_j.
The critical section of the task T_i controlled by the semaphore S_j is denoted by z_{ij}.

2.1 Priority Inversion

$T_0 = \{...., P(S_1), ..., V(S_1), ...\}$
$T_1 = \{.................................\}$
$T_2 = \{.., P(S_1), ..., V(S_1), ...\}$
$P_0 > P_1 > P_2$

1. T_2 is initiated, locks S_1
2. T_1 is initiated, preempts T_2
3. T_0 is initiated, preempts T_1. It then claims S_1 and becomes blocked.
4. T_1 restarts.

While T_0 is blocked, T_1 is running. The priorities are inverted and the duration of T_0's blocking time is unpredictable because T_1 must finish its work. The general rule of priority inheritance is that a task runs at the highest priority from the priorities of itself and of any tasks waiting for semaphores it holds.

2.2 Priority Inheritance

Because priority inversion can cause an unpredictable blocking time, the priority inheritance was invented. According to the above example,

1. T_2 is initiated, locks S_1
2. T_1 is initiated, preempts T_2
3. T_0 is initiated, preempts T_1. Then claims S_1 and becomes blocked. And T_2 inherits the priority of T_0. T_2 restarts.
4. T_1 releases S_1, then T_0 locks S_1.

The delay of T_0 is the time of blocking S_1 by T_2.

2.3 The Priority Ceiling Protocol

Burns [Burns91] described a protocol:

1. all processes have a static priority assigned.
2. all semaphores have a ceiling value defined which is the maximum priority of the processes that may use it.
3. a process has a dynamic priority that is the maximum of its own static priority and any that it inherits because it blocked a blocking higher priority process.
4. a process can only lock a semaphore if its dynamic priority is higher than the ceiling of any currently locked semaphores. (excluding any that it has already locked itself).

The priority ceiling protocol uses static analysis of the dependency between lower priority (L) tasks and the ceiling priority (H) task of the semaphore S. The simple diagram is like this.

$$L \xrightarrow{S} H$$

When the lower priority task locks S, only taskes dynamically higher than H can lock another semaphore.

2.4 The Semaphore Dependency Protocol

In order to keep the delay of the highest priority process to the minimum, the ceiling protocol has to be too restrictive and conservative.

We can approach the problem by considering the case in which semaphores are held for one unit time, and we would like to ensure that high priority tasks are not delayed more than one unit time per set of semaphores used by that task as a result of lock holding by lower priority tasks. We would also like to do this without holding tasks up unnecessarily, like the ceiling protocol does.

This semaphore dependency protocol has the features below:

1. it uses static a dependency list and the dynamic ceiling mechanism,
2. it provides priority preemptive,
3. it provides deadlock avoidance,
4. it will be applied to multiprocessor.

Definition For each held semaphore, and for each priority level higher than the native priority of the holder (i.e. before any increase due to priority inheritance), the runtime system keeps a list of semaphores dependent on the held semaphore, in the following sense.

Definition of Dependence
S_k is dependent on S_i at priority P if there is a task with priority P which will seek to lock S_i while holding S_k. The dependency list of S_i is static, and works while S_i is being held.

The simple diagram is like this.
$$S_i \xrightarrow{P} S_k$$

We call S_i the "parent semaphore" and, S_k as the "dependent semaphore".

The New Scheduling Rule The new rule becomes:

A process running at priority P is not allowed to lock a semaphore S_k if any semaphore already locked has S_k in one of its dependency lists at a priority greater than or equal to P. A process which is unable to lock a semaphore is blocked immediately before the lock call, and released when the condition on which it was blocked disappears. The task locking the parent semaphore inherits the priority P from the blocked task.

Example-1
$$T_0 = \{\ldots\ldots, P(S_0), \ldots\ldots, V(S_0), \ldots\ldots\ldots\ldots\ldots\}$$
$$T_1 = \{\ldots\ldots, P(S_1), \ldots, P(S_2), \ldots, V(S_2), \ldots, V(S_1), \ldots\}$$
$$T_2 = \{\ldots, P(S_2), \ldots P(S_1), \ldots, V(S_1), \ldots, V(S_2), \ldots\ldots\}$$
$$(P_0 > P_1 > P_2)$$
The dependency lists are:
$$DS_1 = [(S_2, P_2)], DS_2 = [(S_1, P_1)]$$

1. T_2 is initiated, locks S_2.
2. T_1 is initiated, claims S_1, but the dependency list:
   $$DS_2 = [(S_1, P_1)]$$
   shows it must be blocked and T_2 inherits the priority.
3. T_0 is initiated, preempts T_2.
4. T_0 locks S_0 and releases S_0.
5. T_2 restarts, locks S_1 and releases it.
6. Then T_1 locks S_1 and proceeds.

This execution process is the same as the ceiling protocol. This example shows how both protocols avoid the deadlock.

Example-2
$$T_0 = \{\ldots\ldots, P(S_1), \ldots, P(S_2), \ldots, V(S_2), \ldots, V(S_1), .\}$$
$$T_1 = \{\ldots\ldots, P(S_2), \ldots\ldots, V(S_2), \ldots\ldots\ldots\ldots\ldots\}$$
$$T_2 = \{\ldots, P(S_1), \ldots\ldots, V(S_1), \ldots\ldots\ldots\ldots\ldots\}$$

$(P_0 > P_1 > P_2)$
The dependency list is: $DS_2 = [(S_1, P_0)]$

1. T_2 is initiated, and locks S_1.
2. T_1 is initiated, preempts T_2, locks S_2.
3. T_0 is initiated, blocked by S_1, T_2 inherits the priority.
4. T_2 restarts, and releases S_1. T_0 claims S_1, but by the dependency list:
   $DS_2 = [(S_1, P_0)]$
   T_0 is blocked and T_1 inherits the priority.
5. T_1 restarts, then releases S_2.
6. T_0 locks S_1 and S_2, then releases them.

Using the ceiling protocol, T_1 is blocked. The delay of T_0 is only T_2's blocking time. In the case of the dependency protocol, the delay of T_0 is T_2 and T_1's blocking time.

Example-3
$T_0 = \{\ldots\ldots, P(S_0), ., V(S_0), ., P(S_1), \ldots, V(S_1)\ldots\ldots\}$
$T_1 = \{\ldots., P(S_2), .., V(S_2), \ldots\ldots\ldots\ldots\ldots\ldots\ldots\ldots\}$
$T_2 = \{\ldots, P(S_2), .., P(S_1), \ldots., V(S_1), \ldots., V(S_2), \ldots\}$
$(P_0 > P_1 > P_2)$
The dependency list is: $DS_1 = [(S_2, P_2)]$

1. T_2 is initiated, locks S_2.
2. T_1 is initiated, blocked by S_2, T_2 restarts and inherits P_1 and locks S_1.
3. T_0 is initiated, locks S_0 and releases it, and claims S_1, then blocked, T_2 inherits its priority.
4. T_2 releases S_1, T_0 claims S_1, locks it and releases it.
5. T_2 releases S_2, then T_1 is initiated.

By the ceiling protocol, T_0 must be delayed to lock S_0 because T_2 locks S_1 as P_0. This is a new form of blocking introduced by the ceiling protocol called ceiling blocking. The ceiling blocking is unnecessary. The semaphore dependency protocol works similarly to the basic priority inheritance protocol.

Example-4
$T_0 = \{\ldots\ldots, P(S_0), \ldots., V(S_0), \ldots\ldots\ldots\ldots\ldots\ldots\ldots\}$
$T_1 = \{\ldots\ldots, P(S_0), .., P(S_2), \ldots, V(S_2), \ldots, P(S_0), \ldots.\}$
$T_2 = \{\ldots., P(S_0), \ldots\ldots\ldots\ldots\ldots\ldots, V(S_0), \ldots\ldots\ldots\ldots\}$
$T_3 = \{.., P(S_2), \ldots\ldots\ldots\ldots\ldots\ldots, V(S_2)\ldots\ldots\ldots\ldots\}$
$(P_0 > P_1 > P_2 > P_3)$
The dependency lists are: $DS_2 = [(S_0, P_1)]$

1. T_3 is initiated, locks S_2.
2. T_2 is initiated, claims S_0, by the dependency lists, becomes blocked.
3. T_1 is initiated, claims S_0, by the dependency lists, becomes blocked.
4. T_0 is initiated, locks S_0.

This example shows that the priority requirement of the scheduling rule is necessary. The ceiling protocol works as well, because the ceiling priority of S_2 is P_1.

2.5 The Properties of the Semaphore Dependency Protocol

The ceiling protocol uses the maximum dependency, it causes unnecessary blockings compared with the semaphore dependency protocol. It is because the semaphores of the higher tasks, which are not related with the current locking semaphores, block the middle tasks or the higher task itself like example3 called the ceiling blocking. In the semaphore dependency protocol, the semaphores in the dependency list (from the current locking semaphore) only block the higher task.

In the dependency diagram, the active locking lines are described like this:
$$S_n \overset{P_i}{\Longrightarrow} S_m$$
Each line never shares S_n and S_m, which represents that this dependency protocol avoids deadlocks.

2.6 Discussion

Both the ceiling protocol and the semaphore dependency protocol use static semaphore dependency lists. What is the essential work of the dependency-based protocol compared with the basic protocol inheritance?

One assumption is that using dependency lists cause an ordering of the priorities of tasks. If it were true, to have more dependencies is better, which means that the ceiling protocol works well, for example:

$T_0 = \{..........., P(S_1), ...V(S_1), ..\}$
$T_1 = \{......, .P(S_2), ...V(S_2), ...\}$
$T_2 = \{...., P(S_2), ...V(S_2), ...\}$
$T_3 = \{.., P(S_1), ..., V(S_1), ...\}$

The dependency protocol cannot find any dependencies in the example. The ceiling protocol assumes the most dependencies, therefore it causes unnecessary blocking called 'ceiling blocking'. The number of dependencies from the semaphore dependency protocol seem low. For example,

$T_0 = \{......., P(S_1)P(S_2)P(S_3), ..., V(S_3)V(S_2)V(S_1)..\}$
$T_1 = \{....., P(S_3), ..., V(S_3), ...\}$
$T_2 = \{...., P(S_2), ..., V(S_2), ...\}$
$T_3 = \{.., P(S_1), ..., V(S_1), ...\}$

The semaphore dependency protocol does nothing at all. It can be said that in the most simple cases, the priority inheritance works well enough, and in the simple cases, the ceiling protocol works well, then, in more complex cases, the semaphore dependency protocol works better, and so on.

3 The Synchronisation in Multimedia

Despite the usefulness of improving performance by less conservative scheduling, one may question how essential it ought to be to solve these problems in multimedia applications. Both the dependency protocol and the priority ceiling protocol implicitly assume that processes struggle through a thicket of conflicting locks, and it is worth considering whether this is a realistic view of a multimedia computation. This section considers the question "Are the circumstances of multimedia computation really like this?"

The characteristic actions of continuous-media systems are copying streams from source to destination, or maybe several destinations, perhaps with a moderate amount of processing on the way. For example from a disc to a network and from a network to a disc called CMFS (Continuous Media File System) which has been developed at the Computer Laboratory [Jardetzky92] , or a network to a screen. It is not easy to see how application level locks are going to cause conflict between different continuous-media operation. While they are going on these are relatively independent of each other and one should not find reference to shared data structures which need protection by locks. To the extent that locks are needed at all they will be between different stages of processing the same stream they should be harmless. For example, suppose that a video stream is being read from a network, each frame is reformatted by software, and then the reformatted frames are displayed by copying them into a frame buffer. The process containing the reformatting program may need to protect its frame by semaphore while it is being worked on, but this will not give rise to any complicated situations.

We conjecture that if applications do encounter lock conflict then it will usually be to do with system data structures, not application data structures. Application programs will need from time to time to make system calls, which will cause system code to be executed which may itself require the use of semaphores. When taken together with the semaphore needs of the application itself, this is capable of giving rise to the kind of complex locking configuration that can require the use of the rather complicated protocols we have discussed above. We further conjecture that these are usually unnecessary, and if they cause trouble it is very likely because of unsuitable system structure, in particular because of two design faults:

1. Such practices as allocating workspace, specifically buffers, from a shared pool which has to be locked rather than from pre-issued pools which do not. This is an example of trying to over-economise on resources. Operating systems in current use tend to date from times when scarcity of resources, in particular of memory, was a serious constraint. They use allocation mechanisms which although perhaps never causing holdups don't become null in circumstances of plenty - they still use locks, and it is in multimedia applications that the unnecessary serialisation resulting from unnecessary locking is noticed.

Correct design will include such features as arranging that a particular data path, for example from network to screen via a software process, has its own dedicated buffer pool so there will never be any conflict. This will have consequences in the design of network drivers, specifically that they will have to demultiplex early; one cannot use the pool associate with a data path until it is established that a particular path is being used.

2. The practice of running system code at the same priority as the user code which called the system - though priority inheritance inside the system should help with this.

4 Conclusion

Rather than ensuring real-time performance in the presence of complicated configurations of semaphore use by erecting complicated structures such as the priority ceiling protocol and the more elaborate ones we propose to succeed it, we believe that one should look carefully at the source of the complexity. It may be that, contrary to our intuitions, the complexity is intrinsic to the application, as it is in some kinds of database transactions. We doubt this; multimedia applications are structurally simple, and should be supported by simple operating system facilities. Perhaps we should not be talking about extending operating systems to handle multimedia material - rather about simplifying or reducing them.

References

[Burns91] Alan Burns, *Scheduling Hard Real-Time Systems: A Review*, Software Engineering Journal, May 1991, pp116-128.

[Jardetzky92] Paul Jardetzky, *Network File Server Design for Continuous Media*, PhD Thesis, Computer Laboratory, University of Cambridge, 1992.

[Khanna92] Sandeep Khanna, Michael Sebree, John Zolnowsky, *Realtime Scheduling in SunOS 5.0*, The proceeding of USENIX 92 Winter, pp375-390.

[Sha90] Lui Sha, Ragunathan Rajkumar, John P. Lehoczky, *Priority Inheritance Protocols: An Approach to Real-Time Synchronisation*, IEEE Transactions on Computers, Vol. 39, No. 9, September 1990.

Session 2: Multimedia On-Demand Services

Chair: Jonathan Rosenberg, Bellcore

The session "Multimedia On-Demand Services" comprised four presentations concerned with systems that provide network delivery of continuous media streams (typically video) to multiple clients on demand. Such systems consist of one or more servers, responsible for the storage and delivery of the media streams, and one or more clients, which request service from the server(s) and receive and present the media. These systems are like networked file servers in basic architecture and functionality, but have significantly different behavioral requirements due to the temporal demands of continuous media streams.

Merely satisfying these requirements presents a challenge. But for economic viability, it is crucial that the system architecture scale appropriately to support multiple clients. The four papers in this session are particularly concerned with this issue of scalability. The first three papers are largely focused on the issues of laying out and retrieving data from disks. These are important issues to resolve since this is where it all begins, in some sense. The final paper describes a series of investigations looking at an entire end-to-end system.

The first presentation was given by Dilip Kandlur from the IBM Watson Research Center ("Design and Analysis of a Grouped Sweeping Scheme for Multimedia Storage Management", Philip S. Yu, Mon-Song Chen and Dilip D. Kandlur). The goal of this work was to investigate disk scheduling schemes that would scale to serve multiple clients, while reducing the worst-case bound on the time required to serve two consecutive requests to the same stream.

Dilip pointed out that much early work on disk layout and retrieval used fixed order scheduling and focused on increasing the size of retrieval units to decrease the effects of seek times. This approach, however, can lead to [reformance degradation or increased buffer requirements. More recent proposals have suggested the use of an elevator-type SCAN algorithm, which orders retrievals so that a single end-to-end disk sweep can retrieve data for all clients. This scheme has the disadvantage of increased buffering requirements and increased startup delay (due to the fact that consecutive retrieval units for a stream can be widely separated in retrieval order).

Dilip and his colleagues suggest a new approach, known as Grouped Sweeping Scheme (GSS). GSS divides the streams into a number of groups, g. Each service cycle then consists of g sweeps across the disk, each sweep retrieving units from one group. GSS is general and subsumes other schemes, with appropriate values of g. For example, the typical round robin scheme is GSS with g=n, the number of streams. The SCAN algorithm can be described by GSS with g=1.

Dilip explained an algorithm that can determine the parameters to minimize the server buffer requirements. A search algorithm can be used to find the optimum number of groups (to minimize buffer requirements). Given a particular g, the algorithm can also find the appropriate number of blocks per stream to retrieve on sweeps.

Dilip also showed that GSS can be used dynamically to determine whether a new stream request can be admitted and to perform the admission. GSS can likewise handle the termination of a stream.

Dilip then summarized the advantages of GSS over existing schemes. These include improved startup time, because a new request can be admitted at the beginning of a group sweep. There is no need to wait for the next cycle, as in the SCAN scheme. This also leads to reduced buffer space requirements. In addition, GSS can handle non-time critical data requests in a convenient manner, by using available time within a group's sweep. In addition, Dilip claimed that GSS may be able to provide improved response for interactive streams. The idea is to assign an interactive streams to more than one group, which should decrease the response time of the server for those streams.

Ongoing work will include generalization of the GSS scheme to handle streams with different service rates (the current scheme only handles streams with the same rate).

The second presentation was by Harrick Vin of the University of California, San Diego ("Admission Control Algotirhms for Multimedia On-Demand Servers", Harrick M. Vin and P. Venkat Rangan). Like the last paper, this work focused on disk scheduling. The goal of this research, however, was to investigate schemes for admission control: how to know whether to admit a new stream request.

An admission policy is, of course, tied intimately with the disk allocation scheme, so Harrick began by presenting a taxonomy of allocation schemes. In unconstrained allocation, consecutive blocks from a stream are laid out without regard to position. While this technique is the easiest to implement, it can lead to large seek delays, making continuous retrieval difficult. Contiguous allocation lays out the blocks of a stream contiguously. Although this makes it efficient to retrieve streams continuously, it is more difficult to implement, can lead to fragmentation and is problematical when streams are edited.

Harrick suggested constrained allocation as a viable alternative. In this scheme, blocks of a stream are placed so as to provide guaranteed seek times. This provides some of the advantages of both of the other schemes.

Harrick also discussed a categorization for block retrieval policies. Deadline based schemes retrieve blocks in order of the times they must be received. Placement based schemes retrieve blocks based on their placement on the disk (to avoid unnecessary seek delays). In addition, a policy can use a local schedule (only considering blocks within a single stream) or global (considering all streams).

The combination of local and global crossed with deadline or placement based leads to four policies. Harrick and his colleague have developed admission policies for all four policies and details are given in the paper.

In addition, Harrick suggested a new service algorithm, QPMS (quality proportional multi-client servicing) that, unlike many other schemes, does not read the same number of blocks per stream on each service cycle. Instead, QPMS reads a number of blocks based on each stream's playback rate.

Harrick then described analyses they had carried out, which revealed several interesting facts:

- If contiguous allocation is used, then a global placement policy incurs lower seek delays and can serve more clients (than a local deadline policy).
- For constrained allocation, a local deadline policy is better (can serve more clients) than global placement. Contiguous allocation can serve more clients than con-

strained allocation (regardless of service policy). But contiguous allocation can lead to fragmentation and copying problems, so Harrick suggested this scheme be used only for read-only servers.

- QPMS can serve more clients than a round robin policy.

The third presentation was given by Phillip Lougher of Lancaster University in the UK ("The Design and Implementation of a Continuous Media Storage Server", Phillip Lougher and Doug Shepherd). Phillip discussed a prototype system, CMSS (Continuous Media Storage Server), built to investigate the storage and retrieval of continuous media. Although Phillip discussed disk layout and retrieval, this paper takes a system-wide view, including discussions of system architecture and real-time scheduling within the system.

CMSS has two interfaces: the client interface, used to allow applications to request and control service, and the data interface, by which data is transferred to and from the system. The data itself is striped onto multiple disks to provide high throughput. CMSS divides this data into two kinds, stored and manipulated separately. The two kinds are media data, the content of the continuous media (typically video), and file metadata, which can be thought of as directory information about the stream (treated as a file).

Real-time retrieval in CMSS is handled by the CM File Service. Phillip discussed the issues that must be addressed in this module, including the granularity of disk blocks, the disk striping method, pre-fetching (and buffering) to maximize concurrency, and deadline calculation.

CMSS handles disk block granularity by normalizing disk blocks across streams, so that each block represents the same amount of playback time (since streams may have different playback rates). This simplifies the real-time scheduler and gives streams with high playback rates more disk space (and, therefore, system resources). This also avoids the problem of clients with slow streams having to deal with increasing amounts of backed-up data.

The CMSS real-time scheduler is a split level scheduler, providing two levels of service: hard guaranteed streams and soft guaranteed streams. Details on the hard guaranteed scheduling can be found found in the paper (soft guarantee scheduling is not covered), which also provides an analysis of the scheduling constraints used.

QMSS is part of a larger system built as a networked multimedia infrastructure. The enclosing system provides a rich environment for studying various network architectures and protocols.

The final presentation was by Kathleen Nichols of Apple Computer ("Performance Studies of Digital Video in a Client/Server Environment", Kathleen M. Nichols). Kathleen presented a series of end-to-end studies of networked video servers to identify potential bottlenecks in video service systems.

The studies were based on Apple's QuickTime architecture and assumed the clients and server used a simple request-response protocol for each data unit. Studies were performed by making system measurements, by analysis and by simulation.

The first study was of a system using three kinds of storage: a hard disk, a SCSI-based silicon disk and a RAM disk. Throughputs were measured at the fileserver, the client and the network. At the frame rates and sizes being used, these measurements showed that the network was not a bottleneck. The system used software decompression and, matching intuition, this turned out to be the bottleneck in the system.

Kathleen used a closed queueing network model to analyze a video service system. The model used two servers: one for the operating system and disk and one for the network (protocol overhead and transmission time). The analysis considered several values for overheads and error rates (handled by retransmissions). These analyses revealed that the service time per request was the system bottleneck. In addition, the analysis also showed that the system could serve more clients by using better I/O (such as DMA, which allows the CPU to execute in parallel).

Lastly, Kathleen used simulation to investigate effects of changing protocol and network parameters (such as packet size and inter-packet delivery times). These simulations showed the relationship among these parameters, server utilization and response times.

Kathleen stated that the bottom line from this work was to show the importance of using an end-to-end system when investigating video servers.

Design and Analysis of a Grouped Sweeping Scheme for Multimedia Storage Management

Philip S. Yu, Mon-Song Chen, and Dilip D. Kandlur

IBM T. J. Watson Research Center
30, Saw Mill River Road, Hawthorne, NY 10532, USA.

Abstract. This paper presents an efficient grouped sweeping scheme (**GSS**) for disk scheduling to support multimedia applications. The GSS scheme provides a formulation for balancing the two conflicting requirements: on the one hand we want to reduce the amount of seek delay, while on the other hand the worst case bound on the time between two consecutive services to the same request stream needs to be minimized. The GSS approach leads to a general formulation covering a family of disk scheduling schemes. Two previously proposed approaches, a fixed order round-robin scheme and a more sophisticated elevator type scheme, are special cases of the GSS formulation. Our analysis shows that by properly setting the grouping factor the optimized GSS scheme can lead to substantially lower buffer requirement, particularly when compared with that of the round-robin scheme. We develop an efficient procedure to determine the optimum value of the grouping factor. We also incorporate the effect of disk arrays in the GSS formulation and discuss how GSS can be used in dynamic settings such as to provide interactive service in an integrated media environment.

1 Introduction

The requirements of a storage management system for multimedia information are substantially different from that of a conventional computer file system. For example, the storage system for a movie-on-demand service system needs to be able to hold a large number of movies at approximately 1.5 GB per two-hour movie and provide many simultaneous retrievals for viewing by subscribers with possibly different time shifts. In another example, a LAN based multimedia server needs to provide archival services for a large set of video clips of varying length (e.g., from 10 seconds to 10 minutes or longer). Such services also have to support the dynamic retrieval of video clips for in-time presentation for the user's multimedia presentations.

As evident from these examples, the two most demanding requirements of a multimedia storage system are capacity and sustained retrievals. The state-of-the-art storage technology that can most effectively satisfy these requirements is still the Direct Access Storage Device (DASD), commonly called hard disk. This is because fast solid-state memory is too expensive when large capacity is required, and large capacity systems, such as tapes and optical jukeboxes, do not have sufficiently high throughput to sustain a large number of simultaneous retrievals.

DASD, however, has the fundamental problem of long disk arm seek time, which can be as large as 30 ms. If not properly managed, the long gaps due to disk arm re-positioning significantly degrades the video quality. This problem can be addressed in three different angles:

retrieval unit Reduce the fraction of disk arm seek time by increasing the amount of read time between seeks, i.e., increasing the size of retrieval unit.

scheduling Reduce the fraction of disk arm seek time by properly scheduling the read sequence.

disk array Increase the throughput of the storage system through disk array technology.

Previous work has generally focused on increasing the size of the recording unit to reduce the effect of disk arm movement while maintaining fixed-order scheduling [1]. In this approach, a request from each stream is served in a fixed order in each cycle of service. Since the blocks of different streams are scattered on the disk, the stream-based fixed-order servicing can incur a significant amount of seek (or disk arm movement) delay, resulting in performance degradation or an increase in the buffer requirement.

To alleviate this problem, a recent work [2] proposed the use of an elevator-type scheduling scheme to further amortize the effect of long disk arm movement. This scheme, which is an adaptation of the classical SCAN algorithm [3], can effectively support a large number of simultaneous streams. In this approach, the ordering of retrievals is arranged according to their position on the disk in such a way that the disk arm picks up data as it scans from one end of the disk to the other end. In this arrangement each service cycle requires only one maximum disk arm movement independent of the number of retrieval streams.

Since the ordering of service of a particular stream may be changed from the first to the last in two consecutive cycles, up to two cycles of retrieved data must be buffered. Playout of a stream can begin only at the end of the first service cycle. The round-robin scheme, on the other hand, requires the buffering of data from only one service cycle, and playout of a stream can begin at the end of its first retrieval. It is noted, however, that the cycle time using the SCAN approach is typically smaller than that of the round-robin approach [2].

The goal of this paper is to examine the effects of scheduling and disk arrays on multimedia storage management systems. For scheduling, we discuss the pros and cons of the two well-known algorithms, i.e., FIFO and SCAN, and then propose a general formulation, which will be referred to as **Grouped Sweeping Scheduling (GSS)**, to provide an optimization framework. We then incorporate the effect of the disk array in the analysis and optimization.

The structure of the paper is as follows. In Section 2, we present the general formulation, called the grouped sweeping scheme, for the homogeneous situation, in which all video streams have the same data rate. We then extend the formulation to include disk arrays, and present the analysis, optimization, and implementation considerations of this scheme. Numerical examples of several realistic cases are presented in Section 3. Concluding remarks are provided in Section 4.

2 The Grouped Sweeping Scheme

In this section, we present the **grouped sweeping scheme (GSS)** for efficient disk scheduling to support multi-media applications to minimize the buffer requirement. We do not make any particular assumption on the layout of the data blocks for each stream on disk. (Since data blocks from different streams need to be interleaved dynamically at playtime, there is no particular advantage of putting the data blocks from each stream on contiguous cylinders. In any case, this would not affect the analysis.) To simplify the presentation, we use the case of homogenous media streams to demonstrate the main ideas.

2.1 The Scheme

The issue here is that on the one hand we want to reduce the amount of seek delay, while on the other hand the worst case bound on the times between two consecutive services to the same request stream needs to be minimized. The basic idea is to strike a balance between these two conflicting requirements. The goal is to design a disk scheduling approach which is general enough to cover the whole range of traffic conditions from a very low request load to high request load by providing a few tuning parameters. We need to serve multiple request streams during each scan to reduce the number of seeks, but the number served must be controlled to prevent the worst case bound on inter-service times from deteriorating significantly. The proposed **GSS** for disk scheduling addresses the above problems, by dividing the number of media streams, n, needed to be served into g groups. Thus g is the tuning parameter whose value is determined below to optimize the buffer requirement. The assignment of streams to groups is arbitrary in the homogeneous case considered in this section. In each cycle of service, we scan g times through the disk tracks from one end to the other. In each scan, at most $\lceil n/g \rceil$ streams per group is served.

The GSS approach provides a general formulation to cover a family of disk scheduling schemes. In muti-media file systems, the traditional approach based on the fixed order round-robin service scheme is a special case of GSS with $g = n$. Similarly, the elevator-type (or SCAN) service scheme is also a special case of GSS with $g = 1$.

Each track contains b blocks of data. In each service cycle, k blocks are read for each requested stream. Let t_r to be the disk rotation time which is also the time to read a track, t_M be the maximum seek time, t_m be the minimum seek time (i.e. seek time between adjacent cylinders), and B_m be the number of bytes in a block. Furthermore, let B_b be the buffer size, T_c be the maximum time to complete a service cycle, and T_p be the duration of playing back one block of data. We also assume that n is chosen such that $nt_r \leq bT_p$.

We now estimate the total buffer requirement. For each stream a total of kB_m bytes of buffer is required. Furthermore, buffer space is needed to temporally store the blocks from each group after they are read out from the disk but before they can be moved into the buffer designated for each stream. This staging buffer needs to accommodate the blocks from $\lceil n/g \rceil$ media streams and has a size of $\lceil n/g \rceil kB_m$ bytes. Thus,

$$B_b = (n + \lceil \frac{n}{g} \rceil)kB_m \tag{1}$$

The maximum time to complete a service cycle is the time to position the disk arm and read k blocks for each stream. For the GSS scheme, the disk positioning time is a sum of the track to track seek time for each stream and the time to make g complete scans over the disk. Define

$$t_s = t_M - t_m. \tag{2}$$

Then,

$$T_c = n(\lceil \frac{k}{b} \rceil + \delta)t_r + nt_m + gt_s \tag{3}$$

Note that $\delta \in [0, 1]$ is a correction factor to handle the case when an additional track may need to be accessed.

The value of k needs to be chosen such that the media stream can be played continuously without interruption

$$k \geq \frac{T_c}{T_p}$$
$$\geq \frac{n(\lceil \frac{k}{b} \rceil + \delta)t_r + nt_m + gt_s}{T_p} \tag{4}$$

For a given g, we can solve for the optimum k value that minimizes B_b in Equation 1 subject to the constraint in Equation 4. Using a simple search procedure we can find the optimum g value that minimizes the buffer requirement. In the rest of this section it is assumed that k is chosen such that $\delta = 0$, so the δ term is dropped from the equation.

2.2 Extension to Disk Array

Disk arrays have been recognized as an effective way of improving the performance of DASD based storage systems. In particular, the concept of **Redundant Array of Inexpensive Disks (RAID)** has attracted a lot of attention in meeting the high performance requirements in, for example, large transaction processing and high-end real-time systems. By employing techniques such as striping the throughput and latency of the storage system can be substantially improved. Sophisticated redundancy and error protection schemes, however, need to be implemented to maintain the same level data integrity and mean-time-to-failure (MTTF) as that of the single disk.

To extend the above formulation to include a disk array environment, let D be the number of disks in an array. Assume that data are striped across disks and the same number of blocks, say k, can be read out from each disk simultaneously in $\lceil \frac{k}{b} \rceil t_r$. Then the service cycle time, i.e., T_c, required in reading out Dk blocks of data for each stream from an array of D disks is the same as the T_c in Eq. 3. Eq. 1 on the buffer requirement becomes

$$B_b = (n + \lceil \frac{n}{g} \rceil) Dk B_m, \tag{5}$$

The constraint on playing continuously without interruption becomes

$$Dk \geq \frac{T_c}{T_p}. \tag{6}$$

It is interesting to note that disk arrays for multimedia storage systems may be in many cases subject to less stringent redundancy requirements than commonly seen in RAID systems. In RAID, redundancy is built-in in such a way that on-line recovery is provided upon any data corruption. In multimedia storage systems, however, on-line recovery is much more difficult, if feasible at all, because of the strict isochronous requirements. Hence, it may be more effective to eliminate this complicated function by providing backup. In movie-on-demand, for example, it is assumed that there is always a master or backup copy of each movie to fall back to whenever necessary.

The same argument is applicable to the consideration of MTTF. As the number of disks is increased in an array and data are striped across disks, MTTF of the storage system decreases, i.e., MTTF is proportional to the reciprocal of the number of disks. In multimedia storage systems, based on the same argument that there is always a master copy backed up somewhere, MTTF is not critical. In combination, disk arrays are an effective technique to increase the performance of DASD based multimedia storage systems. In addition, there may often be the added advantages of trading frequent and reliable back up support for the complexity in the design of redundancy and error recovery mechanisms that are common in RAID systems.

2.3 Solution Scheme

For ease of presentation, the constraint in Eq. 4 can be represented by a function $f(k)$, which must be no smaller than zero, as

$$f(k) = kT_p - nt_r \lceil k/b \rceil - nt_m - gt_s \geq 0.$$

The optimum solution can be computed by the following procedure. For each group value g, find the optimum k which minimizes the buffer requirement while satisfying the playtime constraint. Let k_g^* and B_g^* denote the optimum value of k and corresponding minimum buffer requirement B, respectively. The minimum buffer requirement is then the smallest B_g^* for all possible g between 1 and n.

The procedure of finding k_g^* for a given g is as follows.

1. Compute the initial value of k by

$$k = \lceil \frac{gt_s + nt_m}{bT_p - nt_r} \rceil b$$

2. If $f(k - 1) \geq 0$, $k = k - 1$ and repeat this step.
3. Otherwise, $k_g^* = k$ and stop.

Theorem 1 *For a given g, the above procedure produces the optimum value of k, i.e., k_g^*, which minimizes B_b in Eq. 1 subject to the constraint $f(k) \geq 0$.*

Proof: Since B_b is a linear function of k, k_g^* is essentially the smallest k that satisfies $f(k) \geq 0$. To prove the optimality of the procedure, we first examine the characteristics of $f(k)$ function, which is depicted in Figure 1. When $k = mb$, for integer m, $f(k)$ is linear to m, i.e., $f((m + 1)b) - f(mb) = bT_p - nt_r$. When $k = mb+l$, where $0 < l < b$, $f(k)$ is linear to l, i.e., $f(mb+l+1) - f(mb+l) = T_p$.

The initial value of k, as computed by Step 1 of the procedure, is mb where m is the smallest integer such that $f(mb) \geq 0$. From Figure 1, it is clear that the search in Step 2 finds the smallest value of k which satisfy $f(k) \geq 0$ and hence the optimum value of k, i.e., k_g^*.

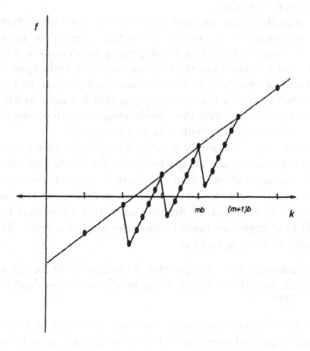

Fig. 1. Plot of the function $f(k)$.

2.4 Dynamic Operation

The description of GSS thus far has been for a *static* case in which there are a fixed number n of multimedia streams to be supported. In addition to minimizing the buffer requirement, GSS has several other advantages. Consider the dynamic case in which requests for streams arrive and depart during the operation of the system. Compared with the SCAN type service scheme, GSS could also reduce the initial delay to start the playback to just the service time for a group. The GSS algorithm can be used to pick an operating point to support n_{max} streams, depending upon the available buffer space and the disk throughput. The dynamic algorithm will operate within these limits and reject requests for streams when the number of streams in the system reaches n_{max}. The actions taken to process a request are:

1. Check whether $n < n_{max}$. If not, reject the request.
2. Assign the stream to the nearest group that has cardinality $< \lceil n_{max}/g \rceil$.

The *response time* for a request is defined as the time from the arrival of a request to the time when the playback starts. In the SCAN algorithm a new request can be admitted only at the beginning of a cycle and playback can begin at the end of the cycle, so the average response time for a request that is admitted into the system is 1.5 times the current cycle time and the minimum response time is the current cycle time.

The GSS algorithm can improve the response time since streams can be admitted at the beginning of a *group*. The minimum response time is T_c/g and occurs when the request arrives at the beginning of a group and is assigned to that group. It can be seen that the response time depends upon the group to which the stream is assigned, so the allocation scheme has to pick the group closest (in playout order) to the current group that has slots available. In order to reduce the response time with this greedy assignment, it is essential to ensure that open slots are evenly distributed among the groups.

When a stream departs from the system, it may be necessary to rearrange the remaining streams among the groups to better distribute open slots. Consider the problem of moving a stream from one group to another, for example, from g_i to g_j. Playout constraints have to be satisfied even during the transfer to prevent any disruption in service. Let T_c^ℓ denote the current cycle that is in progress, and assume that changes are made at the beginning of a cycle. The procedure to move the stream from g_i to g_j is:

1. Allocate the stream to g_j and prefetch k blocks of data for the stream in T_c^ℓ.
2. If $i > j$ deallocate the stream from g_i in T_c^ℓ, otherwise deallocate it in the next cycle $T_c^{\ell+1}$.

Note that the second condition is necessary to maintain continuity in the playback. Prefetching is always done in units of k blocks so as to preserve the read alignment and prevent additional rotational latency (the δ factor). Consequently it is not possible to deallocate all the buffers assigned to the stream in g_i, since only some of the buffers may be played out by the time new buffers are available for g_j. The additional amount of buffer space required for a stream that is moved from g_i to g_j is:

$$B_{add} = \begin{cases} B_m \lceil k(g-j+i)/g \rceil & \text{if } j > i \\ B_m \lceil k(i-j)/g \rceil & \text{otherwise} \end{cases}$$

Hence, in order to reduce the excess buffer requirement, g_j must be chosen to be far from g_i in the playout sequence, i.e., $(j-i) \bmod g$ is close to g. When $k > b$, it is possible to use another technique to reduce buffer requirement. In Step 1 above, the prefetch amount can be changed to an integral number of tracks, h_i, such that $h_i \geq \lceil k(j-i)/g \rceil$. This number of tracks is just sufficient to meet the continuity requirement for the stream. Additional buffers are needed even in this case because $k(j-i)/g$ may not be an integer. The additional buffer requirement in this case is one track.

Integrated Service The grouping concept can be extended to provide better services to the interactive users in an integrated media service environment. Interactive users can get more frequent services in a service cycle by assigning more than one group for interactive service. The worse case bound on inter-service times for the interactive users can thus be reduced.

Data traffic can be accommodated into the workload by assigning it to groups which have open slots. The additional time required to service a request to read k_d contiguous blocks is $T_d = t_m + \lceil k_d/b \rceil t_r$. Assuming that there are n_i streams assigned to the group g_i, the data request can be serviced in g_i if

$$n_i \lceil k/b \rceil t_r + n_i t_m + t_s + T_d \leq T_c/g \tag{7}$$

This analysis assumes that data traffic is handled using bandwidth that is not used by stream traffic. In order to guarantee bandwidth for data traffic, it is possible to allocate some streams (in different groups) for data traffic.

3   Numerical Results

In this section some numerical examples are presented that illustrate the performance of the GSS approach. The examples are based on the characteristics of a state-of-the-art high capacity SCSI disk drive (see Table 1). Given these characteristics, the throughput of the disk is known and the number of streams that can be supported then depends upon the playout rate required for a stream.

The first set of examples considers an MPEG-1 stream which has a playout requirement of 1.5 Mbits/second. In this case, the number of streams that can be supported by the disk is bounded by the disk read rate to 19. Also, in this example k was chosen to be an integral multiple of b, so δ was always zero.

Figure 2 shows a comparison of the buffer requirement for the different algorithms to service n streams. It can be seen that optimum g picked using GSS results in a substantial reduction in the buffer requirement over the fixed order

52

Disk Parameter	Value
t_m	2.0 msec.
t_M	27.0 msec.
t_r	11.1 msec.
B_m	0.5 Kbytes
b	81 blocks

Table 1. Characterisitcs of the disk drive used in the examples.

round-robin scheme as n approaches 19. The round robin scheme performs better than the SCAN scheme only when the number of streams is small. For larger values of n, the buffer requirement increases rapidly thereby limiting the number of streams that it can support. The GSS technique bridges between round robin and SCAN and over the range of n, it has a lower buffer requirement than either of the schemes. The values of g chosen by GSS for this example are shown in Figure 3. When n is large the value of g chosen by GSS, g^*, tends to 1, which is the same as the SCAN scheme.

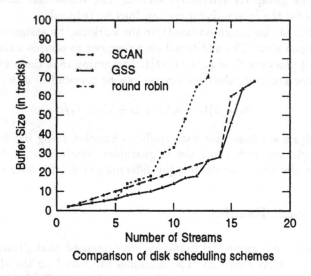

Comparison of disk scheduling schemes

Fig. 2. Buffer requirement for different disk scheduling algorithms.

The second example is used to illustrate the effect of choosing k in terms of blocks, rather than tracks. This has the potential of reducing the buffer requirement because the granularity of the access unit is reduced. In this example the playout requirement is taken to be 768 Kbits/second, which is twice the H0 rate of ISDN. The maximum number of streams that can be supported is 38 in this case.

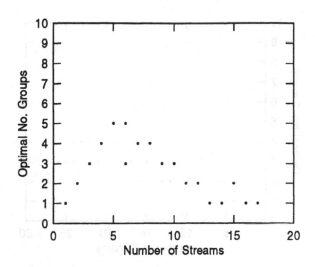

Fig. 3. Variation of g^* with number of streams.

Figure 4 shows the buffer requirement, and it is observed that the requirement for the three algorithms follows a pattern that is similar to the previous example. GSS performs better than SCAN and they both perform much better (by an order of magnitude) than round robin. Figure 5 shows that, as before, g^* approaches 1 as n approaches n_{max}. However, g^* is chosen to be 1 even for some small values of n. This is because k can now be chosen such that T_c is small and B_b is minimized even when $g = 1$.

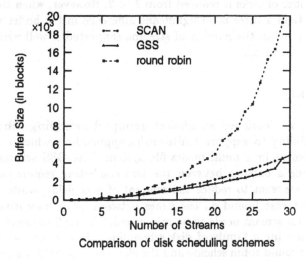

Comparison of disk scheduling schemes

Fig. 4. Buffer requirement for different disk scheduling algorithms.

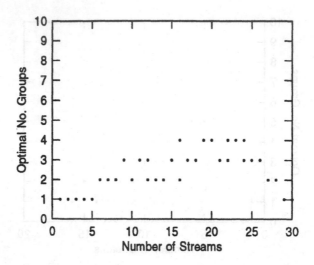

Fig. 5. Variation of g^* with number of streams.

The last example shows the effect of using a disk array on the buffer requirement. The disks in the array have the same characteristics as in the previous examples. Figure 6 shows the results where the number of MPEG-1 streams to be serviced is fixed at 30 while the number of disks in the array varies. The figure shows that the optimal buffer requirement picked using GSS is always lower than the buffer requirement for the other two schemes. When the number of disks is small, the total throughput is small and GSS does not have much flexibility in choosing values for g and k. Hence, there is a jump in buffer requirement when the number of disks is reduced from 3 to 2. However, when the number of disks is more than 3 there is no significant difference in the buffer requirement. This is the case when the number of streams supported is well within the total throughput of the array.

4 Summary

In this paper, we developed an efficient **grouped sweeping scheme (GSS)** for disk scheduling to support multi-media applications which minimizes the buffer requirement in a multi-media file system. The GSS scheme provides a means to strike a balance between the two conflicting requirements that on the one hand we want to reduce the amount of seek delay, while on the other hand the worst case bound on the times between two consecutive services to the same request stream needs to be minimized. The GSS approach is a general formulation covering a family of disk scheduling schemes, and we showed that the fixed order round-robin scheme and the elevator-type SCAN scheme are both special cases of the GSS.

We have presented the GSS for homogeneous request streams and developed an algorithm for computing the optimal number of groups and the buffer require-

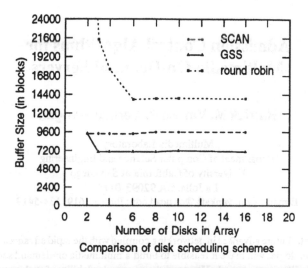

Comparison of disk scheduling schemes

Fig. 6. Buffer requirement for disk array example.

ment. Our analysis and examples showed that the optimal group values chosen by GSS lead to substantially lower buffer requirement as compared to the round-robin scheme. We also presented the applicability of GSS to reduce the initial delay of starting playout of the stream, and for integrating data requests with stream requests. This can be used to provide better support to interactive users in an integrated media service environment.

We are currently working on generalizing our scheme to incorporate heterogeneous media streams with different playout requirements. In this case also, we are investigating a scheme for finding the optimal number of groups and the assignment of streams to groups.

References

1. P. V. Rangan and H. M. Vin. Designing File Systems for Digital Video and Audio. Proc. of 12th ACM Symposium on Operating Systems, 1991.
2. D. D. Kandlur and M.-S. Chen and Z.-Y. Shae. Design of a Multimedia Storage Server. IBM Research Report, June 1991.
3. J. L. Peterson and A. Silberschatz. Operating System Concepts, Second Edition. Addison–Wesley, 1985.
4. R. Graham. Bounds on Multiprocessing Timing Anomalies. SIAM Journal of Computing, Vol. 17, pp. 416–429, 1969.
5. H. M. Vin and P. V. Rangan. Designing a Multi-User HDTV Storage Server. UCSD Technical Report CS92–225, January 1992.

Admission Control Algorithms for Multimedia On-Demand Servers

Harrick M. Vin and P. Venkat Rangan

Multimedia Laboratory
Department of Computer Science and Engineering
University of California at San Diego
La Jolla, CA 92093-0114
E-mail : {vin, venkat}@cs.ucsd.edu; Phone: (619) 534-5419

Abstract. Future advances in networking coupled with the rapid advances in storage technologies will make it feasible to build a multimedia on-demand server on a metropolitan-area network. However, the economic viability of such a multimedia on-demand server is dependent on the ability to amortize its operational costs over a large number of clients. In this paper, we present a taxonomy of policies for servicing multiple clients, and propose round robin and quality proportional *admission control algorithms* for determining whether a retrieval request from a new client can be admitted without violating the real-time requirements of any of the clients already being serviced. We evaluate the performance of various servicing policies, and show that they are an order of magnitude scalable compared to straightforward multiplexing techniques such as servicing one subscriber per disk head.

1 Introduction

Recent advances in digital audio and video processing technologies coupled with the availability of high bandwidth networks and large capacity storage devices will make it feasible to build multimedia on-demand services over metropolitan-area networks [11, 12]. The architecture of such a multimedia on-demand service consists of a multimedia on-demand server connected to display sites belonging to clients via a high-speed network subsystem. The multimedia on-demand server (which we will refer to as *multimedia server*, for brevity) digitally stores media information such as entertainment movies, educational documentaries, advertisements, etc., on a large array of extremely high-capacity storage devices that are permanently on-line. Clients can select a multimedia object, and request its retrieval for real-time playback on their display sites. The multimedia server, if it has the necessary resources (such as service time and buffer space), services the client's request by transmitting the chosen multimedia segment to the client's display site(s). The multimedia server may also permit clients to record or edit media objects, thereby subsuming the functions of VCRs, videotapes, audio recorders, etc.. Furthermore, such multimedia servers can serve varying sizes of clientele: from individual households to entire neighborhoods, and from commercial organizations and educational institutions to national services.

Whereas the above architectural vision of a multimedia server is technologically feasible within the next several years [10], the economic viability of such a multimedia server is critically dependent on the ability to amortize its operational costs over a large number of clients. Consequently, an important question that need to be addressed in designing such a multi-client multimedia server is: how can multiple retrieval requests from multiple clients be serviced simultaneously by the same multimedia server? The design of a high-performance storage server that can satisfy a large number of multimedia retrievals from multiple clients is the subject matter of this paper.

Most of the multimedia services that are being built or proposed make use of analog transmission and storage of video and audio by employing cable-TV style distribution network and crossbar switches for routing [6, 8]. On the other hand, multimedia services that are indeed digital and integrated into distributed computing systems, have focused mainly on still images and/or audio [1, 4, 7]. Recently, Anderson et al. [2] and Gammell et al. [3] have described file system designs for supporting audio playback, but they do not address multi-user video on-demand services. A model for the design of a file system for storing real-time video and audio streams on disks have been presented by Rangan and Vin [9]. A qualitative proposal for a video on-demand service is presented by Sincoskie in [12]. However, admission control algorithms and policies for servicing multiple client requests have remained relatively unexplored.

In this paper, we address the problem of a multimedia server servicing multiple clients simultaneously. We propose a taxonomy of policies for servicing multiple clients, and propose round robin and quality proportional *admission control algorithms* for determining whether a new client can be admitted without violating the real-time requirements of any of the clients already being serviced. We evaluate the performance of various servicing policies, and show that they are an order of magnitude scalable compared to straightforward multiplexing techniques such as servicing one subscriber per disk head.

The rest of the paper is organized as follows: A framework for storing digital continuous media is presented in Section 2. The problem of servicing multiple retrieval requests simultaneously is formulated in Section 3, and admission control algorithms are described in Section 4. Performance of various admission control policies is evaluated in Section 5, and finally, Section 6 summarizes our results.

2 A Framework for Storage of Digital Continuous Media

Digitization of video yields a sequence of frames, and that of audio yields a sequence of samples. We refer to a sequence of continuously recorded video frames or audio samples as a *Strand*. A multimedia server must divide video and audio strands into blocks while storing them on a disk. Continuous playback of media strands requires that the time for retrieving a media block of a strand from disk does not exceed the media block's playback duration.

Most existing storage server architectures employ *unconstrained allocation* of blocks on disk. In such storage servers, reserving computational cycles to meet real-time requirements is not sufficient to support continuous retrieval of media strands. This is because, separations between blocks of a strand may not be constrained enough to guarantee bounds on seek and rotational latencies incurred while accessing successive blocks of the strand.

At the other end of the spectrum, *contiguous allocation* of media blocks guarantee that successive blocks of a strand can be retrieved without incurring any seek or rotational latencies. However, contiguous allocation of media blocks is fraught with inherent problems of fragmentation, and can entail enormous copying overheads during insertions and deletions. Thus, contiguous allocation of media blocks, although suitable for read-only storage systems (e.g., compact discs, CLVs, etc.), is not viable for interactive, read-write storage systems.

Constrained allocation, on the other hand, maintains the access time of media blocks within the real-time playback requirements of strands by bounding the separation between successive media blocks on disk. There are two questions that need to be answered in constrained allocation of media blocks on disk: (1) What should the size of the blocks (i.e. the *granularity*) be? and (2) What should the separation between successive blocks (i.e. the *scattering*) of a strand be? Whereas granularity can be determined using the available buffer space at display sites, scattering can be derived using the requirements of continuous playback of media strands. Table 1 defines the symbols for these parameters for digital video (which is the most demanding medium with respect to performance and storage space requirements), using which, it can be seen that the playback duration of a media block is given by $\frac{\eta_{vs}}{\mathcal{R}_{vp}}$. Retrieval at media playback rates requires that the total delay to retrieve each media block from disk (given by $l_{ds} + \frac{\eta_{vs}*s_{vf}}{\mathcal{R}_{dr}}$) be bounded by the its playback duration:

$$l_{ds} + \frac{\eta_{vs} * s_{vf}}{\mathcal{R}_{dr}} \leq \frac{\eta_{vs}}{\mathcal{R}_{vp}} \tag{1}$$

Symbol	Explanation	unit
$\mathcal{R}_{vp}$	Video playback rate	frames/sec
$\mathcal{R}_{dr}$	Disk data transfer rate	bits/sec
η_{vs}	Granularity of video storage	frames
s_{vf}	Size of a video frame	bits/frame
l_{ds}	Disk scattering	sec
τ	Service time per round	sec
T	Number of tracks	tracks
l_{seek}	Seek latency	sec
l_{rot}	Rotational latency	sec
B	Number of blocks per track	blocks

Table 1. Symbols used in this paper. A *display unit* represents a frame for video, and a sample for audio

Whereas both contiguous and constrained allocation of media blocks of a strand ensure continuous retrieval of that single strand, a multimedia server, in practice, has to process requests from several clients simultaneously. A simple mechanism to guarantee that the real-time requirements of none of the clients are violated is to dedicate a disk head to each client, which, however, limits the total number of clients to the number of disk heads. On the other hand, if the data transfer rate of the disk is higher than the requirements of a single client, then the number of clients that can be serviced simultaneously can be significantly increased by multiplexing a disk head among several clients. However, given the maximum rate of disk data transfer, the multimedia server can only service a limited number of clients. Hence, a multimedia server must employ admission control algorithms to decide whether a new client can be admitted without violating the continuity requirements of any of the clients already being serviced.

3 Formulating the Admission Control Problem

Continuous playback of a media strand consists of a sequence of periodic tasks with deadlines, where tasks correspond to retrievals of media blocks from disk, and deadlines correspond to the scheduled playback times of media blocks. Thus, servicing multiple strand retrieval requests requires the derivation of a real-time schedule, for which the complexity of the best known algorithms show quadratic dependence on the number of tasks [5, 14]. Since strands usually consist of a large number of media blocks (e.g., if each media block contains one video frame, then a five minute clip of a HDTV video strand recorded at 60 frames/s contains 18000 blocks), the number of tasks can be very large. Hence, direct application of traditional real-time scheduling techniques for servicing multiple strand retrieval requests is out of question.

Since each request is periodic, the multimedia server can service them by proceeding in *rounds*. During each round, the multimedia server retrieves a sequence of media blocks for each strand. In order to ensure continuous retrieval of each strand, the total *service time* of a round should not exceed the minimum of the playback durations of the blocks retrieved for each strand during the round. Whereas the playback duration of a sequence of media blocks of a strand is a function of the playback rate of that strand, the retrieval rate of media blocks is a function of their allocation on disk as well as the policy used for their retrieval.

Since unconstrained allocation of media blocks cannot ensure continuous retrieval of individual strands, media strands must be stored on disk using either contiguous or constrained allocation policy. Furthermore, media blocks can be retrieved from disk using either a *deadline based* (which retrieves media blocks based on the earliest deadline first scheduling policy) or a *placement based* (which retrieves media blocks from disk so as to minimize the total seek and rotational latencies incurred during retrieval) servicing policy. These servicing policies can be applied either to the media blocks within a strand (yielding a *local schedule*) or to the global pool of media blocks from all the strands (yielding a *global schedule*), and can be referred to as local and global servicing policies, respectively. Clearly, when servicing policies are applied among media blocks within a strand to derive a local schedule, the multimedia server has to employ ordering techniques (such as, round robin ordering) to switch from one strand to next during each round.

Notice that if successive media blocks are allocated contiguously on disk, then employing a local placement based servicing policy for retrieving media blocks of a strand from disk yields exactly the same order of retrieval as the local deadline based servicing policy. If, on the other hand, media blocks are stored on disk using the constrained allocation policy, then a local placement based servicing policy may incur lower seek and rotational latencies as compared to a local deadline based servicing policy. However, the performance of the local deadline based servicing policy provides a lower bound on the performance of the local placement based servicing policy.

Regardless of the block allocation policy, the global deadline based servicing policy may incur significant seek and rotational latencies during each round if successive media blocks retrieved from disk (in accordance with the earliest deadline first ordering) are widely separated on disk. On the contrary, a global placement based servicing policy can retrieve media blocks from disk so as to minimize the total seek and rotational latencies, thereby yielding an upper bound on the performance of the global deadline based servicing policy. Hence, in this section, we will analyze the performance of the local deadline based and the global placement based servicing policies:

- *Local deadline based servicing policy*: During each round, the multimedia server retrieves a finite number of media blocks k_i of each strand S_i, $i \in [1, n]$ in accordance with the earliest deadline first policy, and then switches to the next strand (chosen using a round robin ordering policy).
- *Global placement based servicing policy*: During each round, the multimedia server retrieves a finite number of media blocks k_i of each strand S_i ($i \in [1, n]$), by employing the SCAN disk scheduling policy to the global pool of ($k_1 + k_2 + \cdots + k_n$) media blocks to be retrieved during each round. At the end of each round, disk head is repositioned at the innermost track that contains a media block to be retrieved in the next round.

Since unconstrained allocation of media blocks cannot ensure continuous retrieval of individual strands, we will analyze the local deadline based as well as the global placement based servicing policies only for contiguous and constrained allocation of media blocks on disk.

For our analysis, let us suppose that a multimedia server is servicing n clients, each retrieving a different media strand (say, S_1, S_2, ..., S_n, respectively). Let $\mathcal{R}_{vp}^1$, $\mathcal{R}_{vp}^2$, ..., $\mathcal{R}_{vp}^n$ be their playback rates, and k_1, k_2, ..., k_n be the number of blocks of each strand retrieved during a round. Let the media strands be stored on a disk whose storage space is divided into T tracks, with each track being capable of holding at most B media blocks. Let the seek time of the disk be given by $l_{seek} = a + b * t$, where t is the distance in terms of tracks and a and b are constants. Finally, let l_{seek}^{max} and l_{seek}^{min} denote the maximum and the minimum seek latencies, respectively, and l_{rot}^{max} denote the maximum rotational latency of the disk.

3.1 Service Time Formulation Assuming Contiguous Allocation

A multimedia server employing contiguous allocation policy stores successive media blocks of a strand into adjacent disk blocks until a track is completely occupied. Then the multimedia server selects an adjacent track, and the procedure repeats. Thus, if k_i blocks of strand S_i are retrieved during each round, then the time for their retrieval is dependent on the number of tracks spanned by k_i media blocks. Since each track on the disk contains B media blocks, k_i successive blocks of strand S_i can span at most $\left(\left\lceil \frac{k_i}{B} \right\rceil + 1\right)$ tracks. Consequently, once the disk head is positioned on the first of the k_i blocks to be retrieved in a round, the disk head may have to be moved to an adjacent track at most $\left(\left\lfloor \frac{k_i}{B} \right\rfloor + 1\right)$ times, each such move incurring a seek latency of l_{seek}^{min}. Furthermore, the total latency incurred while accessing media blocks from a track is bounded by l_{rot}^{max}. Hence, k_i blocks of strand S_i can be retrieved from disk within time:

$$\left(\left\lfloor \frac{k_i}{B} \right\rfloor + 1\right) * l_{seek}^{min} + \left(\left\lceil \frac{k_i}{B} \right\rceil + 1\right) * l_{rot}^{max} \qquad (2)$$

Consider the local deadline based servicing policy for retrieving media blocks during each round: In accordance with this policy, k_i media blocks of strand S_i can be retrieved within the time derived in Equation (2). However, switching from one strand to another may entail an overhead of up to the maximum seek and rotational latencies (since the layout does not constrain the relative positions of two different strands). Hence, the total service time τ for each round is bounded by:

$$\tau = n * l_{seek}^{max} + n * \left[\left(\left\lfloor \frac{k_i}{B} \right\rfloor + 1\right) * l_{seek}^{min} + \left(\left\lceil \frac{k_i}{B} \right\rceil + 1\right) * l_{rot}^{max}\right] \qquad (3)$$

If, however, the global placement based policy is employed for servicing multiple requests, then for each round, the multimedia server retrieves $(k_1 + k_2 + \cdots + k_n)$ media blocks during one sweep of the disk head from the inner most track to the outer most track (or vice versa). During such a sweep, the disk head has to be repositioned at least n times, once for each request being serviced. Furthermore, once the disk head is positioned on the first of the k_i blocks of strand S_i to be retrieved during a round, it may have to be moved to at most $\left(\left\lfloor \frac{k_i}{B} \right\rfloor + 1\right)$ adjacent tracks in order to retrieve all the k_i blocks. Since the total number of tracks that can be spanned during a single sweep of the disk is bounded by T, the total seek latency incurred for each round is bounded by:

$$a * \left(n + n * \left(\left\lfloor \frac{k_i}{B} \right\rfloor + 1\right)\right) + b * T \qquad (4)$$

Since k_i blocks of strand S_i may span at most $\left(\left\lceil \frac{k_i}{B} \right\rceil + 1\right)$ tracks, the total rotational latency incurred while accessing media blocks of all the strands during a round is bounded by:

$$n * \left(\left\lceil \frac{k_i}{B} \right\rceil + 1\right) * l_{rot}^{max} \qquad (5)$$

Finally, since the seek latency to move the disk head to the inner most track, at the end of each round, is within l_{seek}^{max}, the total service time τ for each round is bounded by:

$$\tau = l_{seek}^{max} + n * a * \left(1 + \left(\left\lfloor \frac{k_i}{B} \right\rfloor + 1\right)\right) + b * T + n * \left(\left\lceil \frac{k_i}{B} \right\rceil + 1\right) * l_{rot}^{max} \qquad (6)$$

3.2 Service Time Formulation Assuming Constrained Allocation

Constrained allocation of media blocks on disk is governed by the granularity and scattering parameters associated with each media strand. Let η_{vs}^1, η_{vs}^2, ..., η_{vs}^n, be the granularities of storage, and l_{ds}^1, l_{ds}^2, ..., l_{ds}^n denote the bounds on scattering for all media strands. Since granularity and scattering for each strand is derived using Equation (1), a sequence of k_i blocks of strand S_i can be retrieved from disk in time:

$$\sum_{j=1}^{k_i} \left(l_{ds}^i + \frac{\eta_{vs} * s_{vf}^i}{\mathcal{R}_{dr}} \right) \tag{7}$$

Consider the local deadline based policy for retrieving media blocks during each round: Whereas the retrieval of the first of the k_i blocks of each strand S_i during each round may entail an overhead of up to the maximum seek and rotational latencies (since the layout does not constrain the relative positions of two different strands), the total overhead incurred in retrieving the remaining $(k_i - 1)$ blocks can be derived using Equation (7). Thus, the total service time τ for each round is bounded by:

$$\tau = n * (l_{seek}^{max} + l_{rot}^{max}) + \sum_{i=1}^{n} \sum_{j=1}^{k_i-1} \left(l_{ds}^i + \frac{\eta_{vs} * s_{vf}^i}{\mathcal{R}_{dr}} \right) \tag{8}$$

On the other hand, consider the global placement based policy for retrieving media blocks during each round: Since constrained placement algorithm does not restrict the allocation of successive media blocks to the same track, all the media blocks to be retrieved during a round, in the worst case, may be stored on separate tracks. Hence, the disk head may have to be repositioned onto a new track at most $(k_1 + k_2 + \cdots + k_n)$ times during each round. Furthermore, retrieval of each media block may incur a rotational latency of l_{rot}^{max}. Since the seek latency to move the disk head to the inner most track, at the end of each round, is within l_{seek}^{max}, the total service time for each round is bounded by:

$$\tau = l_{seek}^{max} + b * T + (a + l_{rot}^{max}) * \sum_{i=1}^{n} k_i \tag{9}$$

Using the service time formulations presented in Equations (3), (6), (8), and (9), we will now derive the condition for servicing multiple clients simultaneously, and present algorithms that a multimedia server can employ to determine whether a new retrieval request can be admitted without violating the continuity requirements of the requests already being serviced.

4 Admission Control Algorithms

A multimedia server can service multiple strand retrieval requests simultaneously if and only if the continuity requirements of none of the strands are violated. Since during each round, the multimedia server retrieves a finite number of media blocks k_i of each strand S_i ($i \in [1, n]$), ensuring continuous retrieval for each of strand requires that the service time per round does not exceed the minimum of the playback durations of k_1, k_2, ..., or k_n blocks. We refer to this as the *admission control principle*.

Applying the admission control principle to the local deadline based servicing policy (i.e., Equation (8)) yields:

$$n * (l_{seek}^{max} + l_{rot}^{max}) + \sum_{i=1}^{n} \sum_{j=1}^{k_i-1} \left(l_{ds}^i + \frac{\eta_{vs} * s_{vf}^i}{\mathcal{R}_{dr}} \right) \le \min_{i \in [1,n]} \left(k_i * \frac{\eta_{vs}}{\mathcal{R}_{vp}^i} \right) \quad (10)$$

Thus, the multimedia server can service all the n requests simultaneously if and only if $k_1, k_2, ..., k_n$ can be determined such that Equation (10) is satisfied. Since this formulation contains n parameters and only one equation, determination of the values of $k_1, k_2, ..., k_n$ require additional policies. The simplest policy for the choice of $k_1, k_2, ..., k_n$ is to use the same value for all of them, yielding what is generally referred to as a round robin algorithm with fixed quanta [9]. Formally, if $k_1 = k_2 = \cdots = k_n = k$, Equation (10) yields:

$$k \ge \frac{n * \left((l_{seek}^{max} + l_{rot}^{max}) - \left(l_{ds}^{avg} + \frac{\eta_{vs} * s_{vf}^{avg}}{\mathcal{R}_{dr}} \right) \right)}{\min_{i \in [1,n]} \left(\frac{\eta_{vs}}{\mathcal{R}_{vp}^i} \right) - n * \left(l_{ds}^{avg} + \frac{\eta_{vs} * s_{vf}^{avg}}{\mathcal{R}_{dr}} \right)} \quad (11)$$

Since any media block can be retrieved from disk within time $(l_{seek}^{max} + l_{rot}^{max})$ starting from any other location on disk, it is guaranteed that $\left(l_{ds}^{avg} + \frac{\eta_{vs} * s_{vf}}{\mathcal{R}_{dr}} \right) \le (l_{seek}^{max} + l_{rot}^{max})$. Hence, for k to be non-negative, the denominator must be positive, yielding:

$$n_{max} = \frac{\min_{i \in [1,n]} \left(\frac{\eta_{vs}}{\mathcal{R}_{vp}^i} \right)}{\left(l_{ds}^{avg} + \frac{\eta_{vs} * s_{vf}}{\mathcal{R}_{dr}} \right)} \quad (12)$$

The results of applying the above analysis for each of the other servicing policies (i.e., to Equations (3), (6), and (9)) are illustrated in Table 2.

Notice, however, that the number of clients that can be serviced simultaneously, with the choice of $k_1 = k_2 = \cdots = k_n = k$, is limited by the request with maximum playback rate. This certainly may not be the optimal number of clients, because, whereas the client with the maximum playback rate will have retrieved exactly the number of data blocks it needs for the duration of a service round, other clients whose playback rates are smaller will have retrieved more data blocks than they need in each service round. Consequently, by reducing the number of data blocks retrieved per service round for such clients, it may be possible to accommodate more number of clients. Towards this end, we have proposed a *Quality Proportional Multi-client Servicing* (QPMS) algorithm that allocates values to k_i proportional to the playback rate of the strand S_i. That is, $\forall i \in [1, n]: \ k_i \propto \mathcal{R}_{vp}^i$. Thus, if k is the proportionality constant, we get, $k_1 = k * \mathcal{R}_{vp}^1$, $k_2 = k * \mathcal{R}_{vp}^2, ..., k_n = k * \mathcal{R}_{vp}^n$. Under these conditions, Equation (10) reduces to:

$$n*(l_{seek}^{max}+l_{rot}^{max})+k*\sum_{i=1}^{n} \mathcal{R}_{vp}^i*(l_{ds}^i + \frac{\eta_{vs} * s_{vf}^i}{\mathcal{R}_{dr}})-\sum_{i=1}^{n}(l_{ds}^i+\frac{\eta_{vs} * s_{vf}^i}{\mathcal{R}_{dr}}) \le k*\eta_{vs} \quad (13)$$

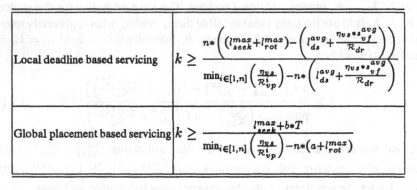

Local deadline based servicing	$k \geq \dfrac{n*\left(l_{seek}^{max}+l_{seek}^{min}+2*l_{rot}^{max}\right)}{\min_{i \in [1,n]}\left(\frac{\eta_{vs}}{\mathcal{R}_{vp}^i}\right)-\frac{n}{B}*\left(l_{seek}^{min}+l_{rot}^{max}\right)}$
Global placement based servicing	$k \geq \dfrac{l_{seek}^{max}+2*n*(a+l_{rot}^{max})+b*T}{\min_{i \in [1,n]}\left(\frac{\eta_{vs}}{\mathcal{R}_{vp}^i}\right)-\frac{n}{B}*(a+l_{rot}^{max})}$

(a)

Local deadline based servicing	$k \geq \dfrac{n*\left(\left(l_{seek}^{max}+l_{rot}^{max}\right)-\left(l_{ds}^{avg}+\frac{\eta_{vs}*s_{vf}^{avg}}{\mathcal{R}_{dr}}\right)\right)}{\min_{i \in [1,n]}\left(\frac{\eta_{vs}}{\mathcal{R}_{vp}^i}\right)-n*\left(l_{ds}^{avg}+\frac{\eta_{vs}*s_{vf}^{avg}}{\mathcal{R}_{dr}}\right)}$
Global placement based servicing	$k \geq \dfrac{l_{seek}^{max}+b*T}{\min_{i \in [1,n]}\left(\frac{\eta_{vs}}{\mathcal{R}_{vp}^i}\right)-n*(a+l_{rot}^{max})}$

(b)

Table 2. Comparison of admission control policies assuming (a) contiguous placement and (b) constrained placement of media strands on disk

Given the granularity and scattering for each strand, Equation (13) can be used to determine k, from which, the number of blocks retrieved during each service round can be obtained as: $k_1 = k * \mathcal{R}_{vp}^1$, $k_2 = k * \mathcal{R}_{vp}^2$, ..., $k_n = k * \mathcal{R}_{vp}^n$. It can be shown that this algorithm always yields values of k_i so as to satisfy Equation (10) whenever a solution exists for the given number of clients [13]. However, the values of k_i's obtained using the QPMS algorithm may not be integral. Since the display of media strands proceeds in terms of quanta such as frames, if k_i is not an integer, then retrieval of a fraction of a frame cannot be used for display, causing the display to starve until the remaining fraction arrives, possibly in the next service round. Such scenarios can be avoided if k_i's are all integers. Given the real values of k_i's yielded by the QPMS algorithm, integral number of blocks of strand S_i to be retrieved during a service round can be set to either $\lfloor k_i \rfloor$ or $\lceil k_i \rceil$, so that on an average, the transfer rate for each strand S_i is k_i blocks/round. However, in doing so, ensuring that Equation (10) is not violated requires that the toggling of $\lfloor k_i \rfloor$ to $\lceil k_i \rceil$ for strands must be dynamically staggered [13]. The QPMS algorithm coupled with such a staggered toggling technique fine tunes the successive number of frames retrieved for each client individually, thereby servicing an optimal number of clients simultaneously.

5 Experience and Performance Evaluation

A prototype multimedia server is being implemented at the UCSD Multimedia Laboratory in an environment consisting of multimedia stations connected to a multimedia server through Ethernet and FDDI networks. Each multimedia station consists of a computing workstation, a PC-AT, a video camera, and a TV monitor. The PC-ATs are equipped with digital video processing hardware that can digitize and compress motion video at real-time rates, and audio hardware that can digitize voice at 8 KBytes/sec. The multimedia server is implemented on a 486-PC equipped with multiple gigabytes of storage.

We have evaluated the relative performance of various servicing policies and the admission control algorithms presented in this paper. Our analysis demonstrates that if media strands are stored on disk is using the contiguous allocation policy, then the global placement based servicing policy incurs lower seek and rotational latencies as compared to the local deadline based servicing policy. Consequently, for a given value of k (which may be derived from the buffer space available at each display site), the global placement based policy can service a larger number of clients simultaneously as compared to the local deadline based servicing policy (see Figure 1(a)).

On the contrary, if the storage of media strands on disk is governed by a constrained allocation policy, then employing global placement based servicing policy for retrieving media blocks during each round may, in the worst case, incur maximum rotational latency for each media block. Consequently, for constrained allocation of media strands on disk, the local deadline based policy can service a larger number of clients simultaneously as compared to the global placement based servicing policy (see Figure 1(b)).

Figure 1 also demonstrates that if media strands are stored on disk using the contiguous allocation policy, then for a given value of k, the number of clients that can be simultaneously supported by the local deadline based servicing policy as well as the global placement based servicing policy is significantly higher than the corresponding values derived for the constrained allocation policy. The performance of the servicing policies, when applied to media strands stored on disk using the constrained allocation policy, improves with decrease in the scattering parameters associated with media strands. Although the contiguous allocation policy yields higher performance as compared to the constrained allocation policy, it is fraught with inherent problems of fragmentation, and can entail enormous copying overheads during insertions and deletions. Consequently, the applicability of the contiguous allocation policy is restricted to read-only storage servers, and the constrained allocation policy is more suitable for interactive, read-write multimedia servers.

Our analysis of the admission control algorithms demonstrates that the QPMS algorithm can service higher number of client simultaneously as compared to the round robin algorithm with fixed value of k. Figure 2 illustrates the gain in the maximum number of simultaneous clients in the QPMS as compared to the round robin algorithm.

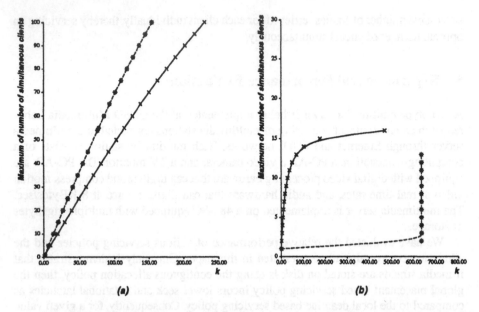

Fig. 1. Comparison of admission control policies: (a) contiguous allocation, and (b) constrained allocation. The solid and the dashed curves denote the local deadline based servicing policy and global placement based servicing policy, respectively

Higher the asymmetry among the playback rates of the client requests, greater is the advantage of employing the QPMS algorithm. When the playback rates of all the clients are the same, the performance of the QPMS algorithm degenerates to that of the round robin algorithm (as depicted by the end-points of the graph shown in Figure 2).

6   Concluding Remarks

In this paper, we have addressed the admission control problem in the design of a multi-client multimedia on-demand server. We have proposed a taxonomy of servicing policies and have illustrated techniques for formulating the admission control problem for local deadline based and global placement based servicing policies. Our performance evaluation demonstrates that, whereas contiguous allocation of media blocks on disk favors the global placement based servicing policy, constrained allocation of media blocks favor the local deadline based servicing policy. We have also demonstrated that the QPMS algorithm can admit higher number of client simultaneously as compared to the round-robin algorithm with fixed quanta. The local deadline based and the global placement based servicing techniques as well as the QPMS admission control algorithm form the basis of a prototype multimedia on-demand server being implemented at the UCSD Multimedia Laboratory.

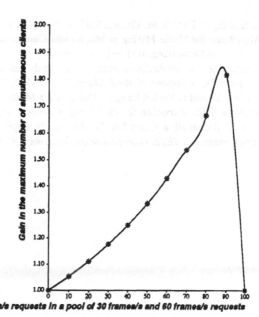

Percentage of 30 frames/s requests in a pool of 30 frames/s and 60 frames/s requests

Fig. 2. Increase in the maximum number of simultaneous clients in QPMS as compared to the round robin algorithm

References

1. C. Abbott. Efficient Editing of Digital Sound on Disk. *Journal of Audio Engineering*, 32(6):394–402, June 1984.
2. D. Anderson, Y. Osawa, and R. Govindan. Real-time Disk Storage and Retrieval of Digital Audio and Video. *To appear in the ACM Transactions on Computer Systems*.
3. J. Gemmell and S. Christodoulakis. Principles of Delay Sensitive Multimedia Data Storage and Retrieval. *ACM Transactions on Office Information Systems*, 10(1):51–90, 1992.
4. S. Gibbs, D. Tsichritzis, A. Fitas, D. Konstantas, and Y. Yeorgaroudakis. Muse: A Multi-Media Filing System. *IEEE Software*, 4(2):4–15, March 1987.
5. J.F. Kurose, M. Schwartz, and Y. Yemini. Multiple-Access Protocols and Time-Constrained Communication. *ACM Computing Surveys*, 16(1):43–70, March 1984.
6. W. E. Mackay and G. Davenport. Virtual Video Editing in Interactive Multimedia Applications. *Communications of the ACM*, 32(7):802–810, July 1989.
7. B.C. Ooi, A.D. Narasimhalu, K.Y. Wang, and I.F. Chang. Design of a Multi-Media File Server using Optical Disks for Office Applications. *IEEE Computer Society Office Automation Symposium, Gaithersburg, MD*, pages 157–163, April 1987.
8. P. Venkat Rangan and D. C. Swinehart. Software Architecture for Integration of Video Services in the Etherphone Environment. *IEEE Journal on Selected Areas in Communication*, 9(9):1395–1404, December 1991.
9. P. Venkat Rangan and Harrick M. Vin. Designing File Systems for Digital Video and Audio. In *Proceedings of the 13th Symposium on Operating Systems Principles (SOSP'91), Operating Systems Review, Vol. 25, No. 5*, pages 81–94, October 1991.
10. P. Venkat Rangan, Harrick M. Vin, and Srinivas Ramanathan. Designing an On-Demand Multimedia Service. *IEEE Communications Magazine*, 30(7):56–65, July 1992.

11. P. Venkat Rangan, Harrick M. Vin, and Srinivas Ramanathan. Communication Architectures and Algorithms for Media Mixing in Multimedia Conferencing. *To appear in IEEE/ACM Transactions on Networking*, 1(1), February 1993.

12. W. D. Sincoskie. System Architecture for a Large Scale Video on Demand Service. *Computer Networks and ISDN Systems, North-Holland*, 22:155–162, 1991.

13. Harrick M. Vin and P. Venkat Rangan. Designing a Multi-User HDTV Storage Server. *To appear in the IEEE Journal on Selected Areas in Communications*, 11(1), January 1993.

14. W. Zhao, K. Ramamritham, and J.A. Stankovic. Preemptive Scheduling Under Time and Resource Constraints. *IEEE Transactions on Computers*, C-36(8):949–960, August 1987.

The Design and Implementation of a Continuous Media Storage Server

Phillip Lougher and Doug Shepherd

Computing Department, Engineering Building, Lancaster University, Lancaster LA1 4YR, UK, E.mail mpg@comp.lancs.ac.uk

Abstract. This paper describes the design and implementation of a file server specially optimised for the storage and retrieval of continuous media. The issues which this paper addresses are disk striping, optimised disk layouts, real-time algorithms, and disk head scheduling.

1 Introduction

With the recent advances in network and workstation technology, it has become possible to develop applications which integrate, store and transmit video, voice, hi-fi audio and graphics, in addition to traditional text. Successful multimedia applications include real-time video conferencing, group working, multimedia electronic mail, computer aided learning and interactive video.

The provision of continuous media support however, imposes immense technical problems for system designers. The transmission of audio and video streams in a distributed system requires *end-to-end support*, in which Quality of Service (QoS) guarantees must be maintained from source device through to destination device. All the intervening components in the communications chain must co-operate to guarantee that the QoS is not broken - this requires negotiated guarantees from the source device, the source device workstation, the interconnecting networks and the destination workstation.

This paper details the design and implementation of an experimental Continuous Media Storage Server (CMSS), which has been built to investigate the techniques required to support the guaranteed storage and retrieval of continuous media. This server forms part of an existing infrastructure of networked multimedia workstations providing a testbed for multimedia application development.

2 Continuous Media vs Conventional Files

Continuous media files differ in many aspects to conventional files. In particular, continuous media files are very large, are highly structured, have complex specialised control operations, have high throughput (180K per second for CD quality audio, and at least 1Mbyte for reasonable quality video) and must have a guaranteed bandwidth. Conventional files in contrast, tend to be very small[1],

---

[1] In an analysis of a BSD Unix system the average file size was found to be 3-4K [7].

are not temporally defined, and are treated as unstructured byte arrays by file servers.

These characteristics have had a large impact on file server design. Conventional file servers are purely demand driven, and all data transfer must be explicitly requested by clients. No interpretation of files is undertaken, and no guarantees are placed on the time taken to execute data accesses - this is a function of varying server load. File server disk layout strategies are usually dominated by the small sizes of conventional files and the desire to minimise file overhead on disk.

A CMSS design has differing objectives. The major objective of a CMSS is the need to retrieve data fast enough from disk to satisfy a medium's QoS. Closely related to this objective is the desire to maximise the number of streams which can be replayed concurrently. This requires the provision of high disk bandwidth and a disk layout strategy optimised to minimise seek time plus hard real-time scheduling to ensure available bandwidth is correctly shared amongst concurrent stream retrievals. The hard real time scheduler requires each continuous media stream to be modelled to enable its throughput requirement to be calculated and to ascertain if enough unallocated disk bandwidth exists to accommodate it (the stream admission test).

This model should include characteristics such as the maximum bandwidth per second, the network packet size, and the rate of packet arrival. In our system, each stream is modelled using the following four parameters:-

T *The constant interval in seconds between generated packet bursts,*
P *The constant size in bytes of each packet,*
R *The rate of packet burst arrival per second,*
N *The number of packets in each burst.*

The above four parameters allow a wide range of packet arrival characteristics to be modelled. The isochronous transmission of full frame rate colour video[2] could be modelled, with T as 1/25, P approximately 1Mbyte, R as 25 and N as 1. In this example, each frame is being transmitted in one packet. More bursty transmissions can also be modelled, for example, each frame could be transmitted in a burst of packets - each packet containing one line. The logical unit of a CM stream can be defined as the N packet burst, with the logical unit size (LU) being defined as $N \times P$. The logical unit refers to the natural structuring of CM data into a sequence of units. In the case of video, this refers to the frame.

A further important objective of a CMSS, is the provision of a user interface that supports high level multimedia abstractions. This, for example, should include the ability to control streams using VCR like operations such as *play*, *pause*, *seek* and *rewind*. In addition, there should exist the ability to retrieve and edit stream data on a non real-time basis.

---

[2] 25 frames per second in Europe.

3 System overview

The logical structure of the CMSS is illustrated in Fig. 1. Externally the CMSS consists of a data interface and a client control interface. The control interface is similar to the interface provided by conventional file servers, and allows clients to set up, close down, and control streams. Clients communicate with the server via this interface using Remote Procedure Calls. The RPCs provide both traditional file oriented commands and stream oriented commands. The file oriented commands include open, close, delete, create, getQos, and getLength. The stream oriented commands include play, pause, seek and step.

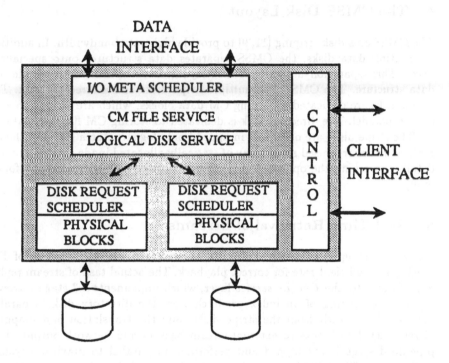

Fig. 1. The logical structure of the CMSS

The data interface provides the interface through which all CM devices communicate. The CM data manipulated through the interface is highly structured and depends on the data's type. At present two types of CM data are supported: video and audio. Additionally, each typed CM stream has an associated Quality of Service, this obviously varies between the types but includes frame rate and frame resolution for video data, and sample size and sample rate for audio.

The receiving and sending of CM data is scheduled by the I/O meta scheduler. It is this layer's responsibility to ensure that data input/output to the network is correctly scheduled to meet all the stream network deadlines. The CM file server layer is responsible for the real-time storage and retrieval of stream data. This layer is multi-threaded, and a separate thread exists for every concurrent stream being serviced. The disk request scheduler layers are responsible for correctly scheduling the requests issued by the CM file service so that data block retrieval/storage deadlines are not missed. Separate disk request schedulers exist for every data disk installed in the system.

The next two sections describe the layout of the data on the CMSS disks, and the real-time retrieval and storage algorithms that implement the server layers. Due to space constraints, the disk layout section is very brief.

4 The CMSS Disk Layout

The CMSS uses disk striping [11, 9] to provide high data bandwidth. In addition to multiple data disks, the CMSS separates data structures onto specialised disks. This separation allows each disk layout to be optimised to handle one data structure. The CMSS has multiple stripe disks and a directory disk. The stripe disks are dedicated to storing CM data blocks, which are striped to each disk in parallel. The directory disk is dedicated to storing CM file metadata.

The stripe disks are organised in append only log structures [8] , and the directory disk is organised as an array of statically allocated inode entries. Further details concerning the operation and rationale of the disk layout can be found in [6].

5 Real Time Retrieval of Streams

The real-time replay of streams from disk is concerned with the retrieval of disk blocks, at a sufficient rate for correct playback. The actual task of stream replay is performed by the *CM file service* layer, which implements a 3 stage retrieval pipe-line, consisting of an index pre-fetch from the directory disk, a parallel data block pre-fetch from the stripe disks, and the translation and mapping of retrieved data blocks to external stream packets and network output. This pipe-lined stage is overlapped and performed in parallel to maximise system performance. The CM file service layer is multi-threaded, and a separate thread exists for *each* concurrent stream. This layer cooperates with the *disk request scheduler layers* which receive multiple concurrent requests for disk block access, and correctly schedules the disks to ensure no stream data deadlines are missed. These layers are now discussed in greater detail.

5.1 The CM file service

The issues which must be addressed by the retrieval algorithm are:

- The mapping of the external streams onto disk, and the granularity of the disk blocks,
- The method of striping the disks,

- The amount of data block *pre-fetch* required to maximise possible concurrency, and thus the amount of buffering needed, and
- The deadline calculation for when a pre-fetched data block needs to arrive from disk.

The disk block granularity calculation is fairly complex, this is due to a number of trade-offs that exist between different factors in the system, i.e. there is a need to provide balanced performance across the CMSS. Firstly, the data block stored on disk should be as large as memory will allow, as this minimises the amount of separate disk read accesses required. Secondly, to ensure efficiency in stream operations, a data block must be a multiple of a stream's logical unit. This is to ensure that data block boundaries are aligned to LU boundaries on disk. Thirdly, and most importantly, all data blocks must be normalised, so that they represent the same playback duration on disk. This normalisation, is attractive for two reasons. Firstly, streams with a high data throughput are allocated larger data blocks than streams with a low data throughput. This results in a distribution of resources ('buffer space etc.) which favours streams with the most stringent data throughput requirements. Secondly, it allows a very efficient hard real time scheduler to be used (discussed later).

The actual calculation is performed using a fixed reference point of one second. All streams are normalised so that their data blocks on disk represent this playout period. This, as shown later, is very important for the disk request scheduler because, for all streams stored on disk, the block size for S_1 (stream one) represents the same playout period as the block size for S_2 and S_3 etc. The mapping is thus:

$$multiplier = \lceil R \rceil$$
$$DLU = multiplier \times LU$$
$$DR = \frac{R}{multiplier}$$
$$DT = multiplier \times T$$
$$DN = 1$$
$$DP = DLU$$

Here $\lceil R \rceil$ represents: if R is non integer $trunc(R) + 1$, otherwise R. DLU and DP represent the size of the data block on disk, DT represents the interval between data block output in seconds, and DR represents the rate of data block consumption per second.

Note $\lceil R \rceil$ is needed where R is not an integer value, in which case the stream is mapped to the multiple representing a playout period closest to (but greater than) 1 second. Intuitively the step is mapping external streams with unconstrained R, T, N and P into streams on disk, with DN constrained to 1 and DR constrained as close to 1 as possible.

The data blocks are striped in sequence to the disks, i.e. the first data block is striped to the first disk, the next data block to the second disk, and the nth

data block to the $n \bmod m$ disk, where m is the number of data disks in the system. On retrieval the algorithm should read all the disks in parallel to ensure maximum performance. Thus to keep all m disks busy the amount of read-ahead needed is at least m buffers. In practice, the minimum read-ahead required to maximise performance is either $m + 1$ data blocks or $2 \times m$ data blocks, this depends on when additional pre-fetching of data blocks is performed. The two alternative methods are outlined below.

One method is to pre-fetch m data blocks in parallel from disk whilst outputting a second bank of m data blocks to the I/O meta scheduler. Once the bank of m data blocks being output to the scheduler has been exhausted the banks are switched. This requires a read-ahead of at least $2 \times m$ data blocks. The other method is to fetch additional data blocks as each data block is exhausted. In this case we have the data block currently being output to the scheduler and m data blocks being retrieved from disk. This requires a read-ahead of at least $1 + m$ data blocks.

The difference in methods extends beyond the amount of read-ahead needed, in particular in the way slack time in the disk accesses is distributed. Assuming no other stream is being serviced concurrently the first method results in the disks being idle at the same time, the second method tends to distribute the idle periods, as it staggers the disk reads. It is important to note that when there exists no slack time in the stream retrieval, i.e. where the time taken to exhaust m blocks is equal to the time taken to read one block from disk, both methods ensure total utilisation of the disks.

To illustrate the above, the pre-fetch algorithm for the second method will now be outlined. On stream start-up an initial pre-fetch of $1 + m$ data blocks is required to 'prime' the buffers. The pre-fetch is to set the pipeline up. Once the buffers are full the parallel output of the pre-fetched buffers and the reading of new buffers is commenced. New data block requests to the disk layers are issued when each data block buffer is exhausted. Assuming sequential playout without logical unit skipping (e.g. frame skipping for video) once the nth data block is exhausted the request for the $n + m^{th}$ is issued to the disk layers, with the new $n + m^{th}$ data block being placed in the newly emptied buffer. Buffer usage is thus cyclic, with disk requests being issued every DT seconds, and each individual disk being accessed every $m \times DT$ seconds. At any one point there can be a maximum of m disk requests outstanding.

From the above it should be clear that the deadline for a disk request to arrive is given by the time to empty all m buffers or

$$deadline = current\_time + m \times DT$$

The schedulability constraint for a single stream is thus given by[3]

---

[3] $Seek_{MAX}$ represents the worst case cost of seeking to a disk block, and D_{MAX} represents the cost of reading one byte from disk. The disk layout has a major influence on the cost of disk accesses, however in this paper, we make no assumptions about layout.

$$Seek_{MAX} + DP \times DMAX \leq m \times DT$$

If this constraint is satisfied, then the stream will be able to be retrieved from disk. The slack in the system for the retrieval of one stream is given by:

$$slacktime = m \times DT - (Seek_{MAX} + DP \times D_{MAX})$$

5.2 The Disk Request Scheduler

The disk request scheduler receives requests for the retrieval and storage of data blocks, index blocks and header blocks. Of these, data blocks are deadline critical.

For single streams the constraints derived in section 5.1 are sufficient for determining schedulability. For multiple streams the constraints are insufficient. In the case of simultaneous stream retrieval/storage the scheduler receives requests from k streams, $x = S_1, S_2, \ldots, S_k$. Each request from S_x arrives with a fixed interval of $m \times DT_x$ and a worst case execution of $Seek_{MAX} + DP_x \times D_{MAX}$. Given this, the schedulability problem can be defined as: how many streams can be serviced without requests missing deadlines, and if the scheduler is currently serving k streams successfully, can another stream be accepted?

The problem therefore is similar to the area of hard real- time scheduling [4, 5]. Traditional real-time scheduling techniques however, cannot be used, because they have been developed for process scheduling, and assume the overhead of descheduling a process for another process is nominal. This is not the case with CM streams, as the overhead incurred in swapping between streams is at least one disk seek.

The scheduler that we have developed is a split level scheduler, which provides two classes of stream service: hard guaranteed streams and soft guaranteed streams. For brevity, only the hard real scheduler will be described.

Scheduling the Hard Guaranteed Streams. Any hard real-time scheduler developed for the CMSS must possess the following characteristics: the feasibility test must be simple and computationally fast, and the scheduler must not impose excessive performance penalties. The scheme used in the CMSS is to use a round robin scheduler (known from here on as RRS), in which stream requests are performed in cycles. If k streams are being processed, a cycle or one round robin iteration will consist of a disk read for $S_1, S_2, \ldots, S_k$. Such a scheme has the advantage that the feasibility test consists of the evaluation of a constraint, and is thus very fast. For the scheme to work efficiently it requires a fairly restricted set of process characteristics. The scheme and the steps taken to ensure the restricted characteristics are detailed below.

A cycle is defined as the sequential retrieval of a data block for stream $S_1, S_2, \ldots, S_k$. The overhead of reading a data block for stream S_j is:

$$Seek_{MAX} + DP_j \times D_{MAX}$$

Thus the overhead of executing one cycle is :

$$k \times Seek_{MAX} + \sum_{j=1}^{k} DP_j \times D_{MAX}$$

Each of the k requests must be read before its deadline expires. The deadlines will be met if the time taken to execute the cycle is less than the minimum request generation interval or:

$$k \times Seek_{MAX} + \sum_{j=1}^{k} DP_j \times D_{MAX} \leq \min_{j=1}^{k}(m \times DT_j)$$

This constraint forms the stream admission test. Slack time in the cycle is given by:

$$\min_{j=1}^{k}(m \times DT_j) - \left(k \times Seek_{MAX} + \sum_{j=1}^{k} DP_j \times D_{MAX} \right)$$

The cycle satisfies one request for each stream. The frequency of cycle execution and the time interval between successive cycles is determined by the most demanding stream. The most demanding stream can be defined as the stream whose data blocks represent the smallest playout period, and thus have the shortest request interval: this is, as shown above, $\min_{j=1}^{k}(m \times DT_j)$.

It should be noted that as the file indexes are stored on a separate disk, the seek and read overheads of reading index blocks do not have to be taken into account by the disk request scheduler.

Restricting the Stream Request Characteristics. The major problem with the RRS scheduler as it has been described is that the k streams must have identical request rates or very similar request rates. This is because all the k streams are scheduled at the rate of the most demanding stream. A less demanding stream will thus have its data blocks read faster than it can consume, and this in turn means it will accumulate data. The rate of the stream's data accumulation will be in proportion to its request rate mis- match (in relation to the most demanding stream's rate). It can be imagined that large mis-matches, for example where the most demanding stream's request rate is double its own, will result in rapid data accumulation.

To prevent accumulation of data, read requests for accumulating streams can be postponed in the RRS cycles. However, if the rate of data accumulation is high then the frequency of read postponement will also be high. The problem arises because the stream acceptance test is based on each cycle being a full read cycle. If read postponement is high then the acceptance test will over estimate bandwidth requirement. This in turn will result in under-utilisation of the disk bandwidth.

Our solution to this problem is to prevent the request rate mis-matches from occurring. We do this by ensuring that all data blocks on disk represent the same playout period. The re-mapping of streams to a fixed rate on disk, discussed in section 5.1, ensures all the k streams' data blocks on retrieval are restricted in playout rate. This re-mapping of streams allows a simple RRS to be used to schedule hard guaranteed streams.

Problems with Data Accumulation. In most cases the k streams on disk will have identical request rates of $\frac{1}{m}$ seconds, thus, $m \times DT_1 = m \times DT_2 = \ldots = m \times DT_k = \min_{j=1}^{k}(m \times DT_j) = m$. In this situation as stream request rates are identical no data accumulation will occur. In certain cases a stream on disk may not be exactly recorded at a $\frac{1}{m}$ second request rate (this is where the rate does not scale up to one second). In this case the stream will have a request rate smaller than $\frac{1}{m}$ seconds, and it will thus accumulate data.

Our solution to this problem is to stop reading data for streams once accumulated data has reached a certain limit. In other words, on cycles where a stream has accumulated data, the read for that stream is postponed to the next cycle. This control of stream reads ensures data accumulation is prevented. The amount of data accumulation allowed before read postponement is determined by the need to prevent stream starvation. If a read for a stream is postponed, the stream must have enough data accumulated to sustain it till the next cycle. Thus the amount of data accumulation allowed is at least the cycle execution interval or $\min_{j=1}^{k}(m \times DT_j)$. The maximum amount of data accumulation needed to prevent stream starvation on read postponement can be shown to be $m \times DT_x$, for each stream, $x = 1, \ldots, k$. The proof for this is outside the scope of the paper.

It is emphasised that when a stream cannot be mapped to exactly one second, the mapping achieved will be very close. Thus the frequency of stream read postponement is small.

6 System Implementation

6.1 Lancaster's Multimedia Infrastructure

We have implemented a prototype CMSS using transputers. This prototype forms part of an experimental multimedia infrastructure consisting of multimedia workstations and a high-speed network emulator. The multimedia workstations consist of standard PC and UNIX computers augmented with a Multimedia Network Interface (MNI). This MNI unit provides each host with high speed video and CD quality audio I/O capabilities. Multimedia data is transmitted, received and processed by each MNI unit independently of the host, without data being streamed over the slow host system bus.

The MNI workstations are inter-connected via a Network Protocol Emulator (NPE). This emulator provides a high speed packet switch through which multimedia data can be transmitted and received. The NPE allows various high speed networks to be emulated, allowing the effects of different network protocols to be studied. Currently, the NPE provides an emulation of FDDI.

Further details of the MNI unit and the NPE can be found in [12, 2, 3].

6.2 The Prototype CMSS

The CMSS is based on the design of the MNI unit, and like the MNI and NPE, is implemented using transputer technology. Transputers were chosen for the units as they offer fairly high performance and easy configurability. The transputer's four links (20 Mbit/s per link) allow high data throughput, and also allow the easy connection of the transputer to commercial cards.

The CMSS is comprised of three dedicated T800 transputers, two transputer based SCSI interfaces, two stripe disks and a directory disk. The inter-connection of the components and the distribution of the CMSS layers onto the transputers is illustrated in Fig. 2. The layers are implemented in OCCAM and communicate using OCCAM channels.

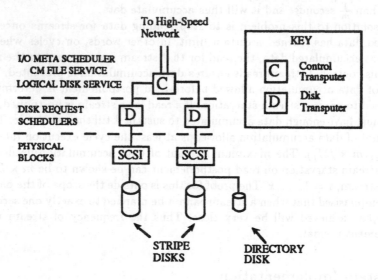

Fig. 2. Schematic of the CMSS

The control transputer implements the upper CMSS layers, and is connected to the NPE via two transputer links. The two links give a maximum total bandwidth to/from the network of 40 Mbits/sec. Each of the data stripe disks have a separate disk transputer and SCSI interface. The use of dedicated disk processors increases performance, and it also allows the disk request schedulers to be implemented correctly without interference from process time slicing.

Each stripe disk is 600 Mbytes in size, giving a total data storage capacity of 1.2 Gbytes. The directory disk is 140 Mbytes.

7 Related Work

Storage servers of interest include work at UCSD [10], UCB [1], and MIT [13]. The UCSD storage server is a single disk system which uses constrained block scattering to ensure media playout constraints are met. The UCB server is a single disk system which has concentrated on scheduling problems. Files in this system must be stored contiguously, with the maximum file size being specified at creation time. The MIT server allows a hierarchy of cooperating servers to be built, with the servers being distributed over a network. This goes some way to avoid bandwidth saturation problems associated with single host servers. This server in other aspects is limited, using a very inefficient disk scheduler.

8 Summary

The increasing use of multimedia in many areas of computing is demanding larger and faster file servers and computer hardware. The demands placed on file servers by continuous media however are not satisfied by current magnetic disk technology or by current file systems techniques. The aim of our research has been to address these critical performance issues.

In this paper we analysed the fundamental differences between CM file storage/retrieval and traditional files. From this analysis it was shown that CM files are characterised by large file size and high bandwidth requirements. This high bandwidth requirement, because of the temporal nature of CM data, needs to be sustained throughout a file's playback from disk even in the presence of multiple requests. Our design concentrated on three major objectives: the provision of high raw disk bandwidth, the use of a disk layout specially organised for CM, and the guaranteed allocation of disk bandwidth on a per stream basis to ensure correct playout of CM data.

We have used disk striping to achieve high raw bandwidth. The bandwidth available in our system for storage/retrieval increases linearly as more disks are added. We have developed a novel disk layout which by splitting metadata and data blocks over specialised disks and by utilising the characteristics of CM files has achieved an efficient layout that exhibits low seek time overhead, efficient allocation and deallocation of disk blocks, and minimal fragmentation. To provide hard guarantees on stream retrieval and storage we designed a split level disk scheduler utilising stream re-mapping and a simple round robin scheduler which provides efficient non pre-emptive hard real time scheduling of disk requests with little overhead.

The transputer based prototype has verified the CMSS design. It is currently being used by a number of applications for the storage of video and CD quality audio. It is also bridged to the ethernet and provides a slow scan video and audio service to networked Sun Sparcstations and Apple Macintoshes.

9 Acknowledgements

The work described in this paper has been carried out in association with the Multimedia Network Interface (MNI) project, funded under the UK SERC Specially Promoted Programme in Integrated Multiservice Communication Networks (grant number GR/F 03097) and co-sponsored by British Telecom Laboratories.

References

1. Anderson D.P., Y. Osawa, and R. Govindan, "Real Time Disk Storage and Retrieval of Digital Audio/Video Data", UCB Technical Report, September 1991.
2. Ball F., and D. Hutchison, "A FDDI Network Protocol Emulator", Proceedings of the Third Bangor Symposium on Communications, pp 287-290, May 1991.
3. Ball F., and D. Hutchison, "A General Purpose Network Protocol Emulator", Proceedings of the Fourth Bangor Symposium on Communications, pp 94-97, May 1992.
4. Cheng S.C., J. Stankovic and K. Ramamritham, "Scheduling Algorithms for Hard Real-time Systems: A Brief Survey", Hard Real-Time Systems, Tutorial, IEEE Computer Society Press, pp 150-173, 1988.
5. Liu C. L., and J. W. Layland, "Scheduling Algorithms for Multiprogramming in a Hard Real Time Environment", Journal of the ACM, pp. 46-67, 20(1), January 1973.
6. Lougher P.K., and D. Shepherd, "The Design of a Storage Server for Continuous Media", To Appear in the Computer Journal's Special Issue on Multimedia, 1993, also available as Technical Report no. MPG-92-27, Department of Computing, University of Lancaster, 1992.
7. Ousterhout J.K., H. Da Costa, D. Harrison, J.A. Kunze, M. Kupfer, and J.G. Thompson, "A Trace-Driven Analysis of the Unix 4.2 BSD File System", Proceedings of the Tenth Symposium on Operating Systems Principles, pp. 15-24, December 1985.
8. Ousterhout J., and F. Douglis, "Beating the I/O Bottleneck: A Case for Log-Structured File Systems", Operating Systems Review, 23(1), January 1989.
9. Patterson D., G. Gibson, and R. Katz, "A Case for Redundant Arrays of Inexpensive Disks (RAID)", Proceedings of the ACM SIGMOD Conference, pp. 109-116, June 1988.
10. Rangan P.V., and H. M. Vin, "Designing File Systems for Digital Video and Audio", Proceedings of the Thirteenth Symposium on Operating Systems Principles, pp. 81-94, October 1991.
11. Salem K., and H. Garcia-Molina, "Disk Striping", Proceedings of the Second International Conference on Data Engineering, pp. 336-342, February 1986.
12. Scott A., F. Ball, D. Hutchison, and P. Lougher, "Communications Support for Multimedia Workstations", Proceedings of the Third IEE Conference on Telecommunications (ICT'91), pp. 67-72, March 1991.
13. Vaitzblit L., "The Design and Implementation of a High Bandwidth File Service for Continuous Media", MIT Thesis, September 1991.

Performance Studies of Digital Video in a Client/Server Environment

Kathleen M. Nichols

Advanced Technology Group
Apple Computer, Inc.
One Infinite Loop, MS 301-4J
Cupertino, CA 95014
nichols@apple.com

Abstract. Networked multimedia services cover a wide spectrum of possible applications. One system architecture of interest is that of a fileserver delivering prerecorded digital movie clips for continuous time playback to clients. This problem is similar to that of delivery of real-time live video to clients, but is more forgiving in that some amount of the movie clip may be prebuffered at both fileserver and client to smooth out variations in the system. On the other hand, such a service requires fileservers capable of maintaining continuous streams of data, or their equivalent. The "fileserver delivering movie clips" is also at the heart of many proposed applications. Although, it is unlikely that all clients would be playing movie clips at once, it is important to understand how many simultaneously active clients a given network architecture can support.

Delivery of stored digital video from a shared fileserver to remote clients is a systems architecture problem with many interacting components. Key items include the physical network, the network protocols, the architecture of both client and server, filesharing and operating system software, and the characteristics of the disk drives used in video storage. If components of the system are not well-matched, we would like to locate the bottleneck components, find how they interact in the system, and explore means of correcting the problem.

This paper presents a study of a commercial system, Apple's QuickTime™ version 1.1, employed in a client/server environment. The measurements studies show clearly the pitfalls of a system that is "out of balance". Bottlenecks and upper bounds are identified, first a simple analytical model is employed to examine small improvements, then simulation is employed to further explore the space of possible system enhancements.

1 Introduction

Networked digital multimedia will require much more than higher bandwidth networks to be feasible. When multimedia implementations are studied, it is apparent the entire systems architecture must be designed to avoid bottlenecks. Some recent work in the literature has focused on previously neglected aspects of these system architectures, like disk scheduling [1,2,3] and at least one company has been presenting solutions that address networked multimedia as a systems problem [4,5]. Some analysis relevant to portions of the systems architecture for delivering continuous media has been presented [3,6], but there is still a lack of systems performance literature on networked multimedia. As commercial and experimental systems become available for study, this is certain to change. However, it is important that researchers become accustomed to thinking of their networked digital video work as part of a complex, interconnected system of components.

In this paper, a client/server environment is used to refer to a number of client personal computers or workstations connected to a remote machine which is acting solely as a fileserver (see figure 1). Client and server use a fileshare protocol to open, browse and read files. The information is exchanged across the network through transport and network protocols. The fileserver stores the digital movie in a disk or a disk array. We are concerned with delivering compressed video clips, or "movies", from a remote fileserver to a client machine over a digital network.

The focus on stored, rather than live digital video, affects the system in a number of ways. Live video may not have the central bottleneck of a single fileserver, unless a compositing or bridge server is used, and can avoid the storage bottleneck as well (unless recording). Retransmission is not possible for live video. Compression and decompression must be fast for live video (or not used). Replaying from storage allows some prebuffering, hence a more constant bit rate may be seen at the network and the server.

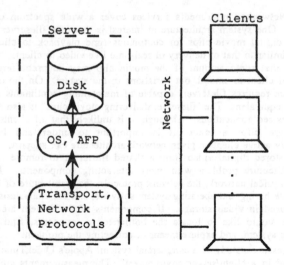

Fig.1. A client/server environment

The system of interest is modeled on the delivery of QuickTime™ movies from a single fileserver over the same network. We assume that the network protocol is request-response, like ATP, or fixed window, where each client may have at most one request, or window, of data outstanding. Requests queue for service at the fileserver and incur operating system overhead, disk access time, and disk transfer time proportional to the size of the data requested. The data is then passed to the network protocol to return to the requesting client. Upon receipt, the client may send the next request through the network to the fileserver.

The number of movie clients that can be supported by a server is affected by two major types of decisions. One is the choice of movie parameters, such as the size of a frame, its frame rate, and its decompression time. The second stems from system architecture parameters: the size of a packet, the average time to service a request, whether network I/O is overlapped with servicing requests to any degree, the interpacket gap, and the speed of the network.

A client playing a movie must issue specific requests through its filesystem over the network to the fileserver. Each of these requests is processed *independently* by the fileserver. This takes system software overhead and seek and transfer time at the disk to retrieve the data. Next, the data must be broken into packets and sent to the client over the network. This incurs an *interpacket gap* time and a packet transmission time for each packet. When the data arrives at the client, it will be buffered until needed (it may represent a portion of a frame or more than one frame). When the client's application is ready for it, each frame is decompressed and copied to the screen. A client has several options for requesting data: it may be requested on a strictly frame-by-frame basis , similar to live transmission, or it may request blocks of data from the movie track. The size of these blocks may be fixed or varied. Blocks may be requested only for the next frame needed or several blocks may be requested ahead, assuming viewing order. A prebuffering scheme may be used that tries to keep a buffer at the client above a low water mark. In the remainder of this paper, we use the term *transaction* to refer to the request and associated data. Note we assume there can be at most one transaction in the system for each client.

This author had access to a working commercial, though relatively low-performance arrangement for networking digital video. Major limitations are imposed by the lack of appropriate network protocols for digital video. In this work, measurement studies of QuickTime^TM 1.1 and more general simulation analysis are used to explore the space of enhancements to the general system architecture. We emphasize that a multimedia system is only as good as its weakest link, not its strongest.

2 Measurement Studies

To analyze the target system we neglect some components and simplify others. The measured system is composed of an application running on a client which displays the video stream ("plays the movie"), a local buffer where data from future frames is stored, the Apple Fileshare Protocol (AFP) and Apple Transport Protocol (ATP) whose overhead is combined here, and a physical connection viewed simply as a pipe of a specified bandwidth. The resulting system is shown in figure 2.

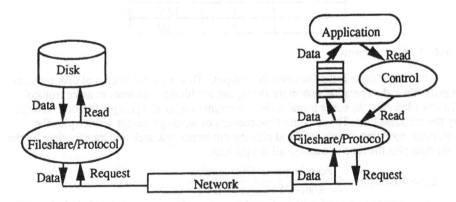

Fig. 2. Model of system

In addition to AFP/ATP overhead, the fileserver has a digital storage medium for the movies. In AFP/ATP, the client sends a short request packet (64 bytes for Ethernet) to the fileserver. The fileserver takes these requests in first-come, first-serve order. A response packet contains up to 4 kilobytes in 8 packets of 622 bytes each. A short release packet is then sent by the client.

For the tests, an eight frame per second variable bit rate compressed video clip with average frame size of 6400 bytes was put into a continuous loop on each client, resulting in about 410 kbps of data sent to each client in separate sessions (no multicast). A Macintosh IIci was used as a fileserver. One of the clients used an instrumented program to determine if frame loss occurred. Measurements were carried out on three types of digital storage for the movie clip: a standard hard disk (HD), a SCSI silicon disk (SD), and a RAM disk (RD). The movie clip was first played from the local hard disk of the instrumented client virtually loss-free. In table 1, the results of the fileserver experiments are summarized. The response time is the time in milliseconds between the fileserver's receipt of a client request and sending the first packet of the response.

The hard disk was not capable of supporting more than one client error-free, while the silicon disk showed minor errors at 2 clients, and was unwatchable at 3 clients. The major difference between the hard disk and the silicon disk is that the silicon disk has no seek time though it transfers at SCSI rates. The RAM disk, with no seek or transfer time, supported 3 clients error-free, and four clients with slight degradation (3% of frames were reported to be below 90% of frame rate, 2% of that below 25% of rate). At five clients, no client performed adequately. Although these results might be improved upon somewhat with a faster hard disk and a faster fileserver architecture, these results were useful in examining system bottlenecks.

Disk type	Number of clients	Response time
HD	1	25.8
	2	55.8
SD	1	6.0
	2	11.7
RD	1	5.2
	2	6.9
	3	12.9
	4	29.0

Table 1. Measurements

The first of these is the fileserver throughput. This is a combination of disk seek and transfer time, the operating system overhead, and the fileshare/network transfer protocol. On any fileserver, the upper bound is the maximum number of bytes per response divided by the response time. The response time consists of waiting time for service plus the minimum operating system overhead plus the minimum disk seek and transfer time for the total data plus the time to transmit all the packets.

$$\text{Server throughput} \leq \frac{\text{Pckts/response} \times \text{Pcktdatasize}}{\text{Wait+OS+Disk+Tx time}}$$

These numbers depend on the I/O architecture and the network bandwidth. For ATP, the maximum transfer block is 4 kbytes in 8 packets. Although the IIci showed an average interpacket gap of about 1.1 milliseconds, a machine with more efficient I/O, like the Macintosh Quadra, can reduce that gap to nearly zero, then the eight packets take a total of about 4 milliseconds on Ethernet. An operating system overhead of about 5 ms can be inferred for the IIci from the RAM disk measurements. Even if this overhead could be reduced to 1 ms, the upper bound on server throughput is 6.5 Mbps for a single client. Of course, with more clients, the waiting time becomes nonzero and the throughput decreases.

Next, we examine the limits to client throughput. A client needs to display one frame every frame time. Prebuffering frames at the client allows the rate to vary in the short term. As an upper bound for synchronous network reads, the interframe time, or spacing between frames, must be large enough to allow time for the average sized frame to be retrieved from the server, decompressed (crucial for software decompression, as in these measurements), and copied to the screen. For asynchronous reads, it is the larger of these two. This can be expressed as:

$$\text{Interframe time} \geq \text{Decompression} + \text{Display} + \frac{\text{Frame size} \times (\text{Server response+transmit time})}{\text{Pckts per response} \times \text{Pckt data size}}$$

or, for asynchronous reads:

$$\text{Interframe time} \geq \text{Max}\{\text{Decompression+Display}, \text{Fileserver return}\}$$

Then the client throughput upper bound is:

$$\text{Throughput} \leq \frac{\text{Average Frame size}}{\text{Interframe time}}.$$

This implies that it is not just the stream rate, but the frame rate which is crucial to the client. In our case of software decompression, the client becomes a serious bottleneck. With hardware to assist the decompression and display steps, the bottleneck moves to the server.

Finally, we turn to network throughput limits. Network throughput is upper bounded by the network capacity or bandwidth, though in practice that upper bound is not achievable. However, simulation studies which neglect fileserver and client overhead predict that we can support 5-7 1.2 Mbps streams on a single isolated Ethernet and 13-14 640 Kbps streams. [7] This is not presently a bottleneck.

These results suggest the possibility of amortizing the overhead of disk seeking and network transfer for a larger block of data. We will explore this approach, though simulation, later in the paper. A fileserver architecture that allows disk and connection scheduling and overlap with network transfer overhead should support more clients and some further simulations are underway in this area.

3 Exploring Bottlenecks through Analysis

As we have seen a large number of system parameters affect performance and many of these are stochastic values of some unknown distribution. For playback of stored video, prebuffering sufficient data at the client can prevent starvation when either the arrival of data from the network or the display rate of the data varies from the average data rate[1], and thus the data rate need only be maintained *on average*. Although this neglects important dynamics of the system, it gives bounds on the number of simultaneous movie reading (or writing) clients that can be attached to the server. For movie playback, it is quite conceivable that buffering can be used to overcome the effects of variance in the movie bit stream, thus the mean values are of primary importance, variance being important in

---

[1]QuickTime attempts to maintain a 64K to 256K buffer at the client.

sizing the buffers. Of course, life is rarely that simple and dynamic factors will affect performance, but results for different sets of system parameters can be compared constructively and ballpark figures for delays and maximum numbers of clients can be determined. In the following, mean parameter values are used to determine whether a system can deliver the required bit rate.

Assumptions and simplifications are made to permit a simple model of the system. The first assumption is that all components of the system operate on one transaction at a time. This assumption means that one client request transaction results in one disk access transaction and one client decompression and display transaction, all the same size. This is not realistic in the case of the client video because the natural transaction size is the frame, but works well for average values. Requests from the client can only be sent after it has received and serviced the previous transaction. This is quite reasonable for average value analysis and for synchronous I/O at the client. Fileserver request queueing and service then take place in parallel with client processing.

Two simplifications in the network architecture are made. First, only requests for data are sent from the client to the fileserver, acknowledge or release packets are ignored. The second is that network usage is low enough that contention on the network can be largely ignored and network transmission can be modeled by the physical transmission time alone.

A closed queueing network can be used to model this client/server system. A single fixed-capacity server is used to model the operating system overhead and the disk access time. A second such server is used to model the fileserver's protocol overhead and the network transmission time. Modeling these as separate servers implies that the server has the capacity to do asynchronous I/O; modeling them in a single server implies no asynchronous capability at all. Contention between server and clients for a shared network transmission medium is ignored, although client request are assumed to queue together. Thus, the disk, fileserver I/O, and the transmission of requests can all be modeled with simple fixed-capacity servers and the clients as simple delays, where the number of clients is equal to the number of transactions in the system. The model includes the case where some of the data must be retransmitted and some of that data must be refetched from the disk. Figure 3 shows the model.

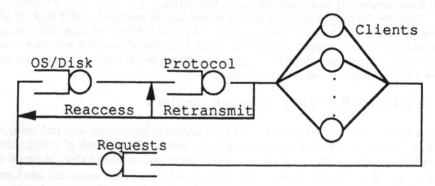

Fig. 3. Queueing model with retransmission for client-server system

Using the model of figure 3, mean value analysis (MVA)[8] was applied to the system and Claris Resolve[TM] was used to create a spreadsheet using average performance numbers. Parameters are entered in the spreadsheet, then a spreadsheet "button" is pushed, running a script which computes the MVA client bounds for those parameters. The button script is programmed to iterate until the mean frame rate is less than some percentage of the desired (input) frame rate. The resultant number of clients can be taken as an upper bound. For studying average behavior, this provides sufficient detail. More detailed studies of system dynamics will require modeling disk transactions whose size may differ from network transactions and client transactions that are actually movie frames which must be processed at a fixed rate, but may be of variable size.

The model was validated against the RAM disk measurements reported in the previous section. Seek and transfer times were set to zero and the OS response time set to the measured value for a single client. The spreadsheet script was set to stop at the last iteration before the frame rate falls below 95% of the nominal movie frame rate. The figure of merit becomes the total time in the system which is the time from issue of a client request until the entire transaction is received. In the measurements this is the request-to-release time, which is the request-to-response time of table 1 plus the time to transmit all eight packets. The spreadsheet showed four clients supported with a request-to-release time of 37.5 ms as compared to the measured value of 38.3 ms.

The primary usefulness of this spreadsheet is for predicting the relative effects of changes in the protocol or I/O architecture very quickly. In this section, two examples of its use are given. First, in figure 4, the number of clients is plotted for a range of average frame sizes and rates. The window size used was 10 packets, and maximum Ethernet sized packets of 1526 (1482 bytes of data) were used. For the first set, 40 ms of client overhead per frame was assumed; for the second set, 20 ms. The movies are assumed to be located in fileserver RAM or efficiently prebuffered so as to incur no access time. It should be noted that these are optimistic values by our current standards.

The plots in figure 4 show that the client overhead has minor effects at the lowest frame rates, 8 fps and 10 fps. As the frame rate increases and the (fixed) overhead becomes a larger proportion of the interframe time, it becomes impossible to support any clients at all. Note the dramatic effects of halving this overhead on the higher frame rates. Presently, client overheads appear to be considerably larger even than 40 ms.

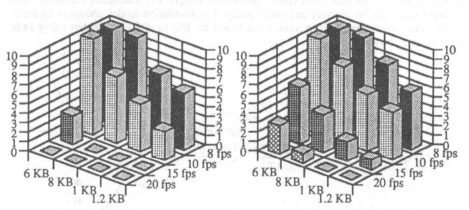

Fig. 4. Maximum number of clients: a) 40 ms client overhead b) 20 ms client overhead

In figure 5, the parameters of figure 4b) are used to carry out an experiment, first with a 10 percent retransmission rate, then with a 20 percent rate. This is a test of the system effects of retransmission alone since movie data is assumed to be located in RAM and thus there is no reaccess cost. Retransmitting ten percent of all transactions causes a reduction of one client in 50 percent of the test cases. Going from ten percent to twenty percent retransmissions causes 70 percent of those values to be further reduced by one client. Again, notice that the lower frame rates and smaller frame sizes are the most robust.

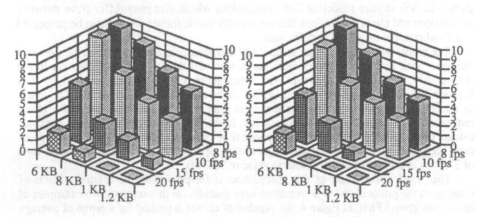

Fig. 5. Maximum number of clients a) 10% retransmissions b) 20% retransmission

This analysis suggests we should consider 1) raising the server throughput by decreasing the service time, 2) making larger requests to amortize server/network overhead, 3) use DMA and other architectural improvements, 3) use decompression hardware at the client, and 4) faster networks and/or hierarchical structured networks.

4 Simulation results

In this section, simulation is used to explore the effects of other solutions such as setting up rate-based sessions with fewer control packets exchanged and scheduling the storage and network resources intelligently and dynamically. The simulation model, shown in figure 6, allows this experimentation. A simulation based on this model was written using Jade Simulations sim++™.

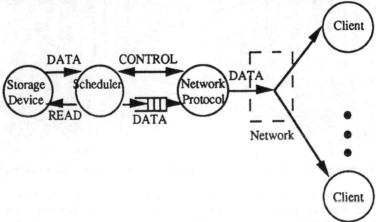

Fig. 6. Simulation model

Client processes first prebuffer a preset amount of data. This can be set to correspond to the prebuffering done by the QuickTime^{TM} data handler. Once the buffer is full, the client removes one frame from the buffer at time at intervals equal to the inverse frame rate, T. At each interval, the frame size, F, is selected by the client from a uniform distribution. After F bytes is deducted from the buffer, its occupancy is compared against a preset low water mark, M. If the occupancy is below the low water mark and there are insufficient server requests pending to bring it up to the low water mark, an additional block of data is added to the server requests pending. In order to create a model similar to the current QuickTime^{TM} data handler, the low water mark is set to 56K bytes and pending requests are incremented in blocks of 8 K bytes This block size is independent of, and should not be confused with, the request/response size permitted by the transport protocol. Block requests are handed to the transport protocol to partition.

First, server access via the AppleTalk Protocol (ATP) is modeled. The ATP model assumes that clients request eight AppleTalk maximum-sized packets (622 bytes, 578 bytes of data) at a time. The client sends a request to the server where it is placed in a FCFS queue (or FIFO). When the server comes to that request, it delays a service time selected from a uniform distribution, then posts each data packet to the client, waiting an interpacket gap between each packet. In the actual ATP , a release packet is posted from the client to the server after all eight packets are received error-free, but this is not included in the model. The average response time is the amount of time before the first packet of the response is generated and includes waiting time in the queue.

To validate the simulation model, measurement conditions were simulated and the results compared to the previously obtained measured results. The measured release-request time is the time between the client's issuing the request and issuing the release of that request. This is comparable to the entire service time at the server, though this does not include the release packet. Hence simulations should yield optimistic values.

Using measurements derived from previous RAM disk measurements, the average frame size was set at 6400 bytes (distributed between 2800 to 10,000 bytes), the frame rate to 8 fps, the service time between 4 and 9 ms, the interpacket gap (including transmission) to 1.5 ms, and the client processing time uniformly distributed between 72 and 93 ms. Each client requires a stream of about 410 Kbps of data from the server. No dropped frames were observed until 4 clients were added where fewer than 1% of frames were dropped by the target client. A five client scenario was completely infeasible. In table 3, we see that there is reasonable, although not perfect, agreement between simulation and measurement and note that both client upper bounds are in agreement.

Clients	measured	simulation
1	17.6	18.7
2	20.2	18.8
3	23.3	26.3
4	38.3	38.3

Table 3. Validation of Simulation Model: System Service Times

This simulation can be used to examine the effects of changing the protocol and network parameters and the fileserver architecture. In these studies, it seemed instructive to use values slightly more optimistic than those measured previously. A target movie is assumed which has an average frame size of 8000 bytes uniformly distributed between 4,000 and 12,000 bytes, a target display rate of 10 frames per second, a request service time uniformly distributed between 3 and 7 milliseconds, and a client processing time uniformly distributed between 40 and 60 ms. First, we experiment with changing packet size.

Using the ATP model described above, 3 clients can be supported. The sever response time from receiving the request until the first packet is sent is recorded in table 4, along with the percentage of the time the fileserver is busy. When the model is changed to use full sized Ethernet packets, and the interpacket gap increased by the added transmission time (0.7 ms), 6 clients can be supported, although some errors are present. This data is also recorded in table 4.

scheme	clients	response	server util
ATP	3	15.2	90%
LP	6^2	97.2	99
ATP(OL)	4	12.7	38
LP (OL)	7	40.4	31

Table 4. Simulation Experiments

The server utilizations are quite high because they include the service time per response and the interpacket gap for all packets in the response. If any of this I/O can be overlapped with OS service time, the utilization decreases. The fileserver CPU needs to be free from request servicing long enough to feasibly spend time prefetching the movie data. If the disk storage is accessed asynchronously, more efficient prefetching is expected. This assumes that clients tend to watch the entire movie, or significant amounts of each movie, at the frame rate without rewinding and that frames are contiguously located (pausing in the movie should not present a problem). These assumptions are clearly not going to be adhered to in all cases, but may apply to a large number of usage styles, and studying the effects of more random access will require more data and experience with such systems. Repeating the experiments with the packet output process completely overlapped with service time gives the results also included in table 4.

Since the longer request blocks yield lower server utilization and longer response times, these would probably work well in cases where the movies can be efficiently "pipelined", that is where our assumptions about no backtracking and contiguously stored frames hold.

Next, we examine changes in the connection protocol assuming some higher performance system parameters. Assume a 100 Mbps network, no interpacket gap and movies of 1.2 Mbps at 15 fps and 10 Kbyte average frame size (variable from 6 to 14 Kbytes). Simple first-come, first-served scheduling is done at the fileserver. Each client takes a randomly selected time between 30 and 50 ms for overhead per frame, but may perform that overhead while waiting for read ahead values to be returned from the fileserver. An erlang distribution time is assumed for the disk and OS overhead is a fixed value.

In table 5, the results show that the rate-based, connection-oriented scenario supported 33-200% more clients, depending on disk speed and transaction size. Results are taken to 95% confidence interval on the queue wait at the server for the transaction-based system and to the same level on the fileserver's total rate for the rate-based connection.

Further simulations are needed to explore the effect of intelligent scheduling at the fileserver, the interaction of the window size and the size of the block read from disk (if any), and tradeoffs in delays and buffer sizes for larger window sizes.

---

[2]A dropped frame rate of about 7% was noted.

window size (Kbytes)	OS time (ms)	disk service (ms)	transaction max clients	rate max clients
4	5	erlang(10)	1	3
7	5	erlang(10)	2	4
9	5	erlang(10)	4	6
4	2	erlang(6)	3	4
9	2	erlang(6)	7	10

Table 5. Simulation results

5 Conclusions

We have shown that a system for delivering stored digital video from a fileserver to clients connected over a network has a number of crucial components. Those most in need of improvement in our measured system were the disk access and the network transport protocol.

We developed some tools for studying client/server digital video and showed how to use them to explore a number of possible changes to our system and what they meant in terms of performance. Analytical tools permit swift evaluation of simple improvements. Simulations allow us to make the most drastic changes in system architecture. James Yee at U.C. Berkeley is working on extensions to the simulation model to compare a number of disk scheduling strategies, including his own [3].

This model is a very simple one, assuming clients view contiguous movie frames without rewinding, and the effects of simultaneous data transfers were not studied. Still, there seems to be a great deal to be learned from this model. Some approaches to scheduling disks for continuous media attempt to allow time for other activities [1] and we may attempt to include this into our simulations at some future date.

Presently, it appears that the greatest gains might be had from using hardware decompression at clients, setting up rate-based sessions for each client at the server, and intelligently scheduling both the fileserver disk and the fileserver's network transmissions.

References

1. D.P. Anderson, Y. Osawa, and R. Govindan, "Real-Time Disk Storage and Retrieval of Digital Audio and Video", Technical Report UCB/CSD 91/646, Computer Science Division, University of California, Berkeley.
2. P.V. Rangan, M.M. Vin, and S. Ramanathan, "Designing an On-Demand Multimedia Service", IEEE Communications Magazine, July 1992, pp. 56-64.
3. J. Yee, "An Analytical Model for Real-Time Multimedia Disk Scheduling", also in these proceedings.
4. J.E. Long, "Perspectives on Multimedia Communications", talk and slides, The Acceleration of Multimedia Communications Symposium, Stanford University, May 13-14, 1992.
5. F.A. Tobagi and J.E. Long, "Client-Server Challenges for Digital Video", Spring Compcon ' 92, pp. 88-91.
6 H.E. Meleis and D.N. Serpanos, "Designing Communication Subsystems for High-Speed Networks", IEEE Network, July 1992, pp. 40-46.
7. K.M. Nichols, Network Performance of Packet Video on a Local Area Network, Proceedings of the IEEE Phoenix Conference on Computers and Communications, April 1-3, 1992.
8. Raj Jain, The Art of Computer Systems Performance Analysis, Wiley, 1991.

Session 3: Media Synchronization

Chair: Riccardo Gusella, Hewlett Packard Laboratory

The Media Synchronization session, one of two events at the Workshop dedicated to synchronization (the other was a panel discussion), featured four papers. The first two papers dealt with modeling synchronization scenarios; the other two address system aspects.

The first paper, "Basic Synchronization Concepts in Multimedia Systems" by Luiz Rust da Costa Carmo, Pierre de Saqui-Sannes and Jean-Pierre Courtiat, proposes a formal notation based on set theory for modeling media synchronization problems. The basic concepts discussed by the authors are: flow, which defines an association between data elements and times; temporal signature, which is the time position of media units within a flow; flow transfer function; bundle, which represents a group of flows; bundle transfer function; and intra- and inter-flow synchronization. The basic concepts and notation are then applied to synchronization services for communication-oriented systems. The authors propose a taxonomy for both intra- and inter-flow synchronization based on the notions of temporal ordering and temporal signatures.

The second paper, "Synchronization in Joint-Viewing Environments" by Kurt Rothermel and Gabriel Dermler, addresses the synchronization issues in an important class of applications–those that provide basic mechanisms for joint manipulation of information by two or more physically separated users. The authors produce several data synchronization scenarios, ranging from the simple case of two users viewing the output of an application in a shared window, to the complex situation of multiple intercommunicating windows produced by various remote applications. The fundamental abstractions introduced by the paper are those of logical clocks, referencing systems, and active and passive bindings. These tools are then used to show how synchronization can be achieved in the various scenarios.

The third paper, "Synchronization of Multi-Sourced Multimedia Data for Heterogeneous Target Systems" by Dick Bulterman, argues that networked multimedia systems must deal with three classes of problems that are not present in single-user multimedia applications. These problems are synchronization control, resource allocation, and heterogeneous system support. The paper focuses on discussing the partitioning of synchronization control among the components of heterogeneous multimedia systems. The CWI Amsterdam Multimedia Framework provides a convenient background for the discussion, which centers around the concept of multimedia document. The synchronization specification provides a user with two major views of a document: a hierarchical view, which shows the logical components of a document and how they interact, and the virtual I/O channel view, which maps the logical components into a collection of virtual channels. Intelligent Information Objects provide resources, representation, and synchronization control.

In the last paper of the session, "An Intermedia Skew Control System for Multimedia Data Presentation", Tom Little and K. Kao observe that in systems that do not employ a channel-establishment resource-reservation protocol, synchronization among several media streams can be achieved by monitoring the skews and the various queue levels and by dropping and replicating media units selectively. The authors claim that this single mechanism is not sufficient and that it must be accompanied by a mechanism for managing queue over- and underflows. The paper offers three algorithms, one for the retrieval and transmission process, one for the playout process, and the last for the monitor and control process, which monitors the queue and skew values of each stream and implements the two intermedia synchronization mechanisms.

Basic Synchronisation Concepts
in Multimedia Systems

Luiz F. Rust da Costa Carmo, Pierre de Saqui-Sannes and Jean-Pierre Courtiat

LAAS-CNRS, 7 avenue du Colonel Roche, 31077 Toulouse Cedex, France
e-mail: {rust, pdss, courtiat}@laas.fr

Abstract. Multimedia systems were so far primarily addressed with communication and performance in mind. It is now acknowledged they essentially capture synchronisation problems, at communication and application levels, respectively. This paper proposes, at first, a terminology that may be convenient for both levels. Communication issues are then emphasised. In particular, multimedia information is presented as bundles of flows characterised by their temporal signatures. Formal definitions are then used to propose dedicated synchronisation services.

1. Introduction

So far, multimedia systems were primarily addressed with communication and performances in mind, particularly discussing the benefits of using high speed protocols [TAW 92] and media-dependent compression techniques. It is now acknowledged that multimedia applications primarily capture synchronisation problems [LIT 90] [STE 90]. Examples include lip-synchro within a teleconference application : the speakers' voice and image need to be transported within "compatible" time intervals and "simultaneously" presented to remote participants.

Such example indicates that both communication and application level issues have to be dealt with. In this paper, we propose a common basic terminology that may subsequently be refined to address each issue separately. Formal definitions are provided accordingly, making temporal constraints explicit. Emphasising on communication issues, multimedia information is presented as bundles of flows characterised by their temporal signatures. Formal definitions are then used to propose dedicated synchronisation services.

The paper is organised as follows: Section 2 briefly introduces a functional architecture for multimedia systems developed within distributed environments. Section 3 presents the formal framework that is used in Section 4 to discuss synchronisation services.

2. Functional Architecture for Multimedia Systems

A two-participant teleconference application will serve as introductory example to discuss synchronisation problems captured by multimedia object exchanges. Such exchange includes the capture of a certain multimedia object at participant 1, the object transmission via the network, and finally, its restitution at participant 2. Along that trajectory, a prime difficulty is to maintain consistency between, e.g., sound and image. Temporal attributes of the emitted object need to be captured, such that the remote participant will be able map the received information onto relevant presentation objects that are accessible to the user. [LIT 90] gives arguments for capturing spatial domain attributes too.

Within a teleconference application, the parallel handling of, e.g., video and voice, is an example of inter-media *composition*, as opposed to *co-operation* synchronisation which, e.g., schedules the right to speak. Apart co-operation, a teleconference poses new requirements in terms of resource reservation and inter-process communication [LIT 92] [JEF 92].

The rest of the section discusses a functional architecture that characterises a wide variety of multimedia systems developed within distributed environments. Five functional blocks are identified (fig. 1).

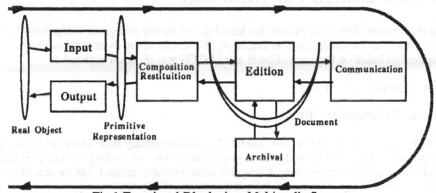

Fig.1 Functional Blocks in a Multimedia System

(1) The *Input/Output functional block* performs multimedia object acquisition and reproduction, including the following two transformations:
- Real object → Primitive (coded) representation : this transformation includes acquisition and binary coding of multimedia information contained by the object.
- Primitive representation → Real object : this opposite transformation restores primitive information on relevant presentation supports and builds up a real object.

(2) The *Composition/Restitution functional block* includes two transformations:
- Primitive representation → Document: this transformation translates the previous primitive representation, extracting temporal and spatial characteristics, so as to build a document. The latter is possibly edited at a later time.
- Document → Primitive representation: this opposite transformation translates spatial and temporal characteristics of a document into a primitive representation that is convenient for the target reproduction presentation support.

(3) The *Editing functional block* has ability to modify both temporal and spatial characteristics of some pre-existing document, so as to compose a new document from pre-existing ones. The opposite transformation decomposes a document into several subdocuments that will be considered in turn as separate documents. By pre-existing document, we intend a document whose composition or reception is underway. If not, that document was supposedly stored in some archival system.

(4) The *Archival functional block* converts a document into a static representation that is convenient for storage.

(5) The *Communication functional block* is intended to transfer documents through a communication service, depending on its offered quality of service.

Within the architecture discussed so far, a multimedia document can be understood as a set of multimedia, possibly synchronised, flows. Each flow consists of a sequence of so-called information units that have to comply with various temporal constraints. These concepts are formalised in next section, at an abstraction level intended to allow their application to any functional block defined so far.

3. Basic Concepts and Terminology

In this section, formal definitions are based on set theory and given independently of any particular formal description technique. We intend to use them as a basis of futher modelings based on Petri nets, Estelle and LOTOS specification approaches.

3.1. Elementary Flow

3.1.1. Information Unit

Definition. An *Information Unit* is defined as an elementary data whose semantics is relevant to the functional block it was output by. According to the previous functional architecture, we may distinguish between Information Units for real objects (IUr), primitive representation (IUp) and documents (IUd), respectively.

No assumption is made on the structure of IUs. The symbol for IUs identity is $\equiv$.

In the sequel, Sr will denote the countable set of information units associated with some real object. Sp et Sd will be used for primitive representations and documents, respectively. The countable set of information units is $Sr \cup Sp \cup Sd$.

3.1.2. Flow

The following notations are introduced: U is a subset of $Sr \cup Sp \cup Sd$, I is an index set and $(u_i)_{i \in I}$ denotes a family, indexed by I, of elements of U.

Definition. Let I be a non-empty finite index set, a *flow* is formally defined as a pair $<(u_i)_{i \in I}, t>$, where $t : I \rightarrow \Re^+$ denotes any total, strictly increasing function, associating a date $t(i) \in \Re^+$ with each index $i \in I$. The following law is therefore verified:
$$\forall i, j \in I, \ i < j \Rightarrow t(i) < t(j).$$

For simplification purposes, a flow $<(u_i)_{i \in I}, t>$ will be considered as a non-empty finite set of timely sequenced information units, noted $\{(u_i, t_i) \mid i \in I\}$ with $t_i = t(i)$.

The exact semantics of date t_i is not important at our abstraction level. For, e.g., the communication functional block, it may correspond to any date between the IU transmission process starting and end dates.

3.1.3. Flow Temporal Signature

Definition. Information units within a flow are characterised by their relative temporal positions, so-called the flow *temporal signature*. Positions are expressed through a sequence of dates w.r.t. the flow starting date. Given a flow $F = \{(u_i,t_i) \mid i \in I\}$, its temporal signature is defined as the family $\sigma = (\sigma_i)_{i \in} I$, the elements of which belong to $\Re^+$ and verify: $\forall i \in I$, $\sigma_i = t_i - t_0$ where $t_0 = \min \{t_i \mid i \in I\}$.

3.1.4. Flow Transfer function

Notation: S being a set, and 2^S being its powerset, we note 2^{S*} the restriction of 2^S to the non-empty subsets of S. S being a set of sets, we note $\cap S = \{x \mid x \in A \text{ for all } A \in S\}$ and $\cup S = \{x \mid x \in A \text{ for at least one } A \in S\}$

Definition. A *flow transfer function* characterises possible alterations suffered by a flow as it is processed by one of the functional blocks mentioned in Section 2. The flow transfer function associates a non-empty output flow subset with a non-empty input flow subset. Formally, be Fin=$\{(uin_i,tin_i) \mid i \in Iin\}$ and Fout=$\{(uout_j,tout_j) \mid j \in Jout\}$ an input and output flow, respectively. A flow transfer function between Fin and Fout is defined as any partial injective function FT: $2^{Fin*} \rightarrow 2^{Fout*}$ verifying the following two properties:

i)   Sout = range(FT) defines a partition over Fout, i.e. $\cup$Sout=Fout and $\cap$Sout=$\emptyset$.
ii) $\forall$sin $\in$ domain(FT), let sout = FT(sin), then
     $\forall(uout_j,tout_j) \subset$ sout, $tout_j > \min \{tin_i \mid (uin_i,tin_i) \in$ sin$\}$.

Note: infra ii) is valid for a global time. This assumption is made in the rest of the paper.

A flow transfer function will presumably not be ideal. In particular, as FT is a partial function, some elements of the input flow may have no mapping by FT, which represents the fact that some information units might be lost during the processing of a functional block.

Definition. A flow transfer function is *unitary* if at most one output flow element is associated with each input flow element. The following properties are implied :
        i) $\forall$sin $\in$ domain(FT), card (sin) = 1   ii) $\forall$sout $\in$ range(FT), card(sout) = 1.

Definition. A unitary flow transfer function is said *ordered* if it preserves the input flow ordering relation, that is:
        $\forall$ sin,sin' $\in$ domain(FT), let sout = FT(sin) and sout' = FT(sin') then
                $tin_a < tin_b \Rightarrow tout_c < tout_d$
        where   sin = $\{(uin_a,tin_a)\}$, sin' = $\{(uin_b,tin_b)\}$,
                sout =$\{(uou_c,tout_c)\}$ and sout' =$\{(uout_d,tout_d)\}$.

Definition. A unitary flow transfer function is said *univocal* if it associates one output flow element with each input flow element. Therefore, card (Jout) = card(Iin).

Definition. A unitary flow transfer function by which all information units are preserved is said to be *conservative*, which is expressed by:

$$\forall \, sin \in domain(FT), \text{ let } sout = FT(sin) \text{ then } uout_j \equiv uin_j$$
$$\text{where } sin = \{(uin_j, tin_j)\} \text{ and } sout = \{(uout_j, tout_j)\}.$$

Definition. A flow transfer function is said to be *ideal* if, being univocal and ordered, it further preserves the flow temporal signature ($\sigma out = \sigma in$).

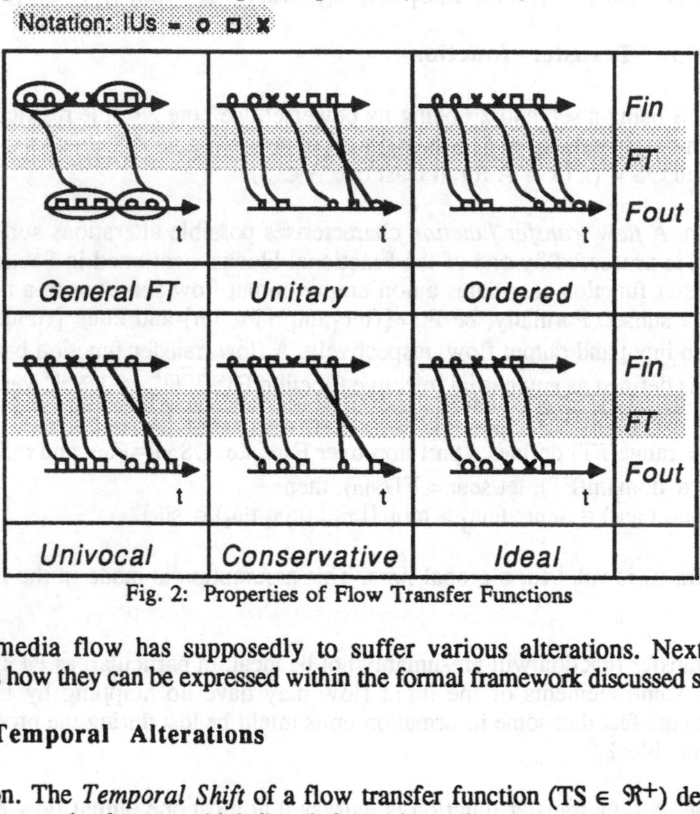

Notation: IUs - o □ x

Fig. 2: Properties of Flow Transfer Functions

A multimedia flow has supposedly to suffer various alterations. Next section discusses how they can be expressed within the formal framework discussed so far.

3.1.5. Temporal Alterations

Definition. The *Temporal Shift* of a flow transfer function (TS $\in \Re^+$) defines the delay between the dates at which flow information units are submitted to a functional block and the date at which they are delivered by the latter, respectively.

Definition. A flow transfer function FT is characterised by its temporal shift TS. Real implementation introduces a *jitter* for each transferred information unit subset. The jitter denotes those temporal drifts which possibly occur on both sides of the expected TS. Be Sout = range(FT). The jitter is formally defined by a function JS: Sout → $\Re$.

Let Fin=$\{(uin_i, tin_i) \mid i \in Iin\}$ be an input flow and Fout=$\{(uout_j, tout_j) \mid j \in Jout\}$ be an output flow obtained by application of a particular flow transfer function. TS and JS together satisfy the following property:

$$\forall \, sin \in domain(FT), \text{ let } sout = FT(sin) \text{ then}$$
$$\min \{tout_j \mid (uout_j, tout_j) \in sout\} = \min \{tin_j \mid (uin_j, tin_j) \in sin\} + TS + JS(sout).$$

Then, the *maximal jitter* (Jmax) of a certain flow transfer function FT is defined as the scalar Jmax = max {IJS(sout)I I sout ∈ range(FT)}.

In previous definitions, we reasoned about information unit subsets. We are now interested in the jitter of a single information unit (Figure 3). We consider an information unit within an output flow Fout obtained applying a unitary flow transfer function FT to an input flow Fin. The jitter is formally defined by a function JIU: Jout → $\Re$.

∀ sin = {(uin_i,tin_i)} ∈ domain(FT), let sout = FT(sin) = {($uout_j$,$tout_j$)}, then JIU(j) = $tout_j$ - tin_i - TS.

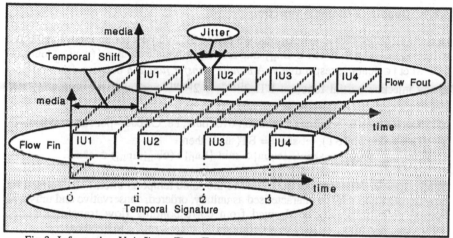

Fig 3: Information Unit Jitter, Flow Temporal Shift and Flow Temporal Signature

3.1.6. Intra-flow synchronisation

Jitter influences intra-flow synchronisation. We assume an input flow Fin={(uin_i,tin_i) I i ∈ Iin}, its corresponding output flow Fout = {($uout_j$,$tout_j$) I j ∈ Jout} obtained by a unitary flow transfer function FT.

Definitions. Synchronisation between Fin and Fout is characterised by a *ratio* : R = (card(Iin)) / (card(Jout)). Be Jmax the maximal jitter for the transfer, Fin and Fout are said *R-synchronous* for Jmax. Further, they are *isochronous* if Fout may be obtained applying an ideal flow transfer function to Fin.

3.2. Bundle

So far, we addressed a multimedia flow in particular. In this section, we discuss the parallel grouping of several flows within a given temporal range. Such a multimedia flow grouping is named a *bundle*.

Definition. Chosen a date t_0, be Φ a set of flows and N a non-empty finite index set. A bundle B is defined as a family, indexed by N, of elements of Φ. Let F_n = {(u_{ni},t_{ni})

$| i \in I_n \}$ be a flow indexed by n. A bundle $(F_n)_{n \in N}$ is denoted by $\{(u_{ni}, t_{ni}) \mid n \in N$ and $i \in I_n \}$ and verifies:

i) $\forall F_n \in B$, $t_n(\min(I_n)) \geq t_0$

ii) $\forall a, b \in N$, $\forall (u_i, t_i) \in F_a$, $\forall (u_j, t_j) \in F_b$ we have $(u_i, t_i) \neq (u_j, t_j)$

Notation. We note β the set of all elements of all the flows that belong to a bundle B, such that $\beta = \cup_{n \in N} F_n$.

As we did for flows, we can define bundle transfer functions along with their properties.

3.2.1. Bundle Transfer function

Definition. Using previous notations, we define βin and βout for an input bundle Bin $= \{(uin_{ni}, tin_{ni}) \mid n \in N \text{ et } i \in Iin_n \}$ and an output bundle Bout $= \{(uout_{mj}, tout_{mj}) \mid m \in M \text{ et } j \in Jout_m \}$, respectively. A *bundle transfer function* between Bin and Bout is any injective partial function BT: $2^{\beta in*} \to 2^{\beta out*}$ verifying the following two properties:

i) Sβout = range(BT) defines a partition over βout, i.e. $\cup S\beta out = \beta out$ and $\cap S\beta out = \emptyset$.

ii) $\forall s\beta in \in$ domain(BT), let $s\beta out = BT(s\beta in)$, then
$\forall (uout_{mj}, tout_{mj}) \in s\beta out$, $tout_{mj} > \min \{tin_{ni} \mid (uin_{ni}, tin_{ni}) \in s\beta in\}$.

Bundle transfer functions are submitted to the same temporal alterations defined for flow transfers and may be characterised as unitary, ordered, conservative and univocal. Formal definitions may be derived from corresponding flow transfer function definitions. Hereafter, we give new definitions that are specific to bundles.

Definition. A bundle transfer function is said symmetric if card(N) = card(M)

Definition. A symmetric, unitary, bundle transfer function is a Single-Flow-Correspondence (SFC for short) if all elements within one output flow are associated with elements of the same input flow. Formally, we have:
$\forall n \in N$, $\exists m \in M \mid ((\forall s\beta in \in$ domain(BT) $\mid s\beta in \subseteq Fin_n)$, $(BT(s\beta in) \subseteq Fout_m))$.

Definition. A bundle transfer function is said *ideal* if, for each input flow, it generates an output flow that preserves the input flow temporal signature. Be $Fin_n \in$ Bin an input flow and $Fout_m \in$ Bout an output flow. Be σin_n and σout_m their respective temporal signature. An ideal bundle function transfer is SFC, ordered, univocal; further, it verifies the following property:
$\forall n \in N$, $\forall m \in M \mid ((\forall s\beta in \in$ domain(BT) $\mid s\beta in \subseteq Fin_n)$, $(BT(s\beta in) \subseteq Fout_m))$,
$\sigma in_n = \sigma out_m$.

3.2.2. Inter-flow synchronisation

Definitions developed for intra-flow synchronisation can be extended to address now inter-flow synchronisation within a bundle.

We assume an input bundle Bin, its corresponding output bundle Bout obtained by a SFC bundle transfer function.

Definitions. Synchronisation between Bin and Bout is characterised by a *ratio* : $R = \Sigma_N \text{card}(\text{Iin}_n)/\Sigma_M \text{card}(\text{Jout}_m)$. Be JBmax the maximal jitter in the Bundle transfer, Bin and Bout are said *R-synchronous* for JBmax. Further, they are *isochronous* if Bout may be obtained applying an ideal bundle transfer function to Bin.

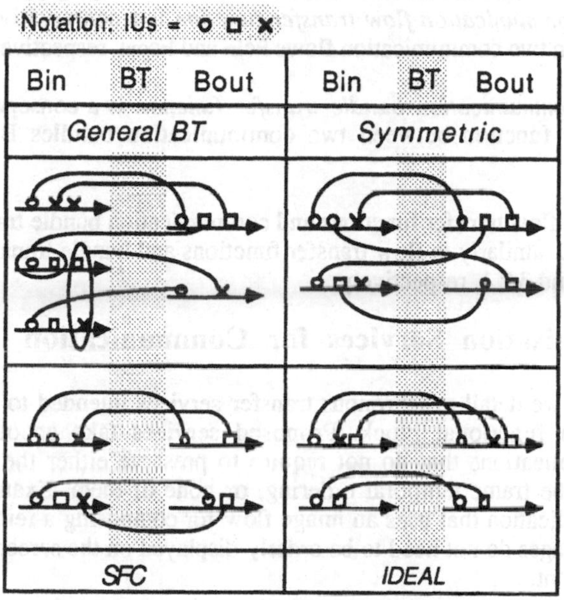

Notation: IUs = o □ ✗

Fig. 4: Properties of Bundle Transfer Functions

3.3. Application to the Communication Functional Block

Formal definitions given so far may be particularised to the communication functional block. For space reason, neither multicast nor broadcast transfer will be discussed.

The communication functional block is characterised as follows:
- Input and output information are of the same type, namely a document,
- Both flow and bundle transfer are always unitary (no multicast), and
- Both flow and bundle transfer functions are always conservative, i.e. the concept of information unit is preserved: a certain residual error rate is assumed acceptable for a IU, without requesting the definition of a new IU.

Definition. An Information Unit within a multimedia document can be encoded using a (Identifier, Type, Length, Value) pattern, enhanced with a so-called *Temporal Duration Interval*. The latter characterises the temporal duration during which the IU should be restituted on the relevant presentation support.

Within a *communication flow* Fc, information units are of document type. A communication flow is defined as $Fc = \{(ud_i, t_i) \mid i \in I\}$ with $ud_i \in Sd$.

Similarly, a *communication bundle* is defined as $Bc = \{(ud_{ni}, t_{ni}) \mid n \in N \text{ et } i \in I_n\}$ with $Fc_n = \{(ud_{ni}, t_{ni}) \mid i \in I_n\}$.

Definition. Be a bundle $Bc = \{(ud_{ni}, t_{ni}) \mid n \in N \text{ et } i \in I_n\}$ and P a set of sites involved in the communication process. A communication bundle is said to be multi-site if there exist a function $D: Bc \rightarrow P$ such that $\forall Fc_n \in Bc, \exists p \in P \mid p = D(Fc_n)$.

Definition. A *communication flow transfer function* is a conservative flow transfer function between two communication flows Fcin and Fcout, respectively.

Definition. A *communication bundle transfer function* is a conservative and SFC bundle transfer function between two communication bundles Bcin and Bcout, respectively.

Communication flow transfer functions and communication bundle transfer functions can be classified similarly to flow transfer functions and bundle transfer functions in Sections 3.1.4 and 3.2.1, respectively.

4. Synchronisation Services for Communication

In this section, we detail synchronous transfer services intended to be used by the communication functional block. Proposed services take account of various multimedia applications that do not require to preserve either the flow temporal signatures, or the frame temporal ordering, or none of them. Examples include a multimedia application that uses an image flow for composing a remote screen layout. Clearly, images do not need to be orderly displayed on the screen; only the final result is important.

4.1. Intra-flow Synchronisation

A basic parameter in synchronous flow transfer services is the maximal temporal shift that is acceptable to the user. The gap between that value and the actual propagation delay may be, indeed, exploited for buffering information. Such facility will be useful for implementing the four synchronisation services proposed in the sequel.

The following notations will be used: ATS denotes the temporal shift allowed by the user. An Information Unit (IU) is characterised by its maximal acceptable jitter *Jdef* and its emission date tin. Additionally, we introduce Rdef, the acceptable ratio between the number of received and transmitted IUs. We call *validity interval* the expected IU reception time interval; it equals [tin+ATS-Jdef, tin+ATS+Jdef].

Definition: Φ being the set of all flows, a synchronous flow transfer service may be characterised by a function SF: $\Phi \rightarrow \Phi$. Relying on formal definitions provided in Section 3 concerning a communication flow transfer function (CFT), each synchronous flow transfer service must verify the following general property:

$\forall$Fcin $\in \Phi$, let Fcout = SF(Fcin) then
$\forall$(udout$_j$,tout$_j$)$\in$ Fcout, $\{(udout_j, tout_j)\} = CFT(\{((udin_j, tin_j)\}) \mid (udin_j, tin_j) \in$ Fcin.

An overview of the proposed four services is given below. Detailed comments follow.

	Service	IU Temporal Ordering	Temporal Signature
1	Fully-loose	Uncontrolled	Uncontrolled
2	Ordered-loose	Guarantied	Uncontrolled
3	Tightly-coupled	Guarantied	Delivery, even outside the IU validity interval
4	Strictly-coupled	Guarantied	Exception raised if the IU cannot be delivered within its validity interval

(1) *Fully-loose service*: neither the IUs temporal ordering, nor the flow temporal signature is preserved. This service looks like a datagram communication service. Formally, a fully-loose service follows the general synchronous flow transfer service definition. Further, there is no assumption on the communication flow transfer function alterations (TS and Jmax).

(2) *Ordered-loose service*: the IUs temporal ordering is preserved, but the flow temporal signature is not. Because of the former property, this service looks like a connection-oriented service. Nevertheless, it is not intended to preserve all the temporal constraints associated with the IUs occurrences within the flow. Formally, an ordered-loose service follows the general synchronous flow transfer service definition and verifies the following properties: i) the communication flow transfer function is ordered ii) there is no assumption on the communication flow transfer function alterations.

(3) *Tightly-coupled service*: both the IUs temporal ordering and the flow temporal signature are preserved. Whether an IU could not be delivered within its validity interval - as expressed by (tin+ Ttrans +/- Jdef) - the IU will be delivered anyway, without caring of preserving the flow temporal signature. Formally, a tightly-coupled service follows the general synchronous flow transfer service definition. Further, the communication flow transfer function is ordered and verifies TS = ATS.

(4) *Strictly-coupled service*: as opposed to the previous service, an IU will not be delivered outside its validity interval. Formally, a strictly-coupled flow service follows the general synchronous flow transfer service definition. Further, the communication flow transfer function is ordered and verifies the following properties: i) TS = ATS ii) Fcout and Fcin are Rdef-synchronous for Jdef.

There is a noticeable progression in the way temporal constraints are dealt with; in particular, the "strictly-coupled" transfer is very strict in regards of the information validity interval. Conversely, the "tightly-coupled" transfer does not exclude that overdelayed information units can be passed, anyway, to the user.

In addition to the four basic services discussed so far, we present a new synchronisation service that we name *memorised-transfer service*. The latter makes it possible for a user to specify that a certain information unit should not be delivered in case previous information units were "lost" during the flow transfer.

Definition : Be M a partial function $M: Fcin \rightarrow 2^{Fcin}$. A memorised-transfer service verifies the following property:

$\forall Fcin \in \Phi$, let $Fcout = SF(Fcin)$ then

$(\forall (udin_i, tin_i) \in domain(M) \mid \{(udin_i, tin_i)\} \subseteq domain(CFC))$,

$(\forall x \in M(udin_i, tin_i), CFT(\{x\}) \subseteq \{CFT(udin_a, tin_a) \mid a \in In \text{ and } a < i\})$

Such service is particularly useful in applications where IUs are presumably semantically related. Examples include compressed video flow transfer, for which IUs are in fact video frames. A frame is often build up considering only changes w.r.t. previous frames, reporting, e.g., only the moving parts of a filmed scenario. In such situation, it may be useful to prevent a frame referring to possibly "lost" frames to be delivered.

4.2 Inter-flow synchronisation

In this section, we investigate how to build up inter-flow synchronisation services relying on the intra-flow synchronisation services discussed so far. In practice, a bundle communication transfer is made up of a set of individual flow transfers, the temporal synchronisation of which may be guarantied by using synchronous flow transfer services. Therefore, the loose and loose-ordered flow services suffice to cover all possible configurations in regards of possible relationships between bundle and flows transfer services.

Any information unit to be delivered within a bundle will be submitted to i) an intra-flow service synchronisation, depending on which flow the IU belongs to, and ii) an inter-flow service synchronisation, depending on which bundle the previous flow belongs to.

Bundle and flow services can be composed in various ways, ranging from an absence of synchronisation requirements up to very strong synchronisation relationships.

We define two bundle synchronous transfer services whose respective characteristics are as follows:

	Bundle Transfer	Temporal ordering of IUs	Temporal signature
1	Fully-loose	Uncontrolled	Uncontrolled
2	Ordered-loose	Guarantied	Uncontrolled

Definition: Ω being the set of all bundles, a synchronous bundle transfer service is formally characterised by a function $SB: \Omega \rightarrow \Omega$. Relying on formal definitions provided in Section 3, we may express the property guarantied by a fully-loose bundle service. Be BTC a bundle communication transfer function with $TS = 0$:

$\forall Bcin \in \Omega$, let $Bcout = SB(Bcin)$ then

$\forall (udout_{mj}, tout_{mj}) \in Bcout$,

$[(udout_{mj}, tout_{mj})] = BTC(((udin_{mi}, tin_{mi}))) \mid (udin_{mi}, tin_{mi}) \subset Bcin$

If BTC is ordered, then the service is said ordered-loose.

Similarly to inter-flow services, we define a *memorised-bundle transfer service* to be applied within a bundle. The main difference is that the information units composing the conditional set for delivering is now made up of all information units of all flows in the bundle.

6. Conclusions

Synchronisation issues in multimedia systems were so far addressed in informal ways. In this paper, we introduced a formal framework related to a five blocks, functional decomposition of multimedia systems developed within distributed environments. Although our prime concern was the communication block, we started discussion on a more general and abstract concept of Information Unit. We viewed a multimedia information as bundle of IU flows. A flow is basically characterised by the concept of temporal signature. Whether the latter and the temporal ordering of IUs are needed to be preserved is the question to be debated in defining synchronous transfer services. Distinction have been made between intra- and inter-flow synchronisation services, for flow and bundle transfer functions, respectively.

The question should now be asked of how the proposed synchronisation services can be used in communication architecture as service primitives. Our first investigations indicate that a connection-oriented service primitives is a good candidate as the basis for further development of a synchronisation layer [YAV 92]. Other directions for future work include the definition of protocol mechanisms intended to provide the proposed synchronous services. Next step will deal with an assessment of existing Formal Descriptions Techniques (particularly Time Petri nets, Estelle and LOTOS) w.r.t. to their ability to cope with those temporal constraints and synchronisation mechanisms newly introduced by multimedia applications.

References

[JEF 92] K. Jeffay, D. L. Stone and F. D. Smith, Kernel Support for Live Digital Audio and Video, Computer Communications, Vol. 15, NO. 6, July/August 1992, pp.388-395.
[LIT 90] T.D.C. Little and A. Ghafoor, Synchronisation and Storage Models for Multimedia Objects, *IEEE Journal on Selected Areas in Communications*, Vol. 8, NO. 3, April 1990, pp.413-427.
[LIT 92] T.D.C. Little and A. Ghafoor, Scheduling of Bandwidth-constrained Multimedia Traffic, Computer Communications, Vol. 15, NO. 6, July/August 1992, pp.381-387.
[STE 90] R. Steinmetz, Synchronisation Properties in Multimedia Systems, *IEEE Journal on Selected Areas in Communications*, Vol. 8, NO. 3, April 1990, pp. 401-412.
[TAW 92] W. Tawbi, S. Dupuy and E. Horlait, High Speed Protocols: State of the Art in Multimedia Applications, Proceedings of the Fourth International Conference on Information Networks and Data Communication, March 16-19, Espoo, Finland.
[YAV 92] R. Yavatkar, MCP: A Protocol for Coordination and Temporal Synchronisation in Multimedia Collaborative Applications, Proceedings of the 12th International Conference on Distributed Computing Systems, June 9-12, Yokohama, Japan.

Synchronization in Joint-Viewing Environments

Kurt Rothermel, Gabriel Dermler

University of Stuttgart
Institute of Parallel and Distributed High-Performance Systems
7000 Stuttgart 80, Germany
E-mail: {kurt.rothermel, dermler}@informatik.uni-stuttgart.de

Abstract

Recent technology advances have made computer-based multimedia cooperation among distributed users feasible. Joint-Viewing (JV) is one such proposed cooperation concept. Correct interaction in it requires synchronization on various levels. In this paper we focus on synchronization aspects on the data stream level. We present several JV scenarios and state for each of these requirements concerning interstream synchronization. Then we introduce basic abstractions of a synchronization scheme, which is powerful enough to model a variety of JV configurations with their various synchronization needs. The presented scheme is based on the notion of a referencing system, which consists of a set of logical clocks, each defining a different time zone in the referencing system. Logical clocks provide timing information for input and output devices, which are used to produce and consume data in a synchronized fashion.

1 Introduction

Advances in the communications and computer technology provide the technical foundation to integrate digital audio and video into today's distributed information processing systems. This is of great advantage for many application areas, especially for those, where users communicate and cooperate to achieve a common task. Joint Viewing (JV) systems are considered to be the basis for many applications in the field of computer-supported cooperative work [ElGi91], such as desktop conferencing [Cr90], joint-editing [FiKr88] and electronic classroom applications [ScCe87]. They provide the basic mechanisms for jointly viewing and manipulating information, user synchronization and user interaction, where the latter may be supported by appropriate video and audio channels.

Cooperative schemes go far beyond classical client/server schemes, where the clients are usually strictly isolated from each other. In cooperative environments, clients cooperate with each other in order to achieve a common task. This implies specific interaction patterns resulting in complex communication and synchronization requirements on various levels of abstraction.

In this paper, we focus on synchronization aspects on the data stream level [CoGa91]. Stream synchronization can be employed between data units within a continuous stream, called intrastream synchronization, and between data units of different streams, which is referred to as interstream synchronization. Various concepts and methods for intrastream synchronization have been proposed [WoMo91], [Fe91], [AlCa91], and most of them can be used for a wide range of applications. In contrast, the requirements imposed on interstream synchronization strongly depend on the corresponding application scenario. In JV scenarios, the sources and/or sinks of streams that have to be synchronized may reside on

This work was supported in part by the EC RACE-II research initiative, within the frame of the RACE project no. 2060 "Coordination, Implementation and Operation of Multimedia Services" (CIO). The views contained in this document are those of the authors, and should not be interpreted as representing official policies endorsed within the CIO project.

different sites, where streams may be of continuous or non-continuous nature. Moreover, streams to be synchronized may originate in different time zones of the same timing system.

The remainder of the paper is structured as follows. In Sec. 2, we introduce some terminology for expressing interstream synchronization requirements and apply this terminology when introducing a number of important JV scenarios. Then, in Sec. 3, we present the basic abstractions of a synchronization scheme, which is powerful enough to model a variety of JV configurations with their various synchronization needs. The presented scheme is based on the notion of a referencing system, which consists of a set of logical clocks, each defining a different time zone in the referencing system. Logical clocks provide timing information for input and output devices, which is used to produce and consume data in a synchronized fashion. Finally, we conclude with a brief summary and an outlook on future work.

2 Synchronization Scenarios in Joint-Viewing Environments

Joint Viewing was conceived as a computer-supported equivalent of a personal meeting, where people convene to jointly work on a work item by developing the work item and by exchanging visual and audible information among them. Joint Viewing allows these users to stay physically at their workstation and travel electronically instead.

A Joint Viewing scenario comprises users, applications and communication relationships among them. Users interacting with an application are presented with the same view of the application output. By introducing the notion of a shared window for the physical representation of the shared output we may alternatively say, the users view shared windows generated by applications.

Users can generate input to an application. The latter either can cope with multiple user input or requires external support of the Joint Viewing mechanism to select one of the users' input. Finally, the users are able to cooperate among each other by exchanging information about the jointly viewed shared windows. For this, they employ user-to-user communication means such as audio or video links or pointing tools, called telepointers, to point at objects in shared windows.

Figure 2.1: A Joint Viewing environment

A typical Joint Viewing scenario is shown in Figure 2.1. It involves joint viewing of an application window, the use of a pointing tool to point within the window and user-to-user communication via audio and video links. The applications may run on either of the depicted machines or on a separate machine connected to the network.

The concept of Joint-Viewing as presented here, was defined within the "Joint-Viewing and Teleoperation Services" (JVTOS) working group of the CIO project. We refer to [DeFr92] for a detailed description of the JVTOS reference model.

In the remainder of this section, we combine the communication elements introduced above (shared windows, user-to-user links) to a set of scenarios with different requirements. We omit considering user input, since it is independent of data stream synchronization. We state synchronization requirements for both continuous and discrete streams.

Continuous streams, such as audio and video streams, comprise a sequence of periodic data units. We do not supply a final definition (or classification) of discrete streams. For this paper, we take the "usual" (text/graphics) output of an application as a typical representant of discrete streams comprising data units occurring at more or less arbitrary time intervals.

2.1 Synchronization Parameters for Joint-Viewing Scenarios

A data stream is generated by an input device (I) and consumed by an output device (O). We refer to a data stream by referring to the corresponding (I,O) tupel. (I,O) is unique since only one stream is attached to an output device[1]. Streams comprise data units. The temporal behavior of these can be observed both at the input and output device. We use I to refer to the input device I as the observation point of the temporal behavior of the generated data and O to refer to the output device as the observation point of consumed data. Temporal behavior is specified by referring to observation points by means of synchronization parameters.

For intrastream synchronization, requirements are expressed in terms of two parameters: delay and jitter. The first one refers to the maximal delay of a data unit of a stream (I,O) between its input and output at the corresponding devices. The second denotes the variance of delay allowed (maximum delay - minimum delay)[2]. Requirements have to state how these values are to be bounded.

The selection of delay bounds considered convenient by users depends on the application used and the user involved. A user requesting fast interactivity with an application, will request low delay bounds for the application output. Similarly, an application requiring frequent input, will be expected to exhibit low output delay. For continuous streams tight jitter bounds are required. Experimental values for audio and video are compiled in [HeSa90]. For discrete streams usually no explicit jitter bounds are specified, thus bounding jitter only to the selected delay bound. Yet, in some cases, as for instance for computer animation sequences, it may be necessary to specify tighter jitter bounds.

For specifying interstream synchronization requirements the skew parameter is used. In order to use it in cooperative environments, we define the following construct:

Skew[(P_1, P_3) -> (P_2, P_4)] < skew_bound.

---

1 We use indices to distinguish between different input and output devices. Input and output device of a stream have the same primary index. Output devices fed by the same source are distinguished by a secondary index.

2 Of course, the values may be defined based on a probabilistic approach.

P_1, P_2, P_3 and P_4 denote observation points of data streams (i.e. input/output devices). The construct above implies that the same data units are observed at P_1 and P_2. The data units can be physically the same, for instance if P_1 and P_2 denote the source and sink device of a data stream. However, it is only required that the data unit content and sequence number is to be the same, as is the case if at P_2 copies of units are visible which were observed at P_1. An analogous relationship is implied for P_3 and P_4.

Let D_1 denote an arbitrary data unit observed at P_1 at time t_1 and D_3 a data unit observed at P_3 at time t_3 (see Fig. 2.2). The same two data units are observed at P_2 and P_4 respectively, at times t_2 and t_4. The skew expression from above simply states, that if between D_1 and D_3 a certain distance in time (d1) was observed at points P_1 and P_3, the distance (d2) between D1 and D3 as observed at points P_2 and P_4 does not differ from d1 by more than skew_bound. More formally:

Skew[(P_1,P_3) -> (P_2,P_4)] < skew_bound
<=> for all D_1, D_3:
    $| (t(P_3,D_3) - t(P_1,D_1)) - (t(P_4,D_3) - t(P_2,D_1)) | <$ skew_bound,
        if $t(P_3,D_3) - t(P_1,D_1) \geq 0$ and
    $| (t(P_1,D_1) - t(P_3,D_3)) - (t(P_2,D_1) - t(P_4,D_3)) | <$ skew_bound,
        if $t(P_3,D_3) - t(P_1,D_1) < 0$.

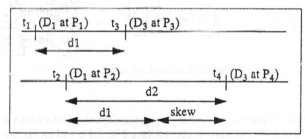

Figure 2.2: definition of skew

As will be shown in section 3, time can be related to different referencing systems. For instance, if time denotes the sequence number of data units of a specific reference stream, all data units of a second stream carry as timestamp the sequence number of the last data unit observed in the reference stream. In such cases, the skew construct is to be augmented by stating which of the observation points observes the reference data units.

The definition given above applies both to live and storage data sources. In the latter case, not creation time is observed initially but the specification of presentation time, as stored together with the data units, for instance as timestamps.

The defined skew construct proves powerful in specifying synchronization requirements between distributed observation points in cooperative environments. For instance, consider an input device generating data units for two users watching their output devices $O_{1,1}$ and $O_{1,2}$. We require that the consumption of the same data units occur at approximately the same time (not more than skew_bound apart), i.e.:

Skew[(I,I) -> $(O_{1,1}, O_{1,2})$] < skew_bound.

It should be remarked that specifying delay and jitter bounds for single streams may automatically bound the skew between them as well. However, for interstream synchronization the primary requirement remains that a skew bound is to be ensured, regardless of how it is done. It is up to the implementation to decide (under consideration of constraints concerning resources, location of sources and sinks, etc.), what requirements are to be derived ultimately.

2.2 Synchronization Scenarios

Scenario 1

In this most simple case shown in Fig. 2.3a, two users view the output of an application through one shared window. There is no communication (i.e no cooperation) between the users. Therefore, synchronizing the contents of the two shared windows is not necessary, i.e.:

Skew[(I_1, I_1) -> $(O_{1,1}, O_{1,2})$] = arbitrary.

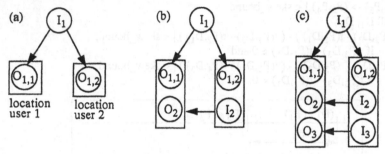

Figure 2.3: scenarios 1 and 2

Scenario 2

The previous scenario is augmented by one-way communication links which are employed by user 2 to refer to the shared window content (see Fig. 2.3 b and c). He may use a telepointer to point at objects in the window, an audio link to comment on them and perhaps a video link to show real objects relating to the shared window content or to support audio communication. The scenario reflects a master-slave relationship where the master gives explanations about the objects jointly viewed. The users may switch their roles, thus making information exchange in both directions possible. However, at any time only one-way communication is possible.

In a first step (Fig. 2.3b), we assume that one communication link exists. One new requirement has to be met: data units generated by user 2 relating to a shared window content are to be received by user 1 while viewing the same shared window content. For example, for a telepointer link, this implies that its movements are visible in front of the same window background at both user sites. Formally, the new requirement is:

(1) Skew[$(O_{1,2}, I_2)$ -> $(O_{1,1}, O_2)$] < skew_bound.

The bound value depends on the characteristics of objects referred to. If they change rapidly, the bound will have to be correspondingly low. At the other extreme, static objects may not require any skew bound at all.

Consider now the case that both an audio and a telepointer stream are employed by user 2 (Fig. 2.3c). The relation between each of these and the shared window content is as required in (1) with possibly different bound values. However, since the two streams may be related to each other as well (e.g. user 2 saying: "I mean the object I point to"), we state a requirement between these streams as well. The same set of requirements is necessary, if user 2 employs audio and video communication. Thus, we require:

(1) Skew[$(O_{1,2}, I_2) \to (O_{1,1}, O_2)$] < skew_bound
(2) Skew[$(I_2, I_3) \to (O_2, O_3)$] < skew_bound.

The skew bound for (2) may be different depending on the involved stream types. We expect that it will be in the range of the jitter bounds of the streams. If the jitter bounds differ, we expect that specifying the skew bound in the range of the largest is a reasonable approach. Own experiments with a telepointer and an audio link have shown, that values up to 0.5 sec are acceptable, if the telepointer is moved at usual mouse pointing speed.

A third case allows communication via telepointer, audio and video at the same time. The specification of the requirements is developed like in the previous case.

Scenario 3

The scenario depicted in Fig. 2.4a differs from the last one in that two-way communication between the users is allowed like in a conference situation. It requires that things occurring at one user location are perceived as occurring at virtually the same time at the location of the other user. For instance, for audio the requirements would have to ensure that the users are able to discuss like in a face-to-face meeting. In the following we only consider communication between users via one medium. Extending the scenario to simultaneous communication via multiple media (like in the previous scenario) leads to analogous requirements. Formally, we require:

(1) Skew[$(I_1, I_1) \to (O_{1,1}, O_{1,2})$] < skew_bound
(2) Delay[I_2, O_2] < nnd.
(3) Delay[I_3, O_3] < nnd.

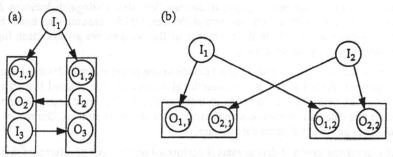

Figure 2.4: scenarios 3 and 4

Requirement (1) states that the content of the shared windows should not diverge more than specified by the skew bound, while (2) and (3) are requirements which are the direct consequence of the conferencing requirement. The delay is restricted in both directions to a value which is not noticeable to humans. For audio and video the jitter bound values of [HeSa90] can be applied. For telepointing a value below 0.4 sec was found satisfactory.

Scenario 4

Joint-Viewing is not restricted to one shared window. An application can generate content for many windows, which may be required to be presented in a synchronized fashion. This may be the case, even if the windows are generated by different possibly remote applications. A good example for this is provided by scientific visualization, where experimental results may be viewed on line on the computer display, and compared with simulation results generated by a computer program and presented on the computer display.

The requirements for such a situation with two users (see Fig. 2.4b) are expressed as:

(1) Skew[(I_1, I_2) -> $(O_{1,1}, O_{2,1})$] < skew_bound
(2) Skew[(I_1, I_2) -> $(O_{1,2}, O_{2,2})$] < skew_bound.

These requirements arise from presenting different shared windows. The scenario can be augmented by communication links, as it was done for scenarios with one shared window. The resulting requirements are identical to those presented in the context of these scenarios.

The skew bounds depend on the applications involved. For instance, for many cases of scientific visualization a very low bound can be expected, since matching simulation and experimental results correctly is of essence. In contrast, for a travel agency, where a still picture of a hotel is presented in one window and text information about cost appears in another, the skew bound can obviously be less strict.

3 Abstractions for Data Stream Synchronization

In the following, we will describe a very simple but powerful model for stream-level synchronization in distributed multimedia systems. This model will serve as a foundation for the programming abstractions our prototype will be based upon. A more comprehensive description of this model will be provided in [DeRo93].

Some of the basic abstractions are very similar to that introduced in [AnHo90]. We also define *logical devices*, such as virtual microphones, speakers, video windows, telepointer devices, that can be mapped to *physical devices*. We also distinguish between *input devices*, which produce data units, and *output devices*, which consume data units. Since physical devices play no role in the remainder of this section, we always mean logical devices when using the notion device.

Information can be conveyed from an input device to one or more output devices by means of *data streams*. A data stream is a unidirectional communication channel that transfers data units. Each data unit may have one or more references (or timestamps), that temporally relate this data unit to other data units of the same and other streams. Our model covers continuous as well as discrete data streams.

Basically, synchronization of data streams is performed on the basis of references that correlate data units of different streams with regard to their relative display time. References can be realized in many different ways, such as implicit (e.g. sequence numbering) and explicit timestamping [AnHo90], or interleaving of data streams [LeBa90]. In our model, we provide the concept of a *referencing system (RS)*, to group a number of data streams that may be synchronized on the basis of a certain referencing method. Consequently, an RS also defines the semantics of skew bounds required between data streams. Of course, multiple referencing systems of different type may coexist in a given MM-application.

More precisely, a reference system consists of a number of data streams and a set of *logical clocks*, which are bound to input and output devices of these streams. A logical clock determines the current time in a certain "time zone" of an RS. Since there may be different time zones in an RS, multiple clocks may be needed, one for each zone.

As already mentioned above, logical clocks may be bound to input and output devices. Conceptually, all devices bound to the same clock operate in the same time zone. In an RS, multiple input and/or output devices may be bound to one logical clock, whereas each device may be bound to at most one clock. In other words, for a given RS a device may operate in at most one time zone.

In the simple scenario depicted in Fig. 3.1, referencing system RS1 comprises a (continuous) audio stream (A) and a (discrete) telepointer stream (T). It contains two clocks, LC1 and LC2, where LC1 is bound to the both input devices and LC2 is bound to the two output devices. That is, both input devices operate in one time zone, the one defined by LC1, and the output devices operate in another time zone, the one defined by LC2.

When describing the semantics of binding a device to a logical clock, we must distinguish between input and output devices. Let's consider input devices first and assume that input device ID is bound to logical clock LC. Whenever ID produces a data unit, the current value of LC is read and assigned to the produced data unit. Therefore, if two devices bound to the same clock produce a data unit at the same time, then both data units carry equivalent timestamps. Now let's consider output devices and assume that output device OD is bound to clock LC. The display of data units at OD is controlled by LC in the sense that a data unit with timestamp t is displayed at OD only when the value of LC does not differ from t by more than a specified skew bound. Consequently, two data units with equivalent timestamps are displayed at (almost) the same time if the corresponding output devices are bound to the same logical clock.

By binding input devices to clocks it can be specified which data streams can potentially be synchronized. The streams originating at input devices connected to a common logical clock carry (implicit or explicit) time information that can be used for synchronization purposes. Which streams are to be synchronized has to be specified in a second step by binding output devices to logical clocks. All data streams that are connected via their output devices to a common clock are synchronized on the basis of this clock and the timing information included in these streams.

In the example illustrated in Fig. 3.1, binding both input devices to LC1 only ensures that the (explicit or implicit) timestamps associated with the data units of the audio and telepointer stream are based on a common logical clock. This specification step says nothing about where and how this time information is exploited for synchronization purposes. This is done in a second step, which binds the both output devices to LC2. As will be seen in more complex examples below, we gain a lot of flexibility by separating specification into these two steps.

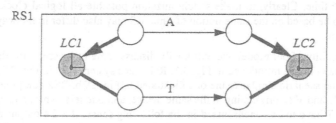

Figure 3.1

In our model, the advance of logical clocks can be driven in different ways. In particular, it can be driven by input devices, output devices or some external devices, such as real time clocks. To be able to capture this fact in specifications, we distinguish between *active and passive bindings*. A device is said to be actively bound to a logical clock if it triggers the advance of this clock, otherwise it is called passively bound. Now, we will consider when and how the different devices cause actively bound clocks to advance:

- *Input devices:* A clock actively bound to an input device "ticks" whenever this device produces a data unit. How far a clock advances when it ticks depends on the type of the RS as well as on the type of the device. For example, if the device produces continuous data, ticking is determined by the (natural) data rate of this device. If an RS implements Lamport clocks [La78] for including (explicit) timestamps into discrete data streams, the clock value advances by one whenever a data unit is produced.

- *Output devices:* Each time an output device that is actively bound to a logical clock consumes a data unit, the value of the bound clock is set to the timestamp of this data unit. Note that no assumptions are made about the type of the device. In the case of a device consuming continuous data, an interrupt of a device-internal clock may cause the logical clock to advance. (This corresponds to Anderson's logical timing systems driven by a master device [AnHo90]). In the case of an output device consuming discrete data, the data arrival on a communication channel may drive the clock advance.

- *External devices:* The advance of a clock can be also triggered by an external device, such as a real time clock. This does not say that the logical clock has the value of the driving real time clock but only means that the logical clock advances with the same rate as the real time clock.

So far, nothing has been said about how and when logical clocks are started. This is subject to start-up synchronization [AnHo90], [LiGh91], which may be a complex procedure, especially in more complicated settings (for example see Fig. 3.2 to 3.4).

In the simple scenario depicted in Fig. 3.1, the audio input device is actively bound to LC1, while the audio output device is actively bound to LC2. That is, LC1 and LC2 advance in the natural data rate of the audio input device and audio output device, respectively. At the input side, this setting enables the audio stream to carry implicit timestamps only. Data units of the telepointer stream have to be associated with timestamps, which - for instance - may be encoded as sequence numbers of audio data units. At the output side, this setting ensures that the audio device displays data units at its natural rate. Display of telepointer data units is synchronized according to LC2, which advances at this rate.

The notion of time may differ from RS to RS. For example, one RS may implement clocks as counters which are incremented whenever an actively bound device produces or consumes a data unit. Another RS, may implement clocks that advance at (approximately) the same rate as real time. Clearly, to make synchronization possible all logical clocks in a given RS have to be based on the same notion of time. RS may also differ in the way they encode timestamps.

To show the flexibility of our model, we will finally discuss some other scenarios that are more complex than the one introduced in Fig. 3.1. Reference system RS2 in Fig. 3.2 models a Joint Viewing scenario similar to the one illustrated in Fig. 2.3b. RS2 comprises two video streams (V1 and V2), originating at the same device, and one telepointer stream (T). Streams V1 and T have to be synchronized in the following sense: if V2's output device and T's input device consume resp. produce two data units at the same time, then T's and

V1's output device must consume these two data units also at (approximately) the same time.

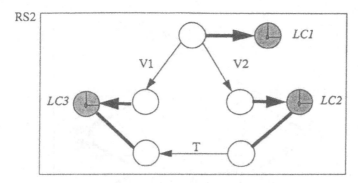

Figure 3.2

RS2 contains logical clocks LC1, LC2 and LC3. The video input device is actively bound to LC1, and V2's output device is actively bound to LC2. In other words, the value of LC2 corresponds to the timestamp of the latest data unit consumed (i.e. displayed) by V2's output device. Since T's input device is (passively) bound to LC2, each telepointer data unit gets associated with a timestamp, which correlates it with the video data unit that has been displayed at the time it has been produced. This timestamp information is used to synchronize streams T and V1. LC3 is actively bound to V1's output device and passively bound to T's output device. That is, the video data units are displayed at the natural rate of the video output device. The telepointer data units are displayed according to the advance of LC3. Remember that the starting of logical clocks is subject to a start-up procedure. LC3 may only be started when enough video and telepointer information is available at both output devices to ensure a synchronized consumption of both streams.

The Joint Viewing scenario illustrated in Fig. 3.3 is modelled by means of the two reference systems RS4 and RS5. In this scenario, two users share a window, in which discrete data are displayed. This is modelled by the two discrete data streams X1 and X2, which originate at the same input device. One user communicates with the other via a telepointer channel and an audio channel. This one-way communication is reflected in our model by audio stream (A) and telepointer stream (T). Two synchronization requirements can be indicated in this scenario: (1) Streams A and T have to be synchronized with each other and (2) stream A has to be synchronized with stream X1 in the following sense: if X2's output device and A's input device consumes resp. produce a data unit at the same time, then A's and X1's output device must consume these two data units also at the same time. RS5 ensures the first synchronization requirement while the latter one is fulfilled by RS4.

The input device of X1 and X2 is actively bound to logical clock LC1. This clock may be realized by a counter for instance, which is incremented whenever a data unit is produced in the bounded input device. Since this input device produces discrete data, the value of such a logical clock has no relationship to real time. Since LC2 and LC4 are in the same RS as LC1, they are based on the same notion of time. X2's output device is actively bound to LC2, which causes the value of this clock to correspond to the timestamp (e.g. sequence number) of the latest data unit displayed by X2's output device. By passively binding A's input device to LC2, the produced audio data units get temporally related to the displayed X2 data units.

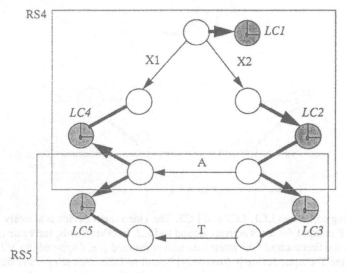

Figure 3.3

RS5 is equivalent to RS1 depicted in Fig. 3.1. RS5 and RS4 are connected by stream A, whose data units are associated with two timestamps, one for each RS. A's output device is actively bound to LC4 and LC5, which reside in different RS. As the advance of these two clocks is driven by A's output device, audio data units can be consumed at the natural rate of this device. The output devices of X1 and T consume their data units in accordance to the advance of LC4 and LC5, respectively. This ensures that streams X1, A and T are synchronized with each other.

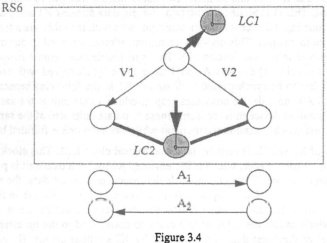

Figure 3.4

The setting depicted in 3.4 corresponds to scenario 3 described in the previous section. Two users share a video window which is modeled by the output devices of streams V1 and V2. Both users communicate with each other via audio channels A1 and A2, which are assumed to have a non-noticeable delay. As discussed in the previous section this scenario requires V1 and V2 to be synchronized.

Referencing system RS6 includes two logical clocks, LC1 and LC2. The former is actively bound to the video input device, while the latter is passively bound to the video output devices. LC2 is actively bound to an external device, such as a global time generator.

Of course, the above examples cannot cover all aspects of synchronization in Joint Viewing environments. A more comprehensive discussion of the concept of referencing systems in the context of JVTOS environments will be given in [DeRo93].

4 Summary and Future Research

In this paper we have first presented several JV scenarios along with their synchronization requirements. In a second part, we have defined a set of abstractions, which combined appropriately, prove powerful in modelling JV configurations with their synchronization schemes. The building blocks of such schemes, were introduced as referencing systems, logical clocks and input/output devices. Examples of several synchronization schemes were provided.

The ultimate goal of the scheme developed is to provide a programming paradigm allowing a programmer to specify all information required for correct synchronization in cooperative environments. So, we will have to look into which information has to be specified by the programmer itself and which information can be derived from this specification by the underlying system. This depends very much on whether synchronization mechanisms prove to be general enough with respect to different application scenarios involving data streams of different characteristics. This applies in particular to the start-up phase of synchronization in a distributed environment.

We will also reconsider the definition of skew, which currently proves to be too rigid, especially with respect to synchronization between discrete data streams. Another open issue is how a global clock can be realized in a distributed environment without a more or less inherently synchronized behavior. Finally, we will investigate what other types of logical clocks and referencing systems are useful in specifying distributed synchronization schemes.

References:

[AlCa91] Almeida N., Cabral J., Alves A.: "End-to-End Synchronization in Packet-Switched Networks", Proc. of the "Second Intl. Workshop on Network and Operating System Support for Digital Audio and Video", Nov 1991, to be published by Springer-Verlag

[AnHo90] Anderson P., Homsy G., Govindan R.: "Abstractions for continuous media in a network window system", Report No. UCB/CSD 90/596, Computer Science Division, University of California, Berkeley, Sep 1990

[CoGa91] Coulson G., Garcia F., Hutchinson D., Shepherd D.: "Protocol Support for Distributed Multimedia Applications", Proc. of the "Second Intl. Workshop on Network and Operating System Support for Digital Audio and Video", Nov 1991, to be published by Springer-Verlag

[Cr90] Crowley T. et al.: "MMConf: An Infrastructure for Building Shared Multimedia Applications", Proc. of the "Third Conference on Computer-Supported Cooperative Work", ACM, New York, 1988

[DeFr92] Dermler G., Froitzheim K.: "Joint Viewing and Teleoperation Services: A Reference Model for a New Multimedia Service", to be presented at the "4th IFIP Conference on High-Performance Networking", Liege, Dec. 1992

[DeRo93] Dermler G., Rothermel K.: "Referencing Systems: A Programming Paradigm for Stream Synchronization", in preparation

[ElGi91] Ellis C., Gibbs S., Rein G.: "Goupware - some Issues and Experiences", Communications of the ACM, Vol.34, No.1, Jan 1991

[Fe91] Ferrari D.: "Design and Applications of a Delay Jitter Control Scheme for Packet-Switching Internetworks", Proc. of the "Second Intl. Workshop on Network and Operating System Support for Digital Audio and Video", Nov 1991, to be published by Springer-Verlag

[FiKr88] Fish R., Kraut R., Leland M., Cohen M.: "Quilt: A Collaborative Tool for Cooperative Writing", Proc. of the "Conference on Office Information Systems", Mar 1988, ACM 1988

[HeSa90] Hehmann D., Salmony M., Stuttgen H.: "Transport Services for Multimedia Applications on Broadband Networks", Computer Communications, Vol. 13, No. 4, 1990

[La78] Lamport L.: "Time, Clocks, and the Ordering of Events in Distributed Systems". Communications of the ACM, vol. 21, July 1978

[LeBa90] Leung W., Baumgartner T., Hwang Y., Morgan M., Tu S.: "A Software Architecture for Workstations Supporting Multimedia Conferencing in Packet-Switched Networks", IEEE Journal on Selected Areas in Communications, Vol. 8, No. 3, April 1990

[LiGh91] Little T., Ghafoor A.: "Scheduling of Bandwidth-Constrained Multimedia Traffic", Proc. of the "Second Intl. Workshop on Network and Operating System Support for Digital Audio and Video", Nov 1991, to be published by Springer-Verlag

[ScCe87] Scigliano J., Centini B., Joslyn D.: "A Real-Time Unix-based Electronic Classroom", Proc. of the "IEEE Southeastcon '87", IEEE 1987

[WoMo91] Wolfinger B., Moran M.: "A Continuous Media Data Transport Service and Protocol for Real-Time Communication in High Speed Networks", Proc. of the "Second Intl. Workshop on Network and Operating System Support for Digital Audio and Video", Nov 1991, to be published by Springer-Verlag

Synchronization of Multi-Sourced Multimedia Data for Heterogeneous Target Systems

Dick C.A. Bulterman

CWI: Centrum voor Wiskunde en Informatica
Postbus 1009, 1009 AB Amsterdam, The Netherlands

Abstract. Accessing multimedia information in a networked environment introduces problems to an application designer that don't exist when the same information is fetched locally. These problems include "competing" for the allocation of network resources across applications, synchronizing data arrivals from various sources within an application, and supporting multiple data representations across heterogeneous hosts. In this paper, we present a general framework for addressing these problems that is based on the assumption that time-sensitive data can only be controlled by having the application, the operating system(s) and a set of active, intelligent information object coordinate their activities based on an explicit specification of resource, synchronization, and representation information. After presenting the general framework, we describe a document specification structure and two active system components that cooperatively provide support for synchronization and data-transformation problems in a networked multimedia environment.

1 Problem Overview

The focus of much of the activity associated with multimedia computing has been on the development of interfaces and software that allow various types of time-based data to be manipulated on a PC or workstation. An obvious evolution of this work is to extend it across a network infrastructure. Such *networked multimedia* offers a number of immediate practical advantages to users: the network provides a convenient means of distributing information to other sites, it provides access to compute servers where special-purpose processing of multimedia data can take place, and last (but not least!) it provides access to file servers that can be used to store the often vast amounts of data required to represent even the most simple multimedia information fragment. Unfortunately, networking multimedia activity also introduces three fundamental problems that don't exist for local-host multimedia; these are network-related *synchronization control*, *resource allocation* and *heterogeneous systems* support. Synchronization control problems are caused by network interfaces, network infrastructures, and remote hosts, all of which introduce various delays while fetching data that are not always obvious to the application requesting that data. Resource allocation problems stem from the fact that the network and its servers are resources that must be shared among various independent applications that have no idea of global network activity. Finally, heterogeneity problems result from a network in which member systems use different underlying data representations or—even worse— on which different underlying multimedia support facilities exist. Taken together, these problems mean that applications which run effectively locally will probably 'break' as soon as they are networked.

While the benefits and problems associated with networked multimedia are not new, the time- and space-sensitive nature of multimedia information provides a new set of constraints that existing infrastructures are not able to handle. This is because the traditional network and operating systems support mechanisms that manage data synchronization and resource allocation do not get enough information from standard I/O requests to intelligently support the resource and data-dependent synchronization needs of multimedia data.

In this paper, we consider the partitioning of control concerns for supporting network-wide synchronization of data in a heterogeneous environment. Synchronization is an important issue because it lies at the heart of allocating resources across the network. Heterogeneous networks (that is, networks in which resource allocation depends not only on the characteristics of the multimedia data, but also on the nature of the systems involved in a multimedia data exchange) are also important because they force a consideration of general techniques rather than any special-purpose solutions that can result from using the particular characteristics of any one support platform.

Our discussion of synchronization control is divided into two parts. We begin by describing a network-wide framework for managing multimedia data manipulation. The purpose of this framework is to clearly partition the synchronization control responsibilities among the "active components" in a multimedia system. We then present a specific synchronization control-sharing model that is based on this framework. This model uses an application-generated multimedia activity specification that is shared by the application, the operating system(s) and a set of intelligent data objects to coordinate activity across the network. As will be shown, the key to our approach is the assertion that the manipulation of multimedia data needs to be controlled by intelligent cooperation among components involved in defining, requesting, supplying and transmitting data; as a result, control issues can not be concentrated within a single layer (as is the case in current operating systems), but it must be shared across layers.

In the sections below, we start by presenting CWI's Amsterdam Multimedia Framework (AMF). We then present our current approaches to partitioning information sharing control among the application, the operating system(s) and an intelligent data storage component. We conclude with a list of problems that need to be solved before true heterogeneous support can be provided for a wide class of multimedia applications.

2 The Amsterdam Multimedia Framework

AMF was developed to organize our study of the problems associated with resource allocation and data synchronization in heterogeneous multimedia networks. This clarification is required because multimedia support has developed on a more-or-less ad hoc basis. For example, since multimedia support has partially evolved from application-based uses of relatively standard I/O devices (such as an audio interface), the applications layer has played a major role in selecting, manipulating and controlling data transfers. On the other hand, operating systems have had the historical responsibility of controlling I/O flow in a computing system. The OS allocates transfer buffers, schedules I/O operations (assigning priorities among applications and devices) and

121

monitors the actual transfer across the system's resources. In addition to the application and the OS, I/O device controllers also play a major control role in managing data transfers. These controllers can consist of multimedia device interfaces or interfaces that control intermediate data transfer channels, such as interconnection networks.

Rather than assign control responsibility to any one layer, AMF consists of a set of active components that cooperate in the control of networked data/information flow. It also defines the "scope of responsibility" that each component can exert before and during the transfer of multimedia information. AMF gives us a foundation upon which to build individual models that address particular synchronization and/or resource control issues.

2.1 Active Multimedia Control Components in AMF

There are four active control components defined within AMF. These are: the application process, a local operating system environment (including I/O controllers), a global operating system and a collection of intelligent information objects. (See figure 1.) The nature of each of these components and their roles with respect to the control of information in a multimedia network is considered in the following paragraphs.

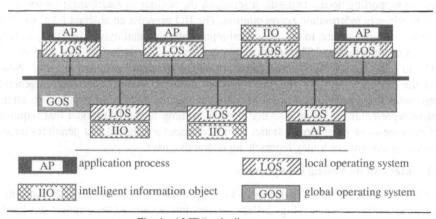

Fig. 1. AMF "active" components.

The application process. The application generates a description of the information objects that will be used (including their location), as well as a definition of how these objects are to be synchronized from the application's perspective. It also provides a control interface to the user to provide high-level interaction with the network. ("High-level" means operations like *start, stop, pause, fast-forward, seek,* etc.—the type of operations that one might expect on, say, a CD-player or home video control panel.) The main control role of the application is to specify the relationships among the information elements and to control the user's production and consumption of information. Given the high performance cost of involving the application at run-time, however, it has only a limited role in *implementing* any dynamic control operations.

The local operating system (LOS). The LOS serves as a scheduling authority that controls access to I/O devices attached to the local workstation. The LOS would typically allocate resources locally based on its knowledge of the local operating environment and an applications specification defined by the application. The LOS has the responsibility for controlling the flow of information into and out of the local environment—including presenting information to and receiving information from the network controller(s)—but it does not participate in global scheduling decisions.

The global operating system (GOS). The role of the GOS is to allocate resources on a network-wide basis. It has a view of network activity that is more comprehensive than the application or the LOS, since it can coordinate activity among independent applications that use the central network but which do interact with all local workstations. As we will describe below, we feel that this is a particularly interesting domain to study distributed operating systems support. Note that it would be possible for a given implementation model to combine the functions of the LOS and the GOS, although from the point of view of the framework, it is important to recognize that the functions served by both abstractions are different.

Intelligent Information Objects (IIO). The IIO is an entity that provides information to applications. It includes the information plus a series of operations on that information. In supporting access requests, it separates the notions of *multimedia information* and *multimedia information representation*. The IIO presents an abstract interface that is used to control access to one of several representations that may exist to implement a block of 'information.' This interface can be used by the application, the LOS or the GOS to select an appropriate representation based on static or dynamic needs. Note that the IIO does not give you something for nothing: it simply provides a general framework that needs to be filled in by data-dependent code and, if appropriate, alternative representations. Note also that there is nothing in the IIO model that requires information to be persistent or static; it may represent a program that generates information on demand, or it may represent an interactive user.

2.2 Interactions Among Components

Within AMF, the control of multimedia is a cooperative process that requires coordination among all components. Fig. 2 illustrates this interaction. At the applications level,

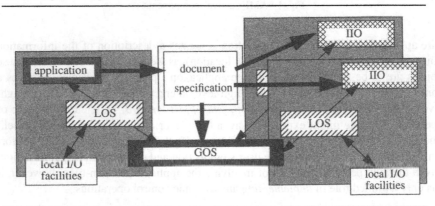

Fig. 2. The document specification.

a *document specification* is generated that describes the logical and physical interactions that are required by the application. The nature of this specification depends on the underlying support model, but its essence is that each application must be able to define the relationships that exist among components with respect to their synchronization needs, their expected resource needs, and their priority in the application. At the applications level, the document specification functions like a programming language that can be used to specify the behavior of the application. At the LOS and GOS levels, it can be used as an I/O specification and as a resource map. (Note that the GOS will have one specification available for each concurrently running application; it can use these for global resource allocation.) At the IIO, the specification can be used to determine when information objects need to be activated, as well as a guide to how the flow of information can be controlled at run-time. In general, the document specification will describe component interactions, but not component data. (This is because data is an active, not passive, component.) Clearly, the more information that is placed in the document specification, the more extensive the level of support that can be provided.

One way of using the document specification is as a guide to resource allocation during the lifetime of an application. Another way is to determine how (and if) the synchronization needs of the application are to be supported at run-time. In both cases, the processing of the document specification could lead to one of three support decisions being made:

- If the support request does not conflict with the present state of the system, a positive support action could be taken by the appropriate level (either the application, the LOS, the GOS or the IIO). This type of support is termed *trivial accept*.
- If the support request produces a fundamental conflict with the present state of the system, a negative support action could be taken by the appropriate level (either the application, the LOS, the GOS or the IIO). This type of support is termed *trivial reject*. (Trivial reject allows for either partial or full rejection.)
- If the support request results in a conflict that is not fundamental, a (light-weight!) process of negotiation could be started to see if support could be offered at a reduced level of service. This type of support is termed *negotiated accept*.

The support for *trivial accept*, *trivial reject* and *negotiated accept* is a consistent theme throughout the AMF. Note that the framework does not dictate any particular means of specifying the activity within a document or the way in which resource decisions are made. Instead, it presents a number of guiding principles that allow individual implementation models to be constructed.

3 Specification of Synchronization in a Heterogeneous Network

In the sections below, we illustrate how synchronization of multi-sourced multimedia information can be supported in a heterogeneous environment. The approach we use is based on AMF; it illustrates how the application, the LOS, the GOS and various IIOs can be architected together to partition and coordinate control activity. In order to put our approach in context, we preface this discussion with a description of the environment we base our research on and its impact on the types of models we investigate in supporting a network of heterogeneous hosts.

3.1 Environmental Impact Statement

The environment used to drive our research consists of a network of heterogeneous host systems that (simultaneously) share one or more multimedia documents with each other. This environment provides the following problems that must be addressed for effective multimedia information sharing:

a) The interconnect bandwidth is a critical resource that must be allocated in an effective manner.

b) Each application is described in terms of a collection of data objects that must be implicitly or explicitly synchronized with other objects. (Synchronization information within an object is specified by the IIO implementing the object.) Each object may consist of composite or single-media data. The information associated with any one object always exists at a single source host, but the collection of objects in a document may be spread across the network. (This means that there may be multiple delays in implementing a document due to the different delay characteristics of each participating source host.)

c) At present, we restrict ourselves to a single destination host for each document, but not a single destination host type. That is, we assume that each destination processes its own copy of a document that is independent of any other documents (or any other instances of the same document) in the network.

d) At present, we restrict ourselves to "static" multimedia documents. That is, we consider only documents whose synchronization specifications have been pre-defined. (For example, we consider "multimedia mail" applications, in which a message is preformatted, but not "multimedia talk" applications, where data is generated on the fly.)

As a consequence of characteristic (a), we use a GOS model that is based on a distributed operating system that is functionally separate from the each LOS at a user or server site. The GOS can be tailored to provide speciality services for resource allocation and synchronization support. As a consequence of (b), the GOS schedules activity in terms of a collection of applications that each may (independently) request one or more data objects simultaneously. As a consequence of (c), each IIO must be able to support data translation operations for the data it "owns" and it must be able to communicate with the application, LOS and GOS to implement demand scheduling requirements. As a consequence of (d), the LOS and GOS can make scheduling decisions that "only" need to adapt to the data and transfer characteristics of the application and the environment; these decisions need to be adaptive, but they can be based on prior knowledge of the document set being processed. Other dynamic concerns (such as instantaneous production of data or the processing of hyperdata jumps) are not considered here.

3.2 CMIF: The CWI Synchronization Specification

The approach we take to defining synchronization is to provide a user with a general document specification model (called CMIF—the CWI Multimedia Interchange Format [2]) that provides a method for collecting data objects into a multimedia document. A CMIF specification consists of two major views of a document: a *hierarchical view* and the *virtual I/O channel view* (or, more simply, the *channel view*). The hierarchial view is used to define the content-based relationship among data objects (that is, it defines which objects are to be presented in parallel and which serially); this relationship defines the implicit synchronization of objects in the document. The channel view maps objects in the hierarchy onto a collection of virtual channels, where each channel represents a collection of similarly-typed information that shares a common resource allocation policy. (See Fig.3.) The channel view consists of a collection of

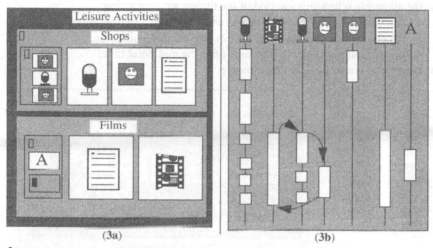

(3a)   (3b)

3a: The *hierarchy view* shows the logical components of a document and how they interact. One block (Leisure Activities) is composed of two sub-blocks; nested blocks are shown by 'nesting rectangles' in the upper corner. This view shows presentation structure, which supplies implicit synchronization information.

3b: The *channel view* maps the logical components to virtual output devices. Each logical block is placed on a channel and is associated with an IIO. The white boxes are self-synchronizing event descriptors. (Note: see figure 5 for a discussion of *synchronization arcs*.) At run-time, channels may be active or inactive, influencing document synchronization.

Fig. 3. CMIF in a nutshell.

event blocks, where each such block is an instance of a data object. Synchronization within a channel can be specified implicitly or explicitly. Implicit synchronization is typically defined by the relationship in the hierarchy view. Explicit synchronization is usually used between event blocks in separate channels; it allows a fine-grained relationship among events in the specification to be defined. Information within an event is usually self-synchronizing, as a result of the underlying data characteristics in the IIO. Explicit inter-event synchronization is required to match content relationships across events or to support presentation on heterogeneous hosts.

When supporting implicit synchronization, the general properties of channels can be used by the LOS for scheduling local I/O activity and by the GOS to schedule interaction among channels. They can also be used by the IIO to determine the filtering (if any) that needs to occur in presenting the data.

When using explicit synchronization, all of the 'clues' available for implicit synchronization exist plus information from explicit synchronization primitives called *synchronization arcs*. (The name comes from the graphical representation used to define relationships among events in different channels; see Fig. 4a.) Each synchronization arc can encode a tuple of information that is used to schedule events in the network. The elements of the tuple, shown in Fig. 4b, are:

a) the source and destination event blocks of the arc;
b) the allowable scheduling interval of the timing arc, containing:
   - the "mid-point" scheduling time as an offset from a starting time;
   - the minimum acceptable start time *before* the mid-point time;
   - the minimum acceptable start time *after* the mid-point time;
c) the type of synchronization relationship.

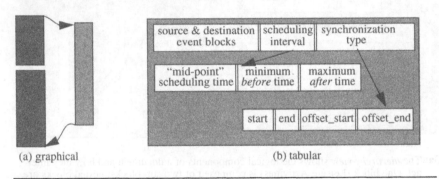

(a) graphical   (b) tabular

Fig. 4. The Synchronization Arc

Information in the synchronization arc is used at each level for different purposes. The GOS, for example, when sensing a synchronization relationship between two events that cannot be supported, can notify the relevant IIOs to stop or retard data transmission. It could also interact with the application, although the cost of doing so may be too high (in terms of scheduling and communications time) to justify this for short-term transient problems.

In addition to synchronization, the channel view also serves as a guide to LOS and GOS resource control. Since each channel is a virtual device, not a real one, similar data types may be specified several times. (For example, there may be two audio channels defined, even if there is only one speaker.) Each time a channel is defined, a different resource allocation policy can be specified for all of its event blocks. This policy can then be passed to the IIO when the corresponding data object is accessed. (If desired, override policies could also be specified with each event block.)

CMIF can be used to provide the application's view of how information should be gathered and presented. These rules are defined in a manner that gives the underlying layers such as the LOS and the GOS a basis for allocating resources and for fetching information out of objects. The CMIF document could be precompiled for static resource allocation, or it could be interactively interpreted by a *player* to provide an application with run-time control.

3.3 Interactive Intelligent Information Objects

The IIO is an active component that manages information. The general structure of the IIO is shown in figure 5. Here we see a single object containing three alternative repre-

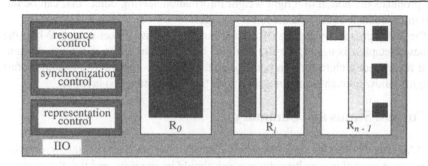

Fig. 5. The intelligent information object (IIO).

sentations; one representation may be a video clip (with composite sound), the second may be an audio track and several still pictures, while the third may contain a text-based description of the information being shown, along with three diagrams and a caption. In all cases, each representation provides the same information, albeit in a different form.

In addition to data, the IIO also contains access control information. Examples of access control may be to support resource allocation (that is, if there is only limited bandwidth available on the interconnect, the IIO select a lean information representation), or heterogeneous mappings (from one display format to another), or synchronization control (as explained above). The control operations provided by the IIO can be divided into three groups:

(1) *resource control*—this interface allows the IIO to control the amount and type of data that is used to represent the information selected. Operations may include conventional activity, such as buffering (local or remote), sub-sampling, and compression, or it may consist of operations such as *aliasing* (where an alternative set of representations may be selected to reduce resource use), *preemption* (where an object can be temporarily suspended), or *optional* (where an object can be totally excluded if necessary)

(2) *synchronization control*—this interface allows the IIO to select a data-dependent synchronization action to meets the type of constraint specified by an explicit or implicit CMIF synchronization request. Policies may include *skipping, sub-sampling* and *forced-delay*.

(3) *representation control*—this interface allows the IIO to map data from one representation to another. It is primarily used for supporting heterogeneous activity in the network. (The use of alternative representations for resource control is discussed above.)

As was noted above, there is nothing inherent in the IIO that would restrict it to supporting static data: the IIO could be an interface into a suite of programs to dynamically generate information representations. The primary value-added benefit of the IIO is that it localizes the policy aspects of supporting synchronization and heterogeneous presentation policies. In providing this support, we delegate any higher-level decision making on the interaction among IIOs to the GOS, LOS or even application. This promotes the notion of a light-weight information serving object that can be tailored to the content of the information it manages.

The IIO shares many aspects of a (distributed) database object model. While the database metaphor is useful, the IIO can also be defined in terms of a more conventional file system architecture. In both cases, the interface into the IIO should be fast enough to allow efficient communication with the GOS (and LOS) layer(s).

4 Current Status and Summary

The AMF and the various support models described, all are based on the assumption that resource control in a multimedia network should be adaptive, and that the adaptive process should be distributed over the application, the local operating system, the global (distributed) operation system and the IIOs involved in a transfer. Each of these layers has a specific insight that is important in controlling multimedia transfers. Although each of these insights are necessary, AMF also attempts to limit the scope of any one layer by giving each layer a specific set of concerns to process.

Support for AMF, the CMIF specification and the IIO are on-going research activities at CWI. At present, we have developed an authoring environment and a run-time player to construct and manipulate CMIF-based multimedia documents in a single-host environment. The CMIF player relies on a commercially-supported LOS, which is a trend that we plan to continue. (The workstation also has to be used for general-purpose work!)

As was reported in [1], we are involved in an on-going project to investigate the development of distributed operating systems at the GOS level that are modelled on a *multimedia co-processor* concept (MmCP). The MmCP acts like an intelligent device controller that can provide distributed resource control operations. It uses the CMIF specification as input, and interacts with other active components to implement the transfer needs of simultaneous, independent multimedia activity.

Support for the IIO is at an early stage. Our present concern is the analysis of the semantic facilities that can be provided to support a wide range of resource, synchronization, and representation control operations. Support for an implementation environment for the IIO will be phased in during the coming year.

All of the activity in the Multimedia Kernel Systems project is aimed at understanding the basic relationships that exist in supporting multiple multimedia applications in a heterogeneous network environment. While the main thrust of this work has been sketched in this article, we are also interested in understanding the impact of interactive generation of CMIF documents for particular applications areas. Our work in this area has begun with the study of user-level hyperdata structures on multimedia systems [3].

129

References

[1] Bulterman, D.C.A. and R. van Liere: "Multimedia Synchronization and Unix," Proceedings of the 2nd International Workshop on Network and Operating Systems Support for Digital Audio and Video, Heidelberg, November 1991.

[2] Bulterman, D.C.A., G. van Rossum and R. van Liere: "A Structure for Transportable, Dynamic Multimedia Documents," Proceedings of the Summer 1991 Usenix Conference, Nashville TN, June 1991.

[3] Hardman, L., D.C.A. Bulterman, and G. van Rossum, "The Amsterdam Hypermedia Model: Extending Hypertext to Real Multimedia," CWI Technical Report (July 10, 1992).

An Intermedia Skew Control System for Multimedia Data Presentation*

T.D.C. Little and F. Kao

Department of Electrical, Computer and Systems Engineering
Boston University, Boston, Massachusetts 02215, USA
tdcl@buenga.bu.edu

Abstract. Delivery of multimedia traffic from distributed sources to a single destination relies on an accurate characterization or guarantee of the behavior of each involved system component. If anticipated bandwidth, delay, and loss characteristics are violated, statistical scheduling approaches fail, resulting in the loss of established source-to-destination synchronization and the introduction of intermedia skew.

We propose an intermedia skew control system for accommodating a class of anomalous behaviors including spurious channel overload, data losses due to corruption or congestion control, and variations in source and destination clock rates. The control system is designed to operate independently from the data source, accommodating the data delivery requirements of an arbitrarily corrupted input stream. Intermedia skew, buffer underflow, and buffer overflow are controlled by regulating the playout rate of each stream through frame drop and duplication. This mechanism is used in conjunction with a statistical source pacing mechanism to to provide an overall multimedia transmission and resynchronization system supporting graceful service degradation.

1 Introduction

One of the difficulties in supporting multimedia data delivery is the synchronization of related media in what we call *intermedia synchronization*. The lip-sync of audio and video is typical, but synchronization to both real-time and to other media can be required in a multimedia application. When data are transmitted (or stored) via different channels (disks) and synchronization is required, a mechanism is necessary to ensure synchronous presentation, or *playout*, to the application. Interleaving, or multiplexing of interrelated media can be used for multimedia storage or transmission. However, it can only be applied to a single storage subsystem or communication channel. Furthermore, the interleaving requires that all media have the same quality of service [3], and prohibits the use of congestion control mechanisms that reduce load by dropping data of a single medium. Initial attempts to satisfy the synchronization problem were achieved by allowing the destination to track the source [6, 13]. In our work we assume a

---

* This material is based upon work supported in part by the National Science Foundation under Grant No. NRI-9211165.

general model supporting multiple media, allowing intermedia synchronization among multiple sources in a multidrop fashion (Fig. 1).

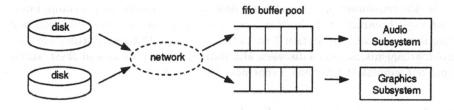

Fig. 1. Data retrieval from multiple independent sources and multimedia playout

Previously, we investigated a session establishment protocol for reservation of channel resources [11]. In our current work we achieve intermedia synchronization after connection establishment by monitoring and controlling skew and queue levels by selective frame drop and duplication. This scheme prevents disruptions in playback caused by unanticipated temporary overloads due to events such as file transfers, and effectively extends the buffering capability of the receiver without increasing memory use or latency. Although this would seem to duplicate the efforts provided by a statistical resource allocation mechanism for bandwidth and delay, we anticipate both short- and long-term guarantee violations that must be accommodated in any real system. In addition, data frames can be lost in the network (e.g., via asynchronous transfer mode congestion control) which will require the source to manage intermedia synchronization in the presence of frame losses. Our system is unique in its ability to recreate a nominal playout sequence in the presence of these aperiodic losses, possibly due to format conversion (e.g., 10 frames/s to 30 frames/s). Furthermore, this system operates independently of the sources in the absence of clock rate mismatches. In essence, a statistical resource allocation mechanism can provide a course reservation of suitable resources, but the fine-tuning required at playout time is provided by our control system.

Drop and duplication of frames has also been used to reconcile mismatches in data playout and generation rates by Anderson and Homsy [1]. Our proposal differs in its provision of source–destination independence, and its generality in providing fine and gross skew correction due to data losses. Furthermore, our solution is designed to correct skew without introducing noticeable discontinuities in data playout. Drop and duplication is also proposed to correct intermedia skew by Rangan et al. [17]. With this approach, synchronization is ensured (in the absence of a global clock) by computing relative timing for a set of distributed sources. In contrast, our system is designed to be easily implementable, independent of the data source, and able to accept data at arbitrary arrival rates. Delay and delay variations are absorbed by buffering while losses or rate conversion are accommodated by the playout correction algorithm.

Other related work in multimedia synchronization includes ths study of real-time operating systems for supporting audio and video synchronization [2, 12, 14, 16, 19], feedback control systems for locking on source delivery rates [4, 5, 6, 15], and rate-based transmission control [7, 9]

In the remainder of this paper we describe our system for providing intermedia synchronization. In Section 2 we define intermedia synchronization and describe its control. In Section 3 we outline the algorithms supporting the control mechanisms. Section 4 discusses alternative control policies and observations from our simulations. Section 5 concludes the paper.

2 Intermedia Synchronization

Synchronization is defined as the occurrence of multiple events at the same instant in time. Intermedia synchronization describes a similar timing constraint among a set of multimedia streams. In this section we characterize intermedia synchronization and introduce a framework for satisfying such timing constraints in a general-purpose computer environment.

2.1 Characterization of Synchronization

Timing parameters can characterize intermedia and real-time synchronization for the delivery of periodic (e.g., audio and video) and aperiodic data (e.g., text and still images). Parameters applicable to aperiodic data are *maximum delay*, *minimum delay*, and *average delay* as measured with respect to real time or with respect to other aperiodic data. For periodic data, maximum, minimum, and average delay are also applicable to individual data elements, but in addition, *instantaneous* delay variation or *jitter* is important for characterizing streams. *Skew*, related to jitter, describes an average jitter over an interval.

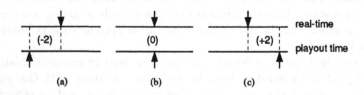

Fig. 2. Skew: (a) lagging, (b) none, (c) leading

For periodic data such as audio and video, data can be lost resulting in *dropouts* or gaps in playout. Such losses cause the stream of frames to advance in time or cause a stream *lead*. Similarly, if a data frame is duplicated, it causes the stream to retard in time or a stream *lag* (Fig. 2). By dropping or duplicating frames it is possible to control the rate of playback. Initiation of frame dropouts is

133

permissible for audio and video which have substantial short-term temporal data redundancy. However, for text and graphics, this is not true as their tolerance to delay is greater.

Synchronization is usually defined as absolute, occurring at an instant in time. To provide a tolerance to timing variations, we adopt the following definition that extends a synchronization instant to an interval (similar to Gibbs' definition [8] and Ravindran's divergence vector [18]).

Definition 1. A composite object with actual playout times $P = \{\rho_i\}$ is *synchronized* with playout reference times $\Pi_i = \{\pi_i\}$ iff $\forall i, |\rho_i - \pi_i| \leq \theta_i$ where $\Theta = \{\theta_i\}$ are the synchronization tolerances between each element and the reference.

Skew can be measured with respect to real-time as an offset to some mutual presentation start time between the source and destination, or can be measured with respect to another stream. Because many streams are possible, we characterize both intermedia and real-time reference skew for k streams using a matrix representation as $skew = sk_{p,q}$, where $sk_{p,q}$ describes the skew from stream p to stream q (q to p is negative) and the $k+1$th element corresponds to a real-time reference. We also define an interstream tolerance $\Theta = \theta_{p,q}$ and target skew matrix $TSK = tsk_{p,q}$ which indicate tolerances and target values between streams and can be interpreted by a skew control function. Related to skew is data *utilization*. Utilization U describes the ratio of the actual presentation rate to the available delivery rate of a sequence of data. Frame drops will decrease utilization whereas duplicates will increase utilization from its nominal unit value. Either skew or utilization can be used as a control variable.

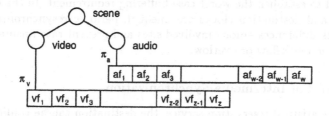

Fig. 3. Blocking for continuous media

Skew best measures intermedia synchronization for continuous media. For characterization of discrete events associated with timed playout of text, graphics and still images, we can apply real-time scheduling terminology as already mentioned (e.g., maximum and minimum delay). However, it is often advantageous to decompose segments of continuous media into *blocks* to permit efficient storage and manipulation [10]. With this decomposition, blocks associate a single start deadline with a sequence of periodic media frames, as illustrated in Fig. 3.

In this example the decomposition is performed on a motion picture and blocks of audio and video are associated with a single logical scene.

2.2 Coarse Synchronization and Session Scheduling

The nature of a packet-switched communications mechanism is inherently asynchronous as is the retrieval of data from rotating storage subsystems. We model the retrieval and transmission of data across a network as an asynchronous system component which must be managed in order to support real-time data delivery required for multimedia presentation. Flow-control (window) protocols can provide feedback necessary to prevent buffer overflow in bulk data transfers. For real-time data streams, rate-based flow-control is more appropriate [7]. However, feedback approaches alone cannot provide consistent presentation quality under wide bandwidth variations due to network delays or multimedia object sizes. For this reason, statistical reservation techniques are used.

Coarse synchronization in a multimedia session is facilitated by reservation of adequate resources. This scheduling process consists of resource reservation, connection establishment, and data transfer initiation. In relation to intermedia synchronization, a scheduler manages data transmission times with respect to the source whereas the destination must provide buffering for delay variations and accommodation for intermedia skew.

We assume the availability of a statistical scheduler [11] that, given the characteristics of a channel and a multimedia data stream (D_v, D_p, D_t, C, S_m, π_i, σ_i, $P(fail)$), corresponding to variable, fixed, and channel transmission delays, channel capacity, packet size, playout deadline, object size, and probability of lateness), a schedule can be constructed. The resultant schedule indicates the times to put objects onto the channel between the source and destination and can be used to establish the worst case buffering requirement. In the event that the source and destination clocks are unequal, periodic resynchronization can correct clock differences among involved sites and prevent rate mismatches that lead to queue underflow or overlow.

2.3 Control of Intermedia Synchronization

By using a statistical reservation service, the destination can be configured with a buffer of sufficient length to accommodate measured or guaranteed channel delay variations. However, changes in the channel delay characteristic in time or other spurious violations in delay and bandwidth guarantees can cause buffer overflow or underflow resulting in anomalous playout behavior. We propose a control mechanism to monitor and control levels of queued periodic stream data as well as intermedia synchronization as specified by a target skew matrix. This mechanism can then provide graceful degradation of playout quality during periods of spurious network or device behavior.

One of our assumptions in the design of a control mechanism is that the playout interval for periodic data streams is constant. For example, video frames have a playout interval of 33 ms. Because this playout interval cannot be changed dynamically, we control the playout rate instead by changing the frame drop and duplication rates. If two streams are skewed we can either drop frames from the lagging one to increase its speed or duplicate frames of the leading one to decrease its speed (or both). The tradeoff between dropping versus duplicating is the loss of data versus the increase in end-to-end latency that affects interactive sessions. Depending on the priority of these two considerations and the current queue levels, the appropriate policy can be applied.

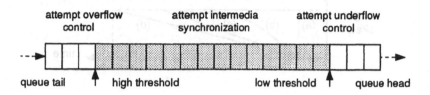

Fig. 4. Three queue states

We propose two related control paradigms. The first is intended to maintain intermedia synchronization between streams, or between a stream and a real-time reference, when operating at nominal queue levels (Fig. 4). The second is for providing graceful degradation when queue underflow or overflow is pending and attempts at intermedia synchronization are abandoned. For intermedia synchronization, we are investigating three policies with respect to intermedia synchronization: (1) minimize real-time skew, (2) minimize inter-stream skew, and (3) minimize session aggregate inter-stream skew. The first policy targets synchronization of playout with a constant end-to-end delay established during connection set-up. This policy is appropriate for streams such as high-quality digital audio. The second policy places priority on maintaining synchronization between streams rather than between a stream and a real-time reference, but could be used in conjunction with the first policy. This scenario is suitable when relative timing between media is more significant than cumulative skew with respect to real-time. The third policy is to minimize skew over a set of streams within a multimedia session. For queue level control, our goal is to provide graceful degradation prior to queue under or overflow. Control takes effect upon reaching either low or high thresholds and results in modification of the playout rate via frame drop or duplication.

To implement intermedia synchronization we use a control mechanism that monitors the queue level and playout skew (at playout time). Control is provided by changing the playout rate (or utilization) using frame drop and duplication. Stability of the system is provided by determination of appropriate time constants for queue level and skew measurement. Utilization, ideally a continuous parameter nominally of unit value, is representative of the playout rate. Duplication causes $U > 1$ while dropping causes $U < 1$. Utilization over an interval comprised of n frames can be determined using the formula, $U = (pass + slip)/pass$, where *pass* and *slip* correspond to played and dropped/duplicated frames. For dropped frames, *pass* indicates the total number of frames evaluated, and *slip*, a positive integer, corresponds to the number to drop from *pass* passing frames. For duplicated frames, *pass* has the same interpretation, however, *slip* indicates the number of duplications of the last frame in the sequence. This formulation leads to a fast evaluation at playout time.

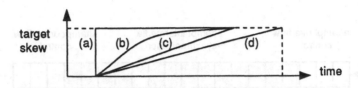

Fig. 5. Correction alternatives. (a) no control, (b) non-linear, (c) constant-rate, (d) constant-time

To set the control variables *pass* and *slip*, we define control functions which are dependent on the tolerable rate of drop and duplication for each medium. These functions can be instantaneous, introducing a large discontinuity in playout, or can be more gradual (Fig. 5). The control algorithm supports any type of selective drop or duplication function. We have initially investigated a constant rate-based correction function, but other functions can be supported as well.

3 Intermedia Synchronization Mechanism

In this section we describe our intermedia control mechanism for providing presentation-time synchronization and quality of service adjustment. The approach relies on an initialization phase, a retrieve/transmit process, a playout process, and a monitor/control process. We have formulated these components using an independent source–destination model with the goal of supporting either a distributed or a local data model. Each component will be described in the following subsections.

3.1 The Initialization Phase

In the course of a multimedia application's execution, the establishment of a session can be requested, requiring the activation of the delivery and playout subsystem. Typical scenarios for this include establishing a video conference or selecting a multimedia motion picture for presentation. Once selected, the session timing requirements must be interpreted for connection establishment and maintenance.

For session establishment, a statistical scheduling framework can be applied based on the characteristics of the requested session and the current system resources. The result is the generation of a set of deadlines which can be used by the retrieve/transmit process as local timing information. Included in this initialization phase are playout time identification [10]; data size characterization; bandwidth, delay and buffer reservation; computation of retrieval/transmission schedules; and initiation of data transfer. The data involved in this initialization phase are summarized in Table 1.

Table 1. Data manipulated during initialization

$object_A$	composite multimedia object
$tree_A$	object temporal representation
$\Pi = \{\pi_i\}$	ordered sequence of object playout times (deadlines)
$\Sigma = \{\sigma_i\}$	component object sizes (bits)
D_v, D_p, D_t	queuing, propagation, and transmission delays for packet of size S_m
S_m	packet size for medium m
C	channel capacity
T_n	control time for nth block
$P(late)$	requested fraction of late packets/blocks
$\Phi = \{\phi_i\}$	ordered sequence of retrieval/transmission times (deadlines)
$\Phi^c \subseteq \Phi$	set of aperiodic deadlines from Φ

3.2   The Retrieval/Transmission Process

The retrieval/transmission process is perceived to be independent of the monitor and playout/receive processes to support both the local and distributed data scenarios. It is therefore an asynchronous process, whereas the transfer of data from the buffer pool to the display mechanism can be synchronous. The **retrieval/transmission** process is outlined as follows. Initially, the multimedia object is characterized in the initialization phase and component object deadlines are determined (Π, Φ). After transfer initiation, data objects are sent to the destination based on their deadlines and sequence number.

1. while $i + j \leq 2m + 1$ do
   (a) if $\phi_i \in \phi_i^c$ and $i \leq m$ then {aperiodic parts of schedule}
       i. if $clock \geq \phi_i + start$ then
          A. {send object i to destination with appended playout deadline π_i}
          B. $i := i + 1$
   (b) if ϕ_i not $\in \phi_j^c$ and $j \leq m$ then {periodic parts of schedule}
       i. if $(clock \leq \pi_j + start \leq clock + T_E + start)$ then
          A. {send object j to the destination with appended playout deadline π_j}
          B. $j := j + 1$

The algorithm operates by evaluating the deadlines of objects awaiting transmission. The first conditional statement in the algorithm tests for impending retrieval deadlines for objects in the culled (Φ^c) retrieval schedule. The second conditional statement identifies the retrieval time of the remaining objects not in the culled schedule. The value of *start* reflects the variation of end-to-end delays for a session that uses multiple channels and allows each source to begin transmission synchronously. After initiation, the retrieve/transmit processes transfer data from sources to the destination with attached deadlines and sequence numbers for playout. At the receiver, the arrived data are queued based on deadlines and sequence numbers inside individual blocks. If no space is available to buffer arriving data frames, they are assumed to be discarded.

3.3 The Playout Process

The playout process must schedule the presentation of aperiodic events such as text and graphic displays and initiate the playout of segments of periodic streams of audio and video. Once established, the ordering of these segments relies on sequence numbers rather than individual deadlines associated with each frame. For example, a motion picture scene can have an initial playout initiation deadline, but its component frames can be played out with respect to this initial deadline and their individual sequence numbers. This mechanism provides less overhead than managing a deadline for each frame. Deadlines are always arranged to be monotonically increasing in time [10], and objects are assumed to arrive in sequence. The **playout** algorithm for managing presentation deadlines is outlined below.

1. if $clock \geq \pi_i + start$ then
   (a) call $playout(object_i)$ {initiate playout of block}
2. for each stream k in *session* do
   (a) if $clock = t_{play}$ then {time to play next frame}
      i. call $playout(rh_k)$ {play frame at head of queue}
      ii. if $slip < 0$ then {drop required}
      iii. elsif $slip > 0$ then {duplicate required}
      iv. else $slip = 0$ {nominal playout}
   (b) $t_{play} := t_{play} + \Delta_k$ {next playout time}

This algorithm performs two critical operations. First, the current time is evaluated against the set of scheduled deadlines. At the appropriate times, their playout is initiated. The second operation performs a similar action on periodic stream deadlines as indicated by sequence numbers. For each stream the current frame is played-out prior to the update of the queue head pointer based on the *slip* and *pass* control parameters set by the monitor and control process. The details of the pointer manipulation are not shown here.

Note that the buffer-to-playout transfer is synchronous. At the appropriate intervals for each medium (e.g., via interrupt), the current frame is evaluated for dropping or duplication. In either case, a frame is always passed to the presentation device for playout. In contrast, data arriving from the source are put in the receiving queue asynchronously.

3.4   The Monitor and Control Process

The monitor/control process serves three functions: monitoring queue and skew values for each stream, providing playout rate control, and providing source timing feedback to initiate time offset correction. The frequency of execution is dependent of the playout rates and block sizes of the individual presentation streams. An outline of the **monitor/control** algorithm is shown below.

1. for each stream k do
   (a) if $qlevel_k < qll_k$ then {low queue level}
       $- (pass, slip) := under(qlevel_k)$ {initiate stream lag}
   (b) elsif $qlevel_k > qlh_k$ then {high queue level}
       $- (pass, slip) := over(qlevel_k)$ {initiate stream lead}
   (c) else {nominal queue level}
       $- (pass, slip) := sync(qlevel_k, skew(k, l))$ {synchronize stream k to l}
2. {update queue level, skew, and lost frame statistics}
3. if $avg\_qlevel_k > threshold$ then
   (a) send $offset$ to source

On each iteration, this algorithm invokes one of the control functions under(), over(), or sync() depending on the queue level. If it is high or low, the algorithm provides service degradation in the form of drops or duplications until service is restored. If the level is nominal, then intermedia synchronization control is applied. In all cases, the values of *pass* and *slip* are manipulated to control the playout rate and to effectively control skew.

4   Discussion

We have exercised our intermedia control mechanism with a number of sequences of simulated audio and video traffic. Correction functions have been selected to provide constant-rate skew correction and underflow/overflow prevention. The results indicate predicted skew correction and demonstrate the validity of our approach. The simulations have also revealed several areas for further study. These include the investigation of appropriate correction functions for each medium, and the development of a predictive solution to prevent underflow and overflow.

When a skew, underflow, or overflow correction function is performed, a resultant reduction in quality of service occurs. Our control mechanism attempts to distribute this degradation over an interval. In the initial simulations, constant-rate functions were used for under(), over(), and sync(). The range of acceptable correction rates depends on tolerated degradation in quality of service. We are currently exploring these ranges and alternate correction functions in relation to specified skew tolerances. Clearly there are differences in correction functions for each medium. Because video data have ample temporal redundancy, they are the target of congestion control schemes, being dropped when the delivery mechanism becomes saturated, and can be subsequently corrected by our scheme.

Audio data are more perceptible when lost, yet can have periods of silence during which skew correction can be achieved without perceivable interruption.

Another observation of our experiments is the need to monitor the rate of queue level change to better predict underflow and overflow. This can be achieved by maintaining additional queue statistics. Furthermore, large instantaneous skew transitions can be eliminated by monitoring frame losses at the time of arrival rather than at playout time.

5 Conclusion

We have proposed a mechanism to enable intermedia synchronization in a distributed multimedia information system. The mechanism is designed to operate independently of a source scheduling mechanism, providing a graceful skew correction for lost frames or other anomalous system behavior. Given suitable correction rate parameters, the algorithm can reconstruct a full-rate playout stream from a stream missing an arbitrary number of elements, as can result from congestion control or when limited bandwidth channels are used.

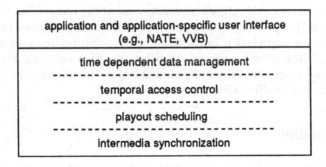

Fig. 6. Layers of timing management

The intermedia synchronization mechanism fits into a larger system framework of managing time dependencies of multimedia data in a distributed multimedia information system (Fig. 6). We are currently incorporating the intermedia skew control system into into our Virtual Video Browser (VVB) application for content-based query and information display as applied to digital movies, and into our News at Eleven (NATE) application which facilitates multimedia information retrieval and presentation.

References

1. Anderson, D.P., Homsy, G.: A continuous media I/O server and its synchronization mechanism. Computer **24** (1991) 51–57
2. Butlerman, D.C.A., van Liere, R.: Multimedia synchronization and UNIX. Proc. 2nd Intl. Workshop on Network and Operating Support for Digital Audio and Video, Heidelberg, Germany (1991)
3. Cambell, A., Coulson, G., Garcia, F., Hutchison, D.: A continuous media transport and orchestration service. Proc. SIGCOMM'92, Baltimore, MD (1992)
4. Chao, H.J., Johnston, C.A.: A packet video system using the dynamic time division multiplexing technique. Proc. Globecom'86, Houston, TX (1986) 767–772
5. Cochennec, J.Y., Adam, P., Houdoin, T.: Asynchronous time-division networks: terminal synchronization for video and sound signals. Proc. Globecom'85, New Orleans, LA (1985) 791–794
6. De Prycker, M., Ryckebusch, M., Barri, P.: Terminal synchronization in asynchronous networks. Proc. ICC'87, Seattle, WA (1987) 800–807
7. Ferrari, D., Verma, D.C.: A scheme for real-time channel establishment in wide-area networks. IEEE J. on Sel. Areas in Comm. **8** (1990) 368–379
8. Gibbs, S., Dami, L., Tsichritzis, D.: An object-oriented framework for multimedia composition and synchronisation. Object Composition, Tech. Rept., University of Geneva (1991) 133–143
9. Gilge, M., Gussella, R.: Motion video coding for packet-switching networks–an integrated approach. Proc. SPIE, Boston, MA (1991)
10. Little, T.D.C., Ghafoor, A., Chen, C.Y.R.: Conceptual data models for time-dependent multimedia data. Proc. MMIS'92, Tempe, AZ (1992) 86–110
11. Little, T.D.C., Ghafoor, A.: Scheduling of bandwidth-constrained multimedia traffic. Comp. Comm. **15** (1992) 381–387
12. Milazzo, P.G.: Shared video under Unix. Proc. Usenix Conf., Nashville, TN (1991) 369–383
13. Montgomery, W.A.: Techniques for packet voice synchronization. IEEE J. on Sel. Areas in Comm. **SAC-1** (1983) 1022–1028
14. Nakajima, J., Yazaki, M., Matsumoto, H.: Multimedia/realtime extensions for the Mach operating system. Proc. Summer 1991 Usenix Conf., Nashville, TN, (1991) 183–198
15. Northcutt, J.D., Kuerner, E.M.: System support for time-critical applications. Proc. 2nd Intl. Workshop on Network and Operating Support for Digital Audio and Video, Heidelberg, Germany, (1991)
16. Pasieka, M., Crumley, P., Marks, A., Infortuna, A.: Distributed multimedia: how can the necessary data rates be supported. Proc. Usenix Conf., Nashville, TN, (1991) 169–182
17. Rangan, P.V., Ramanathan, S., Vin, H.M., Kaeppner, T.: Media synchronization in distributed multimedia file systems. Proc. MM'92, Monterey, CA, (1992) 315–328
18. Ravindran, K.: Real-time synchronization of multimedia data streams in high speed networks. Proc. MMIS'92, Tempe, AZ, (1992) 164–188
19. Szabo, B.I., Wallace, G.K.: Design considerations for JPEG video and synchronized audio in a Unix workstation. Proc. Usenix Conf., Nashville, TN (1991) 353–368

Session 4: Distributed Multimedia Systems

Chair: J.J. Garcia-Luna Aceves, SRI International

First paper: **System Support for Efficiently Dynamically-Configurable Multi-Party Interactive Multimedia Applications,** by *Mark Moran and Riccardo Gusella*

Second paper: **Requirements for Network Delivery of Stored Interactive Multimedia,** by *Darren New, Jonathan Rosenberg, Gil Cruz, and Thomas H. Judd*

Third paper: **Multimedia Processing Model for a Distributed Multimedia I/O System,** by *Rusti Baker, Alan Downing, Kate Finn, Earl Rennison*

Fourth paper: **Enhancing the Touring Machine API to Support Integrated Digital Transport,** by *Mauricio Arango, Michael Kramer, Steven L. Rohall, Lillian Ruston, and Abel Weinrib*

System Support for Efficient Dynamically-Configurable Multi-Party Interactive Multimedia Applications

Mark Moran[1] and Riccardo Gusella[2]

[1] The Tenet Group
Computer Science Division, University of California, Berkeley
and International Computer Science Institute
Berkeley, CA 94720

[2] Hewlett-Packard Laboratories
1501 Page Mill Rd.
Palo Alto, CA 94304

Abstract. An important new class of communication-oriented applications are described as multi-party interactive multimedia applications. In this paper, we examine the programming abstractions, system support and communication services required by these applications; evaluate existing support; and propose strategies for improving support by combining current systems with new designs. We have taken a holistic approach to this problem by considering the local system and communication architecture together in deciding where to implement each service, and in designing interfaces between components.

1 Introduction

Continuing advances in processing power, display capability, and communication capacity are enabling a class of demanding, communication-oriented applications that allow two or more geographically separated persons to interact as if they were at the same location. Escalating time and travel costs and the increasing need, in global markets, of personal and group communication, make these applications highly desirable. Applications that facilitate person-to-person collaboration by providing a variety of communication media—including video, voice, and shared drawing spaces—and support for shared applications—such as a spreadsheet, word processor, or designing tools—are classified as multi-party interactive multimedia (MIM) applications.

Current computer systems and communication architectures do not support MIM applications effectively for two major reasons: they do not provide abstractions and capabilities for MIM communication; and the device access, data copying, and data transport capabilities they do provide are inefficient and could potentially limit both the affordability and scalability of MIM applications. In this paper, we propose a holistic system approach to supporting MIM applications. The local computer system and the communication protocol stack are considered together in order to determine the best place to provide various services for MIM applications and to design interfaces among them. In Section 2 we discuss the requirements of MIM applications. Section 3 analyzes the strengths and weaknesses of current systems in meeting these requirements. Finally, in Section 4 we propose strategies for improving support for MIM applications by designing new abstractions and services to be combined with existing ones.

2 Requirements for Supporting MIM Applications

MIM applications require three types of system and communication support: programming and data abstractions that facilitate application development, a so-called real-time communication service for transmission of time-critical media, and efficient access to supporting services and multimedia I/O devices.

2.1 Programming abstractions

Development of multi-party interactive multimedia applications requires a richer set of programming abstractions and corresponding system support than traditional text- or graphics-based applications. Since MIM applications are inherently distributed, they must be able to operate across heterogeneous computing and communication environments. Therefore, abstractions must be provided to allow the flexible use of a variety of multimedia I/O devices and real-time communication services. We have identified the following abstractions as being fundamental:

Time-based media. Like general multimedia applications, MIM applications require support for time-based media, including device-independent functions to handle the acquisition, presentation, and transmission of time-based media. Presentation functions may perform synchronization between multiple streams.

Multi-party streams. The model for a multi-party stream that we will use is a media stream with one source and multiple destinations. It is important that the source be able to control and *transmit* such a stream without being concerned about the (possibly dynamic) set of participants currently receiving the stream. An abstraction of multi-party streams simplifies programming and enables abstract, efficient access to underlying multicast capabilities.

Receiver control over local presentation devices. Each human user must have the ability to control access to his or her local presentation devices, including the ability to suspend and resume any media stream. Users should be able to determine how media from various streams are multiplexed on an I/O device, and to influence the tradeoffs made between the quality of presentation and the cost of transmission and processing for each stream.

Security abstractions that can be used to protect participants from unwanted eavesdroppers and disrupters are also important. However, we will not consider these abstractions in this paper.

2.2 Communication requirements

The communication requirements of a MIM application are two-fold. First, the communication service must meet certain performance requirements of the application, which we shall denote as Quality of Service (QoS) requirements. The second communication requirement, one of efficiency, is for a multicast service.

Real-time communication service. Because MIM applications support real-time interactions between human users using time-based interfaces, they require a communication service that maintains the QoS requirements designated at the beginning of the conversation, regardless of fluctuations in network load. Without such a service, users may experience difficulty interacting and may notice gaps in the presentation of time-based media. A real-time communication service may offer either *hard guarantees* that have a strict analytical basis (e.g. [14]), or *soft guarantees* that indicate a target quality of service, but lack firm guarantees (e.g. [6]). A real-time communication service must include at least the following elements:

- *Service interface.* The service interface allows a potential network client to describe its expected traffic and to specify its QoS requirements. The expressiveness of the service interface determines the accuracy with which an application may describe its traffic and the quality of service it may request.
- *Support at network switching nodes for guaranteeing QoS.* At the very least, network nodes must be able to distinguish between traffic requiring real-time support, and unprotected non real-time traffic.
- *Admission control.* The network must refuse to accept new conversations when it determines that either (a) the network will (may) not be able to meet the QoS requirements of the new conversation, or (b) that accepting the new conversation will (may) cause the QoS requirements of existing conversations to be violated.
- *Protocol support.* Protocol support consists of the communication protocols that are required to implement the QoS support mechanisms of the internal network, to gather information required for admission control decisions, to establish and manage conversations, etc. Network protocols that actually suspended transmission of data to multicast receivers that have suspended a stream, would enable human users to conserve network and system resources by suspending streams.

Multicast ordering and reliability guarantees. Because most multicast ordering guarantees are incompatible with bounded delays, and because time-based media is usually "stateless", ordering and reliability guarantees are of limited usefulness for the transmission of time-based media. However, ordering and reliability guarantees are useful for MIM communication involving shared workspaces, such as a distributed drawing space or a shared application that allows input from multiple users; and for implementing *floor-control* algorithms that determine which participant of a multi-party conversation is enabled to transmit video and audio stream to the other participants. (e.g. [19]).

2.3 Efficiency requirements

Although some may claim that improvements in processor, memory and network speeds will alleviate efficiency concerns in the future, it seems very likely that efficiency will remain an important concern for MIM applications, since efficiency directly affects the quality, size, and number of simultaneous conversations in which a user may participate. Certainly, in the near future efficient utilization of system and network resources is crucial to ensure the affordability of MIM communication.

Efficiency concerns fall into two major categories: optimization of services, and efficient coordination between services. Optimizations of services includes efficient implementations of a service (e.g. multicast), efficient data handling (i.e. reduction of copying), and dynamic access to available hardware-assist modules, (e.g. a decompression module). Efficient coordination between services includes issues such as eliminating duplication of services between layers and designing interfaces that allow higher layers to access lower-level services efficiently. A holistic system approach is important for improving efficiency.

3 MIM Support in Existing Systems

A number of current systems provide partial support for MIM communication. In this section we will describe a few representative examples and analyze how well each meets the requirements specified in the previous section. We have grouped these systems into those that provide programming and system abstractions and those that provide communication services; efficiency considerations will be addressed in each section. After describing each system, we list its advantages (+) and disadvantages (−) for MIM applications, using **boldface** to indicate a reference to the requirements described in the previous section.

3.1 Programming and System Abstractions

Current systems provide programming abstractions and support for handling time-based media either by extending the client-server paradigm of a window system to handle time-based media and I/O devices (e.g. [1]), or by extending the operating system (e.g. [16] [2]).

Window Systems. Window systems vary considerably in their event-handling architectures, support for graphics, and programming philosophy. However, the features that are of primary importance to MIM applications—abstractions for network communication and for dynamic access to heterogeneous I/O devices— are relatively similar. Therefore, we will consider the X window system [21] as a representative example in discussing the support for MIM applications provided by network-transparent window systems.

The X window system has a client-server architecture: client applications (so-called *X clients*) communicate via a machine-independent protocol with a user-level process (the so-called *X server*) that multiplexes I/O to and from I/O devices (video display, keyboard,...). The use of a machine-independent protocol between client and server allows an application to access a remote display *transparently* without including explicit networking code.

Dynamic access to heterogeneous I/O devices is handled in two ways: device-independent abstractions are defined for accessing I/O devices, and a rudimentary query capability allows clients to determine and adapt to the characteristics of a remote video display. The *Xv* extension provides some support for incorporating *analog* video by providing access to devices that digitize analog video into (or produce analog output from) the framebuffer. Digitized frames may be grabbed out of the framebuffer.

+ Window systems provide an architecture that allows **dynamic, device-independent access to I/O devices** and abstractions for networking.
+ The Xv extensions allow **device independent** access to **video adapters** that enables display and export of analog video signals.
− One important weakness in current window systems is a complete lack of support for **digital time-based media**.
− Another important short-coming is that the networking abstraction does not support access to **multicast** or **real-time communication services**. Just as important, the networking abstraction cannot describe **dynamic**, participation in **multi-party communication.**
− Xv has been designed to support local display of video, not **distribution**. This decision results in two problems: because the signal is digitized directly into the framebuffer, analog video must be displayed locally in order to be distributed; and access to digitized frames through the X server is inefficient.

Although standard window systems do not provide much support for MIM applications, their philosophy of providing abstractions for networking and device I/O can be extended to support time-based media, either by extending the X server itself or by introducing a companion I/O server for time-based media. The next two systems demonstrate these two strategies.

Extending the Window System: SharedX. SharedX[3] is a commercial product that allows windows of existing applications to be shared by multiple, separated users. The application to be shared first establishes a standard X connection with the *SharedX server* (a modified X server). Through a graphical interface provided by the *SharedX client*, the user indicates a list of destination displays for each window that will be shared. The SharedX server distributes X protocol messages regarding shared windows to the remote X servers as if it

---

[3] SharedX is a registered trademark of Hewlett Packard Co.

were an X client. Neither the client application nor the remote X servers need to be changed.

+ The SharedX server provides limited **multi-party communication** between an X client and remote X servers.
+ The SharedX client provides abstractions and user-interfaces for describing **multi-party communication**; thereby simplifying application development and ensuring uniformity of abstractions and interfaces.
− As with X windows, no support is provided for **time-based media**.
− Also as with X windows, the communication abstraction does not afford access to underlying **multicast** or **real-time communication** services, or **dynamic** participation in **multi-party** communication.
− The granularity of **receiver control** is too coarse: in order to share a window, a remote machine must give the machine running the SharedX server complete access to its display.

Separate I/O Server for Time-Based Media: ACME. Anderson, Govindan and Homsey have designed and implemented the ACME I/O server that provides abstractions for device-independent access to continuous media (time-based) I/O devices, and a *logical time system* abstraction that manages timing of, and synchronization between, continuous media (CM) streams [1]. Nicolaou has proposed a similar design in [18]. Govindan and Anderson have also designed and implemented extensions to the UNIX[4] operating system that provide efficient multi-level process scheduling, and *memory-mapped streams* that reduce data copying in order to support efficient implementation of the ACME server as a user-level process [15].

+ Good support for **time-based media**: access to **continuous media I/O devices** and mechanisms for handling the **timing and synchronization** of CM streams.
+ Operating system extensions to make implementation **efficient**.
− ACME supports neither access to **underlying multicast** capabilities nor the description of **dynamic** participation in **multi-party** communication.
− There is no notion of **variable quality,** hence no capability for users to trade off quality against resource utilization.
− Querying capabilities that would allow **flexible access** to I/O devices and hardware processing modules (e.g. video codec) are not provided.
− It does not seem possible to describe and utilize specialized data paths in a machine-independent manner, e.g. a direct data path between a decompression module and a frame buffer.

Operating System Extensions for Time-Based Media: QuickTime. The QuickTime extensions to the Apple Macintosh operating system and associated

---

[4] UNIX is a trademark of USL

toolbox routines support creation, manipulation and display of time-based media as represented by a *movie* abstraction. The movie abstraction associates and describes related audio and video streams, including timing information for playback. It is designed to support fast cut-and-paste editing and efficient sharing of time-based media data between movies. Toolbox support offers flexible and dynamic access to frame grabbers, and hardware or software compression modules.

+ Excellent programming support for generation, manipulation and playback of stored **time-based media**.
+ **Dynamic, device-independent access** to I/O devices and hardware modules that capture and present time-based media.
+ Some support for **user-selectable quality** on playback.
− The current abstractions do not support **network communication** between distributed systems. Although networking capabilities have been announced for the next version of QuickTime, the overall QuickTime paradigm and past Apple networking abstractions suggest that networking will not provide access to **multicast** or **realtime communication** services.
− The *movie* abstraction is not well-suited to describe **distribution** of time-based media, particularly **interactive, multi-party** communication. The authors believe that interactive communication is more naturally represented as a flow of data between devices than as production and immediate playback of a movie to and from a "network device". For example, the *record-playback* paradigm of a movie suggests the use of large buffers in production to prevent loss of data; while interactive applications would prefer to suffer momentary data loss in order to decrease end-to-end delay. An indication of the bias towards a record-playback paradigm is the movie file format, in which the movie *header,* which includes the length of the movie, appears at the end of the file, unnecessarily complicating network transmission.

3.2 Network Communication Services

In this section we describe three proposals to provide communication services for the transmission of time-based media. Because of space considerations, we have selected these three as representatives of a rich area of research. We also describe the IP multicast extensions.

ISI/BBN Protocol Suite. ISI and BBN have collaborated to develop a suite of communication protocols that support N-way video conferencing [5]. The network protocol is ST-II, which provides a connection-oriented service that includes multicast and QoS guarantees [4]. At the transport-level, NVP [7] and PVP [8] support transmission of voice and video streams respectively. The Connection Control Protocol (CCP) [22] manages connections and performs synchronization between streams.

+ These protocols have been used successfully to support video-conferencing across the DARPA Terrestrial Wideband Network [11].

+ ST-II provides 1 ⇒ N **multicast, dynamic addition** and deletion of multicast receivers, and admission control in order to meet **QoS requirements** for the transmission of time-based media.

− Transport-level protocols are *special purpose* for each media type (hence a separate protocol for voice and video). They provide no meaningful **abstractions for time-based media** in general.

− Currently, **performance guarantees** are supported in the Terrestrial Wideband Network via *ad hoc* rather than general algorithms. These techniques do not seem applicable to other networks. Allocation of peak rate bandwidth may artificially reduce the *proportion* of network resources that can be utilized by *real-time* clients. (System resources that are unused by real-time traffic will be available to non real-time traffic.)

− No **receiver control** over transmission of time-based media streams.

Tenet Real-Time Protocol Suite. The Tenet real-time protocol suite provides deterministic and probabilistic guarantees for real-time communication [14]. Analytical algorithms based on worst-case assumptions are used to provide deterministic and probabilistic guarantees. A network-level protocol, RTIP [23], provides simplex, unicast connections with a guaranteed lower bound on bandwidth, and guaranteed upper bounds on end-to-end delay, delay jitter, and buffer overflows within the network. CMTP [17] is a transport-level protocol that provides abstractions to simplify and improve efficiency for the transmission of time-based media. Admission control and resource allocation for both protocols is handled by the Real-Time Channel Administration Protocol (RCAP) [3]. RCAP has a hierarchical architecture that enables it to perform admission control across internetworks of arbitrary topology.

+ RTIP provides flexible **real-time service** to support transmission of time-critical data (including time-based media) according to **QoS requirements** specified by the **application**.

+ Analytical basis of algorithms provides predictable performance that greatly simplifies application development. It also allows admission-control decisions to be made in a *distributed* manner, without requiring global knowledge of network state.

+ RCAP implements admission control algorithms via a hierarchical approach that allows the scheme to work across **heterogeneous internetworks** of **arbitrary topology**.

+ CMTP provides **more efficient** service and **better abstractions** of communication for time-based media than traditional message oriented paradigms [13]. The use of a shared buffer between the sender and the transport service allows data transmission to proceed *periodically* after the start of a stream without further synchronous interactions with the sending applications.

− No **multicast communication**.

- No support for **receiver control** or manipulation of streams.
- Worst-case analysis may artificially reduce the *proportion* of network resources utilized by *real-time* clients. Unused resources may be utilized by non real-time traffic.

Predicted Service. An alternative proposed by Clark et. al is *predicted* service [6]. As with the Tenet suite, a client application describes its traffic and a requested delay bound during an establishment phase. However, instead of basing admission control decisions on analytical models, the current load is measured and used as a predictor of future network load. Predicted service is designed for *adaptive* applications, which try to reduce end-to-end delays (*playback point*) as network load decreases, and which adapt to increases in delays as load increases.

+ Traffic description and **QoS requirements** are **flexible** and can be adapted to **application requirements.**
+ Emphasis on measurement and statistical multiplexing will probably allow a higher-proportion of network resources to be used for *transmission of time-based media* than worst-case analysis.
- Variation in the end-to-end delay *bounds* for a stream, caused by the inaccuracy of using current load as a predictor of future load, requires applications to modify their behavior and may introduce glitches in the display of time-based streams.
- No support for **multicast.**
- No support for **receiver control** for the suspension or manipulation of time-based media streams.

IP Multicast. Deering has designed and implemented extensions to IP to support multicast delivery of IP datagrams to dynamic distribution lists [10]. The assumed network topology is a graph of IP multicast routers connecting broadcast-based leaf networks. Part of the IP address space is designated as logical multicast addresses. Multicast delivery is implemented in a two-level hierarchy: multicast routers distribute messages among themselves and broadcast them to each attached network on which a host is receiving the multicast group. Hosts filter broadcast packets based on IP addresses..

+ **Multicast** service is provided with relatively little additional overhead or modifications to existing protocols.
+ The protocol is virtually stateless: a recovering multicast router is able to fully reconstruct its state by querying attached networks.
+ Hierarchical approach gives reasonable **scalability.**
- The most important weakness is that no provision is made for **real-time performance guarantees.**
- Selection of transient multicast IDs allows for collisions and unpredictable disruptions of service.

– Little or no **receiver control** over transmission of streams. In particular, the time to disconnect from a multicast stream is non-deterministic, causing host resources to continue to be consumed for a longer time than necessary.

4 Strategies for Supporting MIM Applications

We have developed strategies that combine existing systems with new approaches to support MIM communication efficiently and flexibly. An important characteristic of our proposal is a holistic approach to supporting MIM communication. Rather than focusing on a single layer or service, we have considered the entire system and communication architectures together, in order to determine the best place to implement each service, and the proper interfaces between layers to allow abstract, efficient access to such services. The following subsections describe the strategies we have developed for providing programming abstractions and system support, and abstractions of communication services.

4.1 Programming Abstractions and Local System Support

As an overall design architecture, we have decided to provide programming abstractions and local system support through the implementation of a continuous media I/O server co-resident with a standard window server, as in ACME. Our reasons for choosing this approach over media-specific extensions to the operating system are as follows:

1. Applications based on a portable I/O server will themselves be more portable and will be easier to distribute across heterogeneous platforms. This point is demonstrated by current window systems.
2. Applications based on a portable I/O server can run on existing operating systems and can be migrated more easily to future systems that include extensions for time-based media and multi-party communication.
3. By separating the data and control paths, a user-level I/O server can utilize optimized I/O routines and data paths in the operating system and can, therefore, match the performance of operating system extensions.
4. A user-level I/O server will be more flexible and more easily extended than an operating system.

Elements Drawn from Existing Systems. Elements drawn from ACME, QuickTime, and Xv fall into three categories: device-independent access to time-based I/O devices, querying capabilities to allow dynamic utilization of available devices and modules, and support for timing and synchronization of time-based media streams.

Device-Independent Access to I/O Devices. In general, devices are classified as either sources or sinks of time-based media. Routines are provided to move data to and from such devices. As with QuickTime, access to modules that process CM data (e.g. compression/decompression modules) is also supported.

Querying capabilities. Querying facilities (provided in QuickTime, X, or Xv), are required to allow distributed applications to determine the I/O devices and processing modules available at a remote site for processing and presenting time-based media. Since we have chosen an architecture similar to current window systems, we propose query routines that accept a template of the desired device or module.

Timing and Synchronization of Time-Based Media. The *playback* and *capture* routines that provide device-independent access to devices, will also handle the timing of media capture and presentation. An application accessing such devices will need only to initiate and terminate capture or presentation of the stream. Both ACME and QuickTime also provide support for synchronizing the presentation and, in the case of ACME, the capture of related CM streams through the use of logical time systems.

New Designs. In order to support MIM applications, we have developed new abstractions for efficient, device-independent data handling and a richer encapsulation of network communication that enables abstract access to real-time, multicast communication services.

Abstractions for Data Handling. Because of the relatively high data rates required to support MIM communication, we have developed an abstraction for data movement that allows efficient data handling without introducing machine dependencies in the application. The primary strategy of our design is to separate control of data handling from the data path itself by using a *setup* phase. During the setup phase, the application indicates the flow of data between *logical* devices, where a *logical device* may correspond to a physical device (e.g. a framebuffer), or a software module (e.g. an I/O server process that maintains synchronization between several media streams). Data is then moved between logical devices without requiring further intervention by the application. This approach can take advantage of specialized data paths, (e.g. between a video decompression module and the framebuffer) and/or optimized data handling provided by the operating system (e.g. as described in [12]), without requiring modifications to the application.

The interface we have designed uses the abstraction that every device or module describes the location of its own *input buffer*, which may be in system memory or on a physical device. This description is passed as a parameter to the *preceding* logical device during the setup phase. The preceding device is then responsible for directing its output in the most efficient manner into the input buffer of the logical device following it. This abstraction also allows data paths to be altered dynamically, e.g. to switch between input devices.

Abstractions for Network Communication. We have also designed richer networking abstractions to describe dynamic, multi-party communication and to allow system-independent, efficient access to multicast and real-time communication services. This design consists of three components: a logical display to describe multi-party communication; an abstract interface for a real-time communication service; and connection management mechanisms.

- **Logical display.** The networking abstraction supported by the X window system consists of a *display* structure that specifies a connection between a client application and the destination X server. In order to support multi-party communication, we have extended the display structure to a *logical* display that keeps track of participants and destination addresses, and supports dynamic addition, deletion, and suspension of participants.
- **Abstract service interface to real-time communication.** Instead of trying to hide network access, we have developed an abstraction of network communication that enables simple and efficient access to communication services supporting real-time communication and multicast. Access to the network is managed by the I/O server, just as other I/O devices. One advantage to managing network communication in the I/O server is that the same networking abstractions can be applied to a wide variety of networks and internetworks, with the I/O server performing translation of parameters as required.

  Applications describe their communication requirements in terms of a high-level model of multi-party communication that can be mapped precisely onto the network-level models described in [14], [17], [20], [9], and [6].
- **Connection-management mechanisms.** We have also developed connection management abstractions that allow users to join, leave, suspend and resume connections. These capabilities allow a (human) receiver to determine interactively how local system and network resources are proportioned between streams.

4.2 Network Communication Services

The strategies we have developed for supporting the communication requirements of MIM applications combine the real-time communication services and multicast abilities of existing systems with new design features to support scalability by promoting sharing of network resources between related streams.

Elements Drawn from Existing Systems. Real-time communication services will be based upon extensions of, or modifications to, existing systems described in Section 3.2. In addition, we will include the stream-oriented paradigm of time-based communication presented in CMTP and the dynamic membership of multicast groups supported by both ST-II and IP multicast.

New designs. We have targeted two areas for future research in communications support for MIM applications: general algorithms for real-time multicast services, and mechanisms for promoting the sharing of network resources.

Dynamic, Real-Time Multicast. In order to extend a unicast real-time service to support multicast, we are defining new models for traffic characterization and QoS requirements and exploring new routing and resource allocation schemes. Incremental routing mechanisms are being studied in order to support dynamic additions, deletions, and possibly suspensions of participants.

Mechanisms to promote sharing. The real-time communication services we have described in Section 3.2 have each made one of two assumptions about the relationship between streams. Services that provide analytical, worst-case guarantees assume that all streams are independent (e.g. [14]). Other services attempt to promote sharing of network resources by assuming (i) that current measured network behavior is a good predictor of future behavior; and (ii) that statistical multiplexing between streams will mask the burstiness of any single stream (e.g. [6]). However, we propose sharing resources between related streams according to a priori knowledge of the relationship between the streams. For example, a MIM application may use a *floor control* mechanism to serialize audio and video input from participants. Network resources can then be shared between video and audio streams from different sources, even if statistical multiplexing does not apply, e.g., in the host or over any link on which one stream uses a significant portion of network resources.

Conclusion

We have described system and communication support for multi-party interactive multimedia (MIM) applications. In particular, we have shown that a holistic approach is important to ensure that required services are provided at the appropriate place in the system and/or communication hierarchy and to define interfaces that enable abstract, efficient access to underlying systems. We have examined existing systems and proposed several innovations to improve the support given to MIM applications. In particular, we have designed abstractions that enable abstract, efficient access to real-time multicast communication services, multimedia I/O devices and support services.

Acknowledgments

The authors would like to thank many colleagues at Hewlett Packard Labs and the Tenet group for their invaluable discussions, advice, and comments. Walt Hill, Tracy Steelhammer, Srinivas Ramanathan, Audrey Ishizaki, Amit Gupta, Clemens Szyperski, and Giorgio Ventre have been of particular assistance.

References

1. D. P. Anderson and G. Homsy. A continuous media I/O server and its synchronization mechanism. *IEEE Computer*, 24(10):51–57, Oct. 1991.

2. Apple Computer Inc. QuickTime version 1.0 for developers. CD-ROM documentation, 1991.

3. A. Banerjea and B. Mah. The Real-Time Channel Administration Protocol. In *Proc. 2nd Int. Workshop on Netw. & O. S. Support for Dig. Audio and Video*, Nov. 1991.

4. S. Casner, J. Charles Lynn, P. Park, K. Schroder, and C. Topolcic. Experimental internet stream protocol, version 2 (ST-II). *RFC 1190*, Oct. 1990.

5. S. Casner, K. Seo, W. Edmond, and C. Topolcic. N-way conferencing with packet video. In *Proc. 3rd Int. Workshop on Packet Video, VISICOM '90*, Mar. 1990.

6. D. Clark, S. Shenker, and L. Zhang. Supporting real-time applications in an integrated services packet network: Arch. and mech. In *Proc. SIGCOMM '92*, 1992.

7. D. Cohen. A network voice protocol NVP-II. Technical report, USC/Information Sciences Institute, Apr. 1981.

8. E. Cole. PVP - a packet video protocol. Technical report, USC/Information Sciences Institute, Aug. 1981.

9. R. L. Cruz. *A Calculus for Network Delay and a Note on Topologies of Interconnection Networks*. PhD thesis, University of Illinois, July 1987.

10. S. E. Deering. *Multicast Routing in a Datagram Internetwork*. PhD thesis, Stanford University, Dec. 1991.

11. W. Edmond, K. Seo, M. Leib, and C. Topolcic. The DARPA wideband network dual bus protocol. In *Proc. SIGCOMM '90*, pages 79–89, Sept. 1990.

12. K. Fall and J. Pasquale. Exploiting in-kernel data paths to improve I/O throughput and CPU availability. In *Usenix Winter Conference*, 1993.

13. D. Ferrari, A. Gupta, M. Moran, and B. Wolfinger. A continuous media communication service and its implementation. In *Proc. of GLOBECOM '92*, 1992.

14. D. Ferrari and D. C. Verma. A scheme for real-time channel establishment in wide-area networks. *IEEE J. on Sel. Areas of Commun.*, Apr. 1990.

15. R. Govindan and D. P. Anderson. Scheduling and IPC mechanisms for continuous media. In *Proc. of Symp. on O. S. Principles.*, pages 68–80, Oct. 1991.

16. E. Hoffert et al. QuickTime: an extensible standard for digital multimedia. In *Proceedings of CompCon '92*, pages 15–20, Feb. 1992.

17. M. Moran and B. Wolfinger. Design of a continuous media data transport service and protocol. Technical Report TR-92-019, Int. Computer Science Inst., 1992.

18. C. Nicolau. An architecture for real-time multimedia communication systems. *IEEE Journal on Selected Areas of Communications*, 8(3):391–400, Apr. 1990.

19. T. Ohmori et al. Multiparty and multimedia desktop conferencing system: MERMAID. In *Proc. of IEEE Multimedia '92*, pages 122–131, Apr. 1992.

20. C. Partridge. A proposed flow specification. Internet RFC 1363, Sept. 1992.

21. R. W. Scheifler and J. Gettys. The X window system. *ACM Transactions on Graphics*, 5(2):79–109, Apr. 1986.

22. E. Schooler and S. Casner. An architecture for multimedia connection management. In *Proc. of IEEE Multimedia '92 Workshop*, pages 271–274, Apr. 1992.

23. H. Zhang, D. Verma, and D. Ferrari. Design and impl. of the Real-Time Internet Prot. In *IEEE Wrkshp on the Arch. and Impl. of High Perf. Comm. Subsys*, 1992.

Requirements for Network Delivery of Stored Interactive Multimedia

Darren New
Jonathan Rosenberg
Gil Cruz
Tom Judd

Bellcore
Morristown NJ 07962 USA

1 Introduction

This paper gives some insights into the network requirements for delivering stored, interactive, multimedia information. Many applications require the ability to deliver such information. These applications include remote training and education, online access to multimedia reference libraries and electronic shopping. Interest in such applications is at an all-time high. Virtually every large consumer electronics company, computer manufacturer, software producer and entertainment company is actively involved in investigating, trialing or selling products for these applications.

These applications require networks providing appropriate capabilities. Unfortunately, little is known about the network requirements of such applications. There remain questions as to required bit rate as well as the capabilities that are desirable or necessary in a network. For example, what kinds of error handling functionality are needed and where should that functionality reside?

Many of our insights into networked multimedia delivery come from our work on the DEMON (Delivery of Electronic Multimedia over Networks) research project. This project identifies and seeks solutions to the problems caused by attempting to deliver high-quality real-time interactive multimedia information over current and near-term telephone networks.

Our research has shown that it is possible to deliver high-quality interactive multimedia information over networks with bit rates as low as 128 kb/s (such as provided by ISDN). Accomplishing this requires the use of preprocessing techniques we have developed. In addition, we have determined that the network must provide a guaranteed minimum bit rate to enable this preprocessing. We also believe that it is desirable that the network support multiple error handling strategies. The network should apply the strategies to data as specified by the application.

2 Required Functionality

There are a number of functions that an application delivering multimedia over a network must provide that a stand-alone presentation can safely ignore. For example, when a document[1] is local to the presentation devices, bandwidth between the storage and the display is normally not a problem. Format conversion is also rarely needed, since

---

[1] We define a document as a stored collection of multimedia objects along with instructions for presentation.

documents are typically authored on the same platform on which they are presented. Flow control, a necessary part of any network protocol, is trivial when there is only one processor. Finally, error handling is generally ignored during local display of multimedia, since electronic transmissions within a single machine are sufficiently reliable that no particular effort needs to be made.

On the other hand, network delivery must consider and handle all of the above problems: bandwidth allocation and control, format conversion, flow control, and error handling. Our approach to handling these network-induced problems and the implications for network requirements will be presented in this paper.

We first describe the types of multimedia information we wish to be able to deliver and the assumptions we make to enable delivery. We then briefly describe one possible solution. We also describe some of the implications that our work has for networks and for end-user equipment requirements. Finally, we draw some conclusions.

3 Problem Statement

The types of multimedia information we wish to deliver include crisp text and graphics, high-quality still images and motion video, and high-quality sound. We wish to present visual and aural information at the best quality supported by the presentation device, and we wish to present the media at the precise times the author specifies. Documents would be temporal, which means they would display themselves in real-time without additional action from a viewer (unlike for example a "multimedia magazine" with pages to be turned). We assume that variable lossy compression is available, but that it is desirable to compress everything as little as possible to prevent loss of quality.

We make the assumption that we cannot deliver the entire document in advance because the storage requirements are too great or the delivery time would be excessive. Also, delivery of information which is never viewed may be undesirable for economic reasons.

We wish to deliver this multimedia over the public switched network (PSN) in the near future, using the existing copper access facilities (rather than fiber or coaxial cable). The current PSN allows bit rates from 128 kb/s (via Basic Rate ISDN) up to 1.5Mbps (via ADSL [Man91]). While 1.5Mbps is barely adequate for delivering entertainment-quality NTSC video, we require support for high-resolution graphics and high-quality sound. We also wish to support significant numbers of applications at the lower bit rates currently offered by ISDN.

We wish to support interactivity without undue delays. Ideally, we would prefer the viewer to never experience a delay regardless of interaction. This is not possible; however, we can achieve this for a majority of the types of interaction we plan to support. How we do this will be detailed below.

Finally, we would like to require a minimum of support hardware in the network for each subscriber. Details of how we envision supporting such sharing are provided below.

4 A Solution

Having described the problem, we now describe a solution applicable to a limited class of systems. While limited, this class includes a broad range of desirable applications.

Our first simplification is to assume that the Customer Premises Equipment[2] has some amount of local memory, computational power, and media formatting and display capabilities. We do not assume that the memory is large enough to hold an entire document. Hence, incremental real-time delivery will occur during the viewing of a document.

Our second assumption is inherent in our definition of *documents*. That is, all media and structural information is available in advance of the start of delivery.

We also assume that variable, possibly lossy compression is available for some of the media. Without this assumption, the problem becomes trivial, as one can merely compress everything as much as possible without loss and then it is either deliverable or undeliverable. However, since higher compression ratios can be obtained by discarding perceptually insignificant information, using lossy compression without ruining the quality of a presentation can result in many more documents being deliverable at low bit rates.

Given these three simplifications, our solution to the problem of presenting high-quality multimedia over narrow bandwidths involves preprocessing the content extensively before delivery.

5   The DEMON Project

Part of our work in the DEMON project [RKGB92] is developing algorithms to schedule delivery of multimedia objects to allow presentation at good quality, to make efficient use of available CPE memory and network bandwidth, and to support viewer interaction without delays. The offline preprocessing to perform this scheduling is a vital step in making these documents deliverable, and it will be referred to repeatedly below.

To build such algorithms, we need to make two more restrictions. The first is a requirement that an author specify in advance all interactions to which the system should respond without delay. Clearly, it is unreasonable to expect all possible interactions, including forward and reverse search, to be specified by an author. However, those which the author specifies in advance are scheduled and delivered in such a way that any path taken through the document can be followed without delays. If too many interactions are specified, the quality of the presentation may be degraded or excessive memory may be required, so some care must be taken by the author.

Our second restriction is that the documents we can preprocess must be authored in such a way as to include more structural information than is often present in multimedia documents, even interactive ones. This allows compression and delivery to occur on semantically meaningful objects. For example, films do not have the required structural and semantic information since they have already been reduced to a sequence of frames.

There are a large number of applications that fit these restrictions naturally. For example, electronic "yellow pages," "infotainment" documents such as travelogs or documentaries, interactive maps, interactive training and educational presentations, music videos, board games, and interactive role-playing games, as some examples.

The preprocessing is complex, and cannot be presented in detail in a brief paper; but in general, it consists of applying three types of compression to a document. The first compression method relies on an author specifying the semantic roles of various media objects. For example, if an author specifies that an image is to be a background, the

---

[2] Customer Premises Equipment (or CPE) is equipment not inherently part of the network, allowing access to network services. For example, a telephone handset is simple CPE, while a workstation with an ISDN Terminal Adaptor is complex CPE.

image can be transmitted once at the beginning of the presentation and stored locally, to be refreshed without retransmission when overlaying imagery is erased. As another example, text can be sent as ASCII strings rather than as bitmaps representing the characters, and graphical objects (lines, circles, and so) on can be expressed as brief commands rather than bitmaps.

The second type of compression uses standard single-media compressions, such as JPEG, MPEG, ADPCM, and so on. These are applied as appropriate to the individually-specified semantic components of the document being processed. We do not investigate the development of new medium-compression techniques, preferring instead to leverage off the work of others in this field.

The third type of compression is time-shifting, which consists of delivering media objects significantly in advance of their use along with instructions about the conditions under which they are to be presented. This is the technique that allows us to support interaction without delays and to make efficient use of narrowband networks. Obviously, some amount of memory is needed in the CPE to hold this buffered information, but our studies show that something between one and eight megabytes should be sufficient for most documents.

Where delays are explicitly introduced into a document, for example to give a viewer time to admire an image, the bandwidth of the network is available to deliver components scheduled for later presentation, and these may then be presented with better quality than if that bandwidth were wasted. Alternately, if a point at which a viewer may change the flow of the document is approaching, the beginnings of the alternate path can be transmitted before the choice is offered. Along with the media for each path, instructions as to how to handle user interactions, in terms of how to play the beginning of each path and how to decide which path a viewer has chosen, are delivered in advance of being needed. By the time an option to choose an alternate path is presented, sufficient media are available to present the path for long enough for the remote server to respond and begin delivering the remainder of the chosen path. This avoids interaction delays even in high-delay networks.

These three types of compression work together. For example, knowing that an image only appears briefly allows us to compress it more than an image which will be studied over a long duration by a viewer. They also interact with error handling techniques. For example, if an error is detected in an image, but there are insufficient free bits on the network to retransmit it before the image is presented, then there is little point in adding error-detection codes to the information.

6 Network Implications

We have made significant progress on applying these compression techniques to multi-media documents. In so doing, we have learned of a number of implications that these multimedia services will have for the network. Our techniques, for example, assume that a minimum guaranteed bit rate will be available, and we handle all allocation of that bit rate in the preprocessor. This also implies that variable-bit rate networks are not needed for these types of services. As another example, format conversion and compression should be done in the preprocessor, since we need to know in advance how many bits will be required to transmit various media objects. Since the preprocessor is cognizant of all possible paths through the document and also knows how much memory the CPE has and how much various media objects take, flow control in the usual sense is unnecessary. We simply stop transmitting when we know the CPE's memory is full, and restart transmitting when we know the CPE has consumed enough media (i.e., when the CPE sends back a signal indicating that the presentation has progressed to where media objects have been consumed and the memory has been freed).

On the other hand, error correction must be done in the network, since errors are introduced during transmission.

There are three levels of error correction that may be desirable. First, it may not matter whether some information arrives intact. For example, background audio may have corrupted bits or even entire brief segments dropped without unacceptable quality loss.

Second, some items may be able to be retransmitted if they arrive in error. For example, if an image is received in error but will be visible for some time, displaying the image with corrupted pixels is acceptable, but retransmitting it when bandwidth is available is preferable, in order that the image can be repaired. Error detection such as cyclic redundancy checks, with retransmissions during spare bandwidth availability, could support the type of data that can be delivered with errors but would preferably be error-free.

Third, there will be some protocol data which must arrive intact, especially when in-band signalling is used. For example, protocol information detailing packet sizes and media types, and even some media like text, where a single bit error could be unacceptable. For networks with only short error burst lengths, forward error correction could be used for small pieces of reliable transport. These would include protocol data (i.e., headers) and perhaps text or graphic commands.

It is interesting to note that the preprocessor will have available a complete map of the possible bandwidth utilization patterns. It will know, for each media object, whether there will be time to retransmit a second copy before the object is played or displayed. Due to timing of the presentation, it may be that the same image is error-checked at one point in the presentation and transmitted without possibility of correction at another because of lack of bandwidth at the time of the second presentation.

For networks which tend to have long burst errors, perhaps even dropping entire strings of bytes, it may be preferable to reserve some bandwidth and partition the stream into small packets, rather than use forward error correction codes. Each packet would be error-checked and retransmitted as needed. This technique would require a faster turn-around time between transmitter and CPE, perhaps using a specialized intelligent buffer within the network, close to the subscriber.

Ideally, for the class of applications which DEMON supports, we would like the following kinds of functionality available to applications:

If the connection from the device holding the documents to be delivered (the "server") to the edge of the network is a high-speed connection, such as an optical ATM link, the application should be able to deliver a large chunk of bits, with some bits marked as "critical" and some bits marked as "inform of errors" and the remainder marked as "no efforts to correct." For the "inform of errors" bits, some attempt could be made to correct the errors, but these efforts should not take longer than an application-specified length of time lest other data misses its deadline.

The network should deliver the bits to the user as indicated, packetizing or forward-error-correcting as appropriate to the network technology. The more buffer space dedicated to each subscriber within the network, the longer the server machine can take before processing another chunk of information for that subscriber.

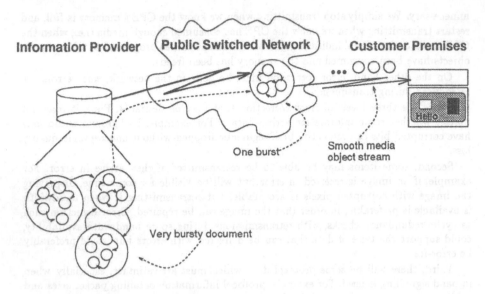

Fig. 1. Proposed delivery architecture

Since we know exactly when the various components of a multimedia stream must arrive, we believe it is possible to engineer a server which can keep up with any given number of subscribers. The preprocessed documents naturally structure themselves into relatively large pieces which get transmitted, separated by delays while the presentation catches up and the server waits for feedback from the CPE. This would imply a useful structuring for bursty transmission from the information provider to an intelligent but smaller buffering device closer to the subscriber. (See Figure 1.) By decreasing the responsiveness required from the server using this technique, it is expected that a larger number of documents can be delivered without a significant increase in hardware costs.

7 CPE and OS Implications

Since the preprocessing makes almost all decisions in advance, the CPE itself needs very little sophistication in the CPU. However, the media capabilities of the CPE need to be fairly powerful, since compression and coding of media needs to be reversed upon presentation. Since so many media bits will flow through the system, it is likely that the hardware will need multiple buses between various media and presentation hardware. The OS, of course, will need to be real-time, and will need to handle multiple independent I/O streams concurrently. In particular, where the network bandwidth is sufficient to support motion video, a direct pipe from the network interface, through the decompression hardware, and onto the screen would be desirable; any additional CPU intervention or bus limitations are likely to reduce the quality of the video and would certainly require additional memory in the CPE.

8 Conclusion

It is possible to design systems capable of delivering high-quality interactive multimedia documents over networks of surprisingly low bit rates. High-quality interactive multimedia documents can even be delivered over bit rates as low as 128 kb/s. Clever preprocessing can be used to schedule and compress multimedia documents for presentation without delays, even in the face of large network delays. Most functionality can be handled outside the network, including flow control and bit rate allocation; the only network requirements are application of specified error handling and guarantee of available bit rate. Finally, our techniques should scale to large user populations, but more research is necessary.

References

[Man91]   Earl Manchester. New uses for residential copper. *Telephony*, June 1991.

[RKGB92] Jonathan Rosenberg, Robert Kraut, Louis Gomez, and C. Alan Buzzard. Multimedia communications for users. *IEEE Communications*, 30(5), May 1992.

Multimedia Processing Model for a Distributed Multimedia I/O System*

Rusti Baker, Alan Downing, Kate Finn, and Earl Rennison
Information, Telecommunications, and Automation Division
SRI International, Menlo Park, California, USA

DooHyun David Kim and Young Hwan Lim
Electronics and Telecommunications Research Institute, Korea

Abstract

This paper describes a data-processing model providing the synchronization, integration, and presentation of multimedia data. This model consists of three layers: (1) stream, (2) multimedia presentation, and (3) hyperpresentation. This paper describes the data abstractions associated with each layer. These data abstractions provide a framework for defining the services provided by each layer, and describe the mechanisms that provide those services. A sample scenario is presented to illustrate the use of this model. Implementation issues, and future directions are also discussed.

1. Introduction

Most vendor multimedia solutions address the problem of attaching various media devices, such as VCRs and camcorders, directly to computers. This approach to multimedia processing limits computers to being only I/O control boxes. Similarly, system developers and researchers have defined abstractions for multimedia processing that support specific vertical applications (e.g., computer-based training), specific topics (e.g., synchronization), or specific media types (e.g., music, video). This approach limits the richness of applications that can be supported by native services. A more general approach to multimedia computing is required to support a variety of applications such as hypermedia [Brondmo et al. 1990; Ogawa et al. 1990; Zellweger 1992], videoconferencing [Watanabe et al. 1990; Ishii and Miyake 1991], multimedia authoring, archiving, and collaboration [Poltrock and Gudin 1992].

In this paper, we discuss a multimedia data-processing model that supports a wide variety of applications. This model supports network-transparent access to stored multimedia data, real-time multimedia input devices, and multimedia processing. The model addresses real-time multimedia data switching and delivery, as well as acquisition, processing, and output. Media translation, compression, and synchronization services are integral to the model.

A fundamental difference between our model and others [Little 1990; Nicholaou 1990] is that our model is based on a full complement of generalized data abstractions of multimedia objects, namely streams, multimedia presentations, and hyperpresentations, rather than focusing primarily on data processing and synchronization. In our model, the data abstractions provide a framework for defining relevant and necessary processing and synchronization services and the mechanisms for providing those services. This leads to a model that is more intuitive to application developers and end-users, while still being powerful enough to accommodate real-time multimedia scheduling and integration services across a network of cooperative processors. A result of this approach is that much of the implementation details are hidden from the application developers, allowing them to focus on application-specific issues.

*This work is sponsored by the Electronics and Telecommunications Research Institute, Korea.

Our model comprises a stream layer, a multimedia presentation layer, and a hypermedia presentation, or hyperpresentation, layer. The *stream layer* provides services similar to a video switch used in video production studios, where sources and destinations are local or remote files or devices (such as microphones, musical instruments, video cameras, displays, and audio speakers). The *multimedia presentation layer* is based on the concept of a programmable media multiplexer or media mixer, such as an audio mixer used in concert productions, or a switch or video editor used in a video production studio. In this layer, the media multiplexer takes input from a variety of sources, mixes them according to controllable parameters, and directs the result to an output port or destination. The *hyperpresentation layer* is a generalization of the links used in a hypermedia document, where the document may include time-based media such as audio and video, and the links are dynamic and vary over time.

We are developing a multimedia I/O system that is based on this model. This system, called MuX, consists of a multimedia I/O server, a presentation manager, an application programmer's interface (API), and a scripting language. The MuX system provides real-time support for streams (e.g., audio, video, mouse, and graphics). It also provides the capability to interactively define, edit, and review a multimedia presentation via the API. Once a presentation has been edited, its structure can be stored for later retrieval, editing, and playback using the scripting language. While a multimedia presentation is being presented, it can be controlled by either the application or the presentation manager. These multimedia presentations can be linked together to form hyperpresentations. Ideally, the time-critical services provided by the MuX server would be implemented at the system level.

In the following sections, we discuss each layer of the model, including its data abstractions, services, and mechanisms. A separate section describes how all these layers are integrated into an example application. Finally, we discuss implementation issues and future work.

2. Stream Layer

The base layer of the model is the stream layer. The abstraction of a stream represents the data associated with a particular medium. Examples of media include audio, video, images, graphics, text, and animation, as well as input from other devices such as mouse/keyboard, pen, and musical instruments. These streams may originate from a file, a device, or a connection to a remote site, or from the higher layers. Streams representing these media can be classified into the following categories:

1. *Digitally sampled continuous media streams:* The medium stream represents a set of samples with a constant sampling rate and pattern.

2. *Synthesized continuous media streams:* These streams are not originally generated by sampling a device. Rather, the output samples are synthesized from a data model to form a continuous stream.

3. *Event-driven streams:* These streams are interrupt- or event-driven and therefore have a nondeterministic sample rate. Although the streams are not continuous, the stream data is time stamped at the time of each event. These streams often correlate to human interaction, such as mouse movement or keyboard input.

Streams within each of these categories can be further classified as real time (i.e., generated at the time of execution) or playback (i.e., prestored and played back from a storage device).

2.1. Stream Services

The services provided by the stream layer include the following:

- Accessing multimedia data from a file, a device, or a connection, or from the higher layers.
- Delivering multimedia data to a file, a device, or a connection to a remote site, or to the higher layers, in a timely manner
- Processing of an individual stream (e.g., compression)
- Selecting an input from one or more streams and distributing it to one or more destinations
- Time-stamping or marking [Shephard 1990] stream data for downstream synchronization.

2.2. Stream Layer Abstractions and Mechanisms

To provide these services, several abstractions and mechanisms have been defined, including a stream, source, destination, filter and filter pipe, and switch. These mechanisms are illustrated in Figure 1 and are described in more detail below.

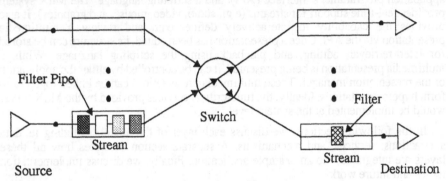

Figure 1. Stream Layer Mechanisms

2.2.1. Stream

A stream is a flow of data through a conduit [Northcutt 1991] that reads data from a source, perform data type conversion, and deliver data to a destination. Source and destination mechanisms provide access to multimedia data in a file, device connection to a remote site or another component in the system. Sources and destinations are similar to transducers, described by Northcutt and Kuerman [1991]. Data from a source can be digitally sampled, synthesized, or event driven, as noted above. For synchronization purposes, the source is responsible for marking data or time stamping data with a system clock time value. For streams that originate from a remote site, the time stamp is corrected, within a margin of error, for differences between the remote site and the local site. A level of performance and quality of service between the destination and the sources may be specified for each stream.

2.2.2. Filters*

Before a stream delivers data to a destination, a filter can perform one of several typen of processing operations on it, including format conversion (e.g., *RGB* images to *YUV* images); data compression and decompression; and data type conversion (e.g.,speech to text). Varying degrees of quality of service and performance can be achieved by having alternate filters for these operations.

---

*Although stream mechanisms are generally media independent, filter mechanisms are, by nature, media dependent.

The basic elements of a filter include an input, an output, control parameters, and a processing program. Filters can be combined to form filter pipes, or collections of filters. If a filter does not have any control parameters, or if the control parameters are provided at the time of processing (such as the quantization table for JPEG* compression [Wallace 1991]), then it is said to be *context free*. A *context-dependent* filter operates within a context that can be specified and controlled independently of the data stream.

2.2.3. Switch

A stream switch provides two main functions: selecting an output stream from multiple input streams, and directing the selected media stream to multiple destinations. Selecting from multiple input streams can be used to support chalk-passing protocols [Crowley et al. 1990]. Directing to multiple destinations allows the input stream to be tapped and tailored to form separate streams. For example, a *YUV* input stream originating from a video device can be output to a stream that converts to an *RGB* video to be displayed locally and to an MPEG† [LeGall 1991] compressed video stream to be sent to a remote location.

3. Multimedia Presentation Layer

The multimedia presentation layer builds on the stream layer. A *multimedia presentation* is a collection of streams that are coordinated with respect to time and space. Streams within a presentation are synchronized and have shared presentation control. There are logical groupings of media streams for integration and media-specific presentation control. The streams of dissimilar media (e.g., aural and visual) are synchronized and presented in parallel. Streams of similar media can be cut, reordered, processed, and mixed to form a new stream. These streams are grouped together as a channel for presentation or further processing. Example presentations are movies, videoconferencing, and collaborative work spaces.

The model's basic elements are derived from the video production model, but almost any media could be accommodated. Concepts from the video production model can be represented as follows: "tracks" are time-ordered streams; "channels" are a group of tracks associated with a mixer; and a "presentation" is the complete set of synchronized channels. The conceptual model for the multimedia presentation layer is illustrated in Figure 2.

3.1. Multimedia Presentation Layer Services

The services provided by the multimedia presentation layer include the following:

- Interstream (i.e., parallel streams) and intrastream (including integrated streams and single-stream order) synchronization [Yavatkar 1992]. Whereby the synchronization timing can be specified using hierarchical relationships, a time line, and reference points [Blakowski, Hubel, and Langrehr 1991].
- Integration of synchronized multimedia data, such as blending two video signals, mapping a video signal onto a graphics surface, and mixing multiple audio streams
- Presentation-specific processing of a stream, such as chroma keying or warping a video signal.

3.2. Media Presentation Layer Abstractions and Mechanisms

To provide the media integration and synchronization services outlined above, we have defined the following mechanisms: logical time system, cue, presentation context, mixer, track, channel, and presentation. These mechanisms are described below.

---

*JPEG: Joint Photographic Experts Group of ISO/CCITT.
†MPEG: Motion Picture Experts Group of ISO/CCITT.

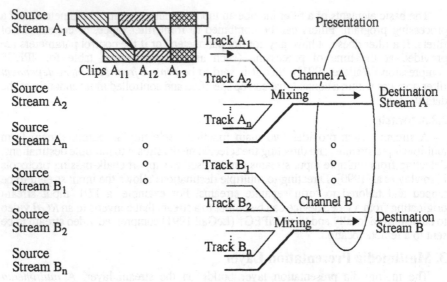

Figure 2. Media Integration Concept

3.2.1. Logical Time System

The logical time system (LTS) is the basic timing specification mechanism for synchronizing streams. Streams are synchronized by specifying their timing relative to the LTS. For example, channel times can be related to presentation time, and track times can be related to the channel times. Through the LTS, presentation of tracks, presentations, and channels can be scaled in a relative manner. An LTS clock consists of a logical start time and tick interval (specified in a real-time measure such as milliseconds). Given these two quantities, it is possible to transform between logical time and real time. Intrastream time can be scaled simply by scaling the intertick time [Anderson and Homsy 1991; Rennison et al. 1992].

3.2.2. Cue

A cue is a synchronization mechanism that allows the specification of timing relationships between streams. It binds one timing event to another, e.g., the start of one stream is bound to the end of another. A timing event can be offset by a delta or a delay value. Given that the duration of the two related streams is known, complex temporal relationships [Steinmetz 1990] can be specified using a cue; including *before*, *after*, *meets*, *overlaps*, *during*, *starts*, *finishes*, and *equals* [Little and Ghafoor 1990].

3.2.3. Presentation Context

The presentation context (PC) provides the mechanisms to specify parameters that define how media streams are mixed, integrated, and presented to the end user. PCs are specific to the medium associated with a stream. For example, a PC for an audio stream would specify the volume at which that stream is presented. The presentation context value associated with a PC varies over time, offering considerable flexibility in defining how media streams are integrated and mixed.

3.2.4. Mixer

The primary element for stream integration and intrastream synchronization is the mixer. Multiple integrated streams are synchronized at the mixer input. The mixer's main function is to take frames of data from the input streams and compose them into a single output stream. Internal to each mixer, streams can be hierarchically integrated [Rennison et al. 1992; Little and Ghafoor 1991] where a group of streams is mixed with other streams, and so on.

3.2.5. Track

A track mechanism facilitates the synchronization, reserialization, and media processing specification of a single-medium stream. Synchronization is achieved by specifying a track's start time relative to the logical time system of a channel. There are three timing-specification methods: (1) a priori specification of the track start time relative to the channel's time domain, (2) specification of track start time with a cue, and (3) interactive direction from the application or user via a *start* command. A track also reserializes streams through the use of *clips*. Clips define cut-in and cut-out (start and stop) times.

Each track has a presentation context associated with it. The presentation context is time based, relative to the track's LTS. When a track receives data from a stream, it converts the data's time stamp (based on real-world time) to logical time, and derives the presentation context values based on the logical time, to be used by the mixer for media integration.

3.2.6. Channel

A channel groups a set of tracks that are integrated by a mixer. It defines a common logical time domain for specifying the timing relationships of the set of streams being integrated. A channel also defines the interrelationship between the streams, including stream ordering and processing within a mixer. A channel has a time-based PC associated with it, which is applied in a hierarchical fashion to each of the tracks associated with the channel.

3.2.7. Presentation

A presentation groups together a set of channels that are synchronized in parallel. It provides a master time domain for specifying the time relationships of the presentation. As with the other mechanisms, it has a PC, which applies to all channels comprised in the presentation.

4. HyperPresentation Layer

The highest level of the three-layer model is the hyperpresentation layer. The abstraction of the hyperpresentation layer is that of a dynamic network of multimedia presentations connected by links, forming multidimensional (time and space) multimedia presentations. Links that connect multimedia presentations together are dynamic in that they change over time and have fixed lifetimes during which they can be activated [Brondmo et al. 1990; Halasz 1988; Michon 1992; Ogawa et al. 1990; Zellwech 1992]. Users can interactively access and manage information contained within the hyperpresentation network by traversing links. In addition, related information can automatically be triggered for presentation when time-sensitive preconditions are met.

The information content of a presentation changes over time. One portion, or *segment*, of a presentation may address one topic, while another segment may cover another topic. These segments may overlap. A specific point in time can be denoted by a *milestone*. Segments and milestones correspond to the visual or aural content (media specific), or correspond to all media that are being presented. Links between one presentation and other presentations come and go with their associated segments and milestones. For example, a segment of a presentation might generally cover "topic X"; and during the presentation of that segment a user might hit a "tell me more" button to see a related presentation on the same topic X.

4.1. HyperPresentation Layer Services

The services provided by the hyperpresentation layer include the following:

- Time-based linking of multimedia presentations, based on the content of the presentation defined by milestones and segments that may overlap
- Triggering of parallel presentations when time-sensitive preconditions are met.

4.2. HyperPresentation Layer Abstractions and Mechanisms

To provide theses services, the hyperpresentation layer supports the following mechanisms: *conditions, control streams, actions,* and *comparators*. Using these fundamental components, flexible, user-definable links and triggers can be constructed. Each of these mechanisms is described below.

4.2.1. Condition

A condition describes an event upon which one presentation is associated or linked with another presentation. This condition is directly related to the information content of a segment or milestone of the originating presentation. The condition is only valid for the duration of the segment or instant of the milestone, which is defined relative to the logical time system of the originating presentation. A condition can be defined spatially with regions or hotspots [Michon, 1992] that are compared to mouse input, aurally with tones or phrases that are compared to audio input, and gesturally with gestures compared to pen input, data glove input, or video input. The condition can also change over time. For example, a hot spot can move and change size along with a graphic object that is moving in a presentation. A presentation context associated with the condition defines how the condition is represented (e.g., visible or invisible button, <x,y> coordinate) to the user.

4.2.2. Control Stream

A control stream is the mechanism that can activate a link. A link is activated when a control stream meets a condition (defined above). A control stream can consist of, but is not limited to, input from the following: mouse, keyboard, data glove, video camera, or pen.

4.2.3. Comparator

A comparator is the primary mechanism for detecting a link or trigger condition and executing the actions necessary to link the presentations together. To perform this operation, a comparator has a control stream and a condition associated with it. During operation (between the start time and the end time), a comparator continuously compares the input from the control stream to the associated condition that activates a link or trigger. This implies that the clock or time system of the comparator must be synchronized with the clock or time system of the multimedia presentation. When a condition is met by the control stream, the comparator executes an action to link the presentations.

4.2.4. Action

When a comparator determines that a condition is met (e.g., a button was mouse selected at the location of a link), the comparator executes a set of programmatically defined actions that perform a link traversal or trigger. For example, if the semantic is a *jump* (i.e., a sequential link), the action pauses the current presentation, pushes that status of that presentation onto a stack, and plays a specified segment of the destination presentation, and upon returning from the destination presentation, resumes the original presentation, using the status from the stack. Since the actions are defined programmatically, an almost limitless variety of links and triggers can be defined.

5. Sample Scenario

A sample scenario, illustrated in Figure 3, shows how streams, switches, tracks, channels, presentations, comparators, and compositors can be interconnected. This example demonstrates the manner in which tracks are aggregated into channels and channels are grouped into presentations, and how multiple presentations can interact through hyperlinks.

This scenario shows a collaborative hyperpresentation application in which two users manipulate independent mouse input devices. Their devices alternate in controlling the cursor through a user-defined chalk-passing protocol that controls the input selection of the switch. Two video sources are played in the first presentation. One video source is read from a file and the second comes from a network connection to a remote device. These streams request data from their input sources, process the data, and deliver to their respective tracks before their assigned deadline. The track converts the data's real-time time stamp into logical time and queues the data in a time-based queue until it is accessed by the mixer. At the appropriate time, the mixer accesses the data, using a synchronized data-extraction technique for both the video tracks and the mouse track (whose cursor icon is defined in the mouse track's presentation context). The mixer then integrates the data and writes the output to a stream that delivers the data to the output display device.

When the user with the cursor control (as determined by the switch) selects a hot spot, the comparator triggers a second presentation to begin presentation of an audio stream, which is played back from a sound file. To perform this operation, the comparator continuously checks conditions against the input control stream (in this case <x,y> coordinates of mouse key presses) against the location of the hot spot. When a condition stream matches user input, a trigger message is sent to the second presentation, notifying it to start playing at an indexed time into the presentation. This message propagates down to the channel, track and stream.

Filtering functions are illustrated in several places in the first presentation. Filter A for the video file input stream performs scaling of data, reducing the frame rate from 30 to 15 fps. The second video source, from the network connection, has filter B attached to its stream. Filter B decompresses data from MPEG encoding. Filtering is also performed on the input mouse streams. In filter C, the input events are transformed from global <x,y> coordinates into window coordinates. In filter D, events are propagated forward, based on whether or not they match the window id of the current focus.

Timely delivery of data to the compositor, and from the compositor to the output destination, uses time constraints specified for the stream objects.

6. Implementation Issues

The abstractions of this model hide many difficult implementation issues from the application developer and end users. For example, real-time data access and flow control issues are abstracted by the stream layer, synchronization issues are abstracted by the multimedia presentation layer, and presentation management issues are abstracted by the hyperpresentation layer. Some of the more important issues in providing an efficient implementation of this model are discussed below.

6.1. Synchronization

Synchronization is a fundamental issue with this model. There are two main factors concerning synchronization: specification of synchronization or timing relationships between media streams, and execution of those synchronization specifications. With regards to specification, some approaches only define methods for specifying

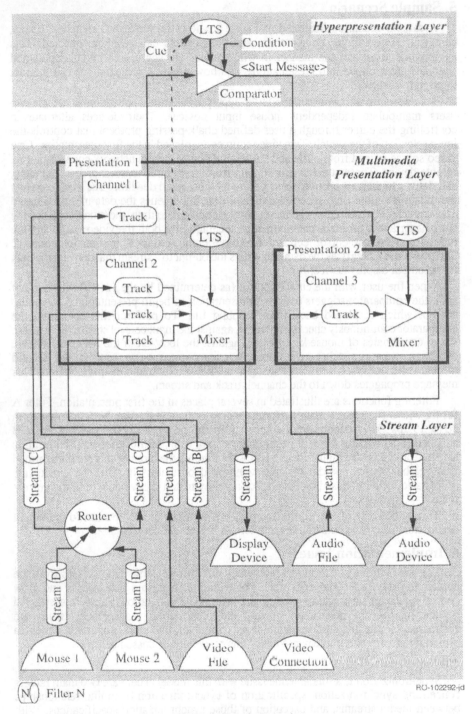

Figure 3. Sample Scenario

synchronization with respect to a time-line [Anderson 1991], but these approaches do not support synchronization under conditions of interaction that are nondeterministic and do not support specification of complex relationships between media. To address this issue, our model provides four methods for specifying synchronization timing, including hierarchical structures consisting of presentations, channels, and tracks; logical time systems for specifying timing relative to a time line; reference points using a cue mechanism to define direct timing relationships between streams; and direct user interaction. These timing mechanisms are used to determine stream integration timing and playout either a priori or dynamically.

The model defines two points of synchronization: input of the mixer for integrated stream synchronization, and the playout point of parallel streams of a presentation. Considering the three methods for synchronizing multimedia data streams with a network (at the source, at the network, and at the receiver [Leydekkers 1991]) it is clear that the model requires synchronization at the receiver. However, this method alone is not sufficient. The delay constraints for each stream in the synchronized presentation must be considered to minimize buffering requirements. Communication between the receiver and the source is necessary during session establishment, to ensure timely and reasonably predictable delivery of data with minimum buffering requirements for interstream synchronization and jitter control [Little and Ghafoor 1991]. This operation must be controlled by the *presentation* mechanism because it has knowledge of all the streams associated with a multimedia presentation.

During operation, several mechanisms play key roles in executing synchronization. In our implementation, synchronization begins at the source where data is marked [Salmony and Shepherd 90] or time stamped which is specified in real time or relative time using SMPTE* time code. The time stamp is carried with the data to the point of synchronization. When data is exchanged between sites, the time stamp is corrected, within a margin of error, relative to the site where the data is being synchronized. After a stream delivers the data to a track, the track mechanism converts the time stamp into logical time relative to its channel's logical time system and places the data in a time-based queue where buffering occurs. At this point, the mixer synchronously extracts the data from the time-based queues and mixes the data to form a new integrated stream, thus performing intrastream synchronization. Interstream synchronization is achieved through flow control using clocking parameters specified by the channel timing relative to the presentation's logical time domain.

6.2. Media Integration

Media integration is performed by the mixer, which defines the domain in which the media is composed. This domain provides a relative basis upon which the individual streams can specify how they are integrated within that domain. In a visual domain, still images, video images, and 2- and 3-D graphic objects are rendered to form a final *rgbαZ* image [Duff 1985] using such techniques as warping, chromakeying, morphing, alpha blending, compositing, and texture mapping. Alternatively, a *voxel* space can be defined to facilitate composition and visualization of volumetric data [Drebin, Carpenter, and Hanrahan 1988].

Since the model's data transport mechanisms, such as streams and tracks, are generic and do not directly address media-specific presentation aspects, a presentation context is associated with each track, channel, and presentation. A presentation context defines parameters for composing a media stream within the composition domain described abov and presenting it to the output destination.

---

*Society of Motion Picture and Television Engineers.

6.3. Data Access in HyperPresentations

The hyperpresentation layer provides the capability to traverse links in response to user actions; therefore, it must provide interactive control over presentation initialization and must minimize latency in accessing new tracks of data. Prefetching presentations for active links can reduce latency, but also incur overhead. Since the end user determines the order of link traversal, access patterns for data can be difficult to predetermine. Furthermore, the central concept of hyperpresentations is nonsequential viewing; and the location of the data for one presentation may not be colocated with data for a linked presentation and may have dramatically different access characteristics. In order to provide more efficient retrieval of this data, services are needed for intelligent memory-fetching operations and network retrieval. A facility such as madvise or a network version of madvise for distributed data sources is appropriate [Loeb 1992].

6.4. Other General Issues

In general, we believe the services provided by the model should be implemented at the system level on an object-oriented microkernel, with high-speed messaging to support the interactive real-time demands of time-critical media. Barring this ability, these services should be implemented via a single server located on each machine to provide integrated and central control of multimedia devices within a site, and efficient handling of multimedia data [Rennison et al. 1992]. Efficiency considerations include support for copy-on-write and optimizations for efficient data management, whereby the system provides real-time data flow control between buffers located on devices and in system memory. The application programmers' interface to the model services should support distributed object messaging, to allow object messages to transparently cross the boundaries of threads, processes, and the network. The multimedia I/O server lends itself to a multithreaded implementation that utilizes a best-effort real-time scheduler [Northcutt 1991; Downing 1992] to provide timely access, processing, and delivery of data to the compositors and devices.

7. Conclusion

We have defined a model that is based on the data abstractions of medium streams, multimedia presentations and hyperpresentations. These data abstractions provide a framework in which to define multimedia data processing and synchronization services in a fashion that is intuitive to application developers and end-users alike.

In the future, we plan to continue investigation into real-time communication support within the model, including multicasting in heterogeneous environments [Shacham 1992], and adaptive file access. We are exploring the use of hierarchical encoding techniques to meet this goal. In addition, we are exploring adaptive resource management policies that are tailored to users' preferences [Downing and Davis 1992] in order to provide a personalizable multimedia environment.

8. Acknowledgments

This work was conducted with the cooperation and support of the Electronics and Telecommunications Research Institute, Korea, which is sponsoring the "Integrated Multimedia User Interface for Intelligent Computing" project. The authors would like to acknowledge the following people who contributed to preparing this paper: Mike Davis, Jose Garcia-Luna, Joshua Nicholas, Nachum Shacham, and Lou Schreier.

References

Anderson, D.P., and G. Homsy. 1991. "A Continuous Media I/O Server and Its Synchronization Mechanism," *Computer*, Vol. 24, No. 10 (October).

Blakowski, G., J. Hubel, and U. Langrehr. 1991. "Tools for Specifying and Executing Synchronized Multimedia Presentations," Second International Workshop on Networking and Operating System Support for Digital Audio and Video, Heidelberg, Germany (November).

Brondmo, H.P., and G. Davenport. 1990. "Creating and viewing the Elastic Charles - a hypermedia journal," in *Hypertext: State of the Art*, R. McAleese and C. Green, eds., Intellect, Ltd., Great Britain.

Crowley, T., P. Milazzo, E. Baker, H. Forsdick, and R. Tomlinson. 1990. "MM Conf: An Infrastructure for Building Shared Multimedia Applications," *Proceedings of the Conference on Computer-Supported Interactive Work, October 7-10 1990*, Los Angeles, CA, pp. 329 ff.

Downing, A., and M. Davis. 1992. "System Resource Management for Distributed Real-Time Systems: Sample Scenario," Technical Report ITAD-2655-TR-92-123, SRI International, Menlo Park, CA (August).

Drebin, R.A., L. Carpenter, and P. Hanrahan. 1988, "Volume Rendering," *Computer Graphics*, Vol. 22, No. 4, (August).

Duff, T. 1985. "Compositing 3-D Rendered Images," *Computer Graphics*, Vol. 10. No. 3 (July).

Halasz, F. 1988. "Reflections on NoteCards: Seven Issues for the Next Hypermedia Systems," *Communications of the ACM*, Vol. 31, No. 7, pp. 836-852.

Ishii, H., and N. Mikaye. 1991. "Toward an Open Shared Workspace: Computer and Video Fusion Approach of Team WorkStation," *Communications of the ACM*, Vol. 34, No. 12, pp. 36-50 (December).

LeGall, D. 1991. "MPEG: A Video Compression Standard for Multimedia Applications," *Communications of the ACM*, Vol. 34, No. 4 (April).

Leydekkers, P. 1991. "Synchronization of Multimedia Data Streams in Open Distributed Environments," Second International Workshop on Networking and Operating System Support for Digital Audio and Video, Heidelberg, Germany (November).

Little, T.T.D., and A. Ghafoor. 1990. "Synchronization and Storage Models for Multimedia Objects," *IEEE Journal on Selected Areas of Communications*, Vol. 8, No. 3 (April).

Little, T.T.D., and A. Ghafoor. 1991. "Spatio-Temporal Composition of Distributed Multimedia Objects for Value-Added Networks," *Computer*, Vol. 24, No. 10 (October).

Loeb, S. 1992. "Delivering Interactive Multimedia Documents over Networks," *IEEE Communications Magazine*, Vol. 30, No. 5.

Michon, B. 1992. "Highly Iconic Interfaces," in *Multimedia Interface Design*, M. Blattner, M. Meera and R. B. Dannenberg, eds., ACM Press.

Nicolaou, C. 1990. "An Architecture for Real-Time Multimedia Communications Systems," *IEEE Journal on Selected Areas of Communications*, Vol. 8, No. 3 (April).

Northcut, J.D., and E.M. Kuerman. 1991. "System Support for Time-Critical Applications," Second International Workshop on Networking and Operating System Support for Digital Audio and Video, Heidelberg, Germany (November).

Ogawa, R., Harada, H., and A. Kameko. 1990, "Scenario-based hypermedia: A model and a system," in *Hypertext: Concepts, Systems and Applications*, A. Rizk, N. Streitz, and J. André, eds., Cambridge University Press, Great Britain.

Poltrock, S., and J. Gudin. 1992. "Computer Supported Cooperative Work and Groupware," Tutorial Notes, CHI92, Monterey, CA.

Rennison, E., R. Baker., D.H. Kim, and Y.H. Lim. 1992. "MuX: An X Co-Existent Time-Based Multimedia I/O Server," *The X Resource*, Issue 1, pp. 213-33 (Winter).

Shacham, N. 1992. "Multipoint Communication by Hierarchically Encoded Data," presented at INFOCOM92, Florence, Italy (May).

Shepard, P., and M. Salmony. 1990. "Extending OSI to Support Synchronization Required by Multimedia Applications," *Computer Communications*, Vol. 13, No. 7, pp. 399-406 (September).

Steinmetz, R. 1990. "Synchronization Properties in Multimedia Systems," *Journal on Selected Areas of Communications*, Vol. 8, No. 3 (April).

Wallace, G.K. 1991. "The JPEG Still Picture Compression Standard," *Communications of the ACM*, Vol. 34, No. 4 (April).

Watanabe, K., S. Sakata, K. Maeno, H. Fukuoka, and T. Ohmori. 1990. "Distributed Multiparty Desktop Conferencing System: MERMAID," *Proceedings CSCW '90 Conf. on Computer-Supported Cooperative Work*, Los Angeles, CA, pp. 27-38 (October).

Yavatkar, R. 1992. "Issues of Coordination and Temporal Synchronization in Multimedia Communication," *Multimedia '92*, Monterey, CA (April).

Zellweger, P. T. 1992. "Toward a Model for Active Multimedia Documents," in *Multimedia Interface Design*, M. Blattner, M. Meera. and R. Dannenberg, eds., ACM Press.

Enhancing the Touring Machine API to Support Integrated Digital Transport

Mauricio Arango, Michael Kramer, Steven L. Rohall,
Lillian Ruston, Abel Weinrib

Bellcore, 445 South Street, Morristown, NJ 07962-1910

Abstract. The current version of the Touring Machine™ software platform for multimedia communications uses analog transport of voice and video media streams. The next iterations of system design will, among other enhancements, support digital (packet-switched) transport of all media streams, placing new requirements on the system. This paper describes proposed changes to the Touring Machine Application Programming Interface in support of integrated digital transport. We identify new abstractions in two areas: network access control and specification of transport topology. These new abstractions will provide flexible control of multiple streams sharing the underlying integrated transport (for instance, separate packet-video streams each being displayed in separate windows on a multimedia workstation) and will support heterogeneous terminal equipment and networks by transparently converting media-stream formats within the network.

1 Introduction

The Touring Machine project at Bellcore is concerned with developing a research software platform to support multimedia communications applications, with the goal of learning about key technical questions important to realizing a public network infrastructure to support such applications. The Touring Machine software infrastructure provides to application programmers an Application Programming Interface (API) that facilitates the development of multimedia communications applications by supporting a rich set of abstract capabilities of the system. The project includes an experimental multimedia communications testbed composed of a network of desk-top video and audio devices and workstation-based shared workspaces controlled via users' workstations. The testbed provides the infrastructure for communication tools used daily by 150 users in two Bellcore locations 50 miles apart; these tools include multimedia, multipoint conferencing and information services as well as point-to-point communications. For a more complete description of the Touring Machine project and API, see, e.g., [1]

The current version of the Touring Machine software is the second iteration of system design. It uses analog transport to provide the voice and video media, controlling analog audio and video switches and other specialized hardware devices (such as bridges to create multi-party communications sessions) which

---

™ Touring Machine is a trademark of Bellcore.

connect monitors, cameras, microphones, and speakers on the users' desk tops. The signaling messages between applications and the Touring Machine infrastructure, and between different components of the Touring Machine software architecture, are carried "digitally" on our local internet, as is the "data" media type used, for example, to transport text or X Window System^{TM} protocol messages. We are presently in the design phase for a third version of Touring Machine, which will include many enhancements, among them support for integrated digital transport for the different media streams controlled by the system. The choice of off-the-shelf analog voice and video hardware for the current version has allowed us to support a significant user community, but in the next version we want to broaden the transport technologies we support.

Moving from analog to digital transport places a number of new requirements on the system and raises many interesting questions. This paper focuses on some of the changes to the Touring Machine API and its underlying abstractions required to support a hybrid analog and digital network infrastructure. In *digital* we include packet-switched transport that supports dynamic bandwidth allocation as well as circuit-switched transport such as ISDN; by *hybrid*, we mean that the system will support heterogeneous transport of media (analog as well as different formats of digital) providing an integrated, media-format-independent set of abstractions across the API. A hybrid system will transparently convert between different media formats when needed to facilitate communications between terminal equipment with different capabilities. Thus, a single *terminal* can comprise a mixture of both analog equipment and workstation-supported digital audio and video, allowing an application the flexibility to, for example, display a video stream wherever the user prefers. The goal is for an application programmer, using the API, to be able to write an application independently of the specific hardware that will be used, whether analog or digital.

2   API Changes

The current Touring Machine API separates the control of media streams into two components: network access control via mapping of logical endpoints to physical ports, and definition of a transport topology connecting the logical endpoints. This separation is viewed by programmers who have created applications to run on the Touring Machine system as one of the strengths of the API because, for example, it supports management of, and sharing of network access resources between, multiple concurrent multi-party sessions; we want to preserve this separation when moving to digital transport. We first discuss the new network access abstractions of the Touring Machine API that allow an application programmer to take advantage of the flexibility of integrated digital access to the Touring Machine controlled network. Then, we will turn to specifying the transport topology.

---

^{TM} X Window System is a trademark of MIT.

2.1 Network Access

In the current analog-only system, network access is described by ports, where each port provides access for a single stream of a given media type to the system. In our analog system, the port abstraction is appropriate: a port corresponds to a cable that connects desk-top equipment to the Touring Machine-controlled network, such as a coax cable used to carry a single stream of NTSC video that terminates on a video switch within the network. The ports are static, changing only when a new cable is installed. The current port abstraction could be applied to a digital network which emulates circuit-based network access using pre-established virtual circuits reserved to carry only a single media stream of a single type. However, this approach would not allow the application programmer to utilize the flexibility of integrated digital transport, where a more dynamic port abstraction is appropriate.

With digital transport, the same network access link (e.g., an Ethernet[TM] tap, or a fiber carrying ATM cells) can be used to carry a variety of different information streams. Thus, we extend the current network access abstractions to include two new types of objects in addition to ports: access links and access channels; see Figure 1.

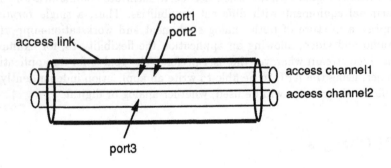

Fig. 1. Links, channels and ports.

An access link provides "raw" connection into the network, for instance a fiber carrying ATM cells, and is relatively static, changing only when a terminal's access into the system is modified. A link is described by the basic transport it provides (e.g., ATM cells or NTSC analog video) and the link resources (for instance, bandwidth) it contains. While an access link may correspond to a single physical connection into the network, the link abstraction is more general: a link may also correspond to a portion of a fiber's bandwidth set aside to carry ATM cells, or to a set of coax cables that all carry analog video.

[TM] Ethernet is a trademark of Xerox.

An access channel is created on a particular access link; its type is the same as that of the link. Creation of a channel reserves a subset of the link resources (bandwidth, etc.), and fails if there are not enough resources available. The channel abstraction provides a mechanism to the application programmer for managing separate portions of the bandwidth on a link, thus supporting quality of service guarantees for multiple applications (perhaps serving different users) on a single desk-top terminal. This abstraction also gives the application programmer explicit mechanisms for multiplexing different streams into limited portions of the total link bandwidth. (Few available packet-network technologies support the bandwidth reservation and "fire walls" required to fully implement channels at this time, but work in this area is active. Furthermore, the channel abstraction can be useful even if only to control link usage in software.) After creation, the channel can be deleted or modified to change the link resources allocated to the channel.

Once channels have been created, an application can create a typed port on a particular channel. The type of a port fully specifies the media stream that is going to be transported over the channel. The port type must be consistent with the type of the channel, and serves as a complete description of everything that the system needs to know to transport the stream. Thus, for example, attempting to create an analog video port on an IP packet network will fail, while a fully specified video port that could be successfully created might be "JPEG-encoded video with frame rate f and quality parameter q in IP packets, UDP transport protocol." The specification of the port is used for two purposes: it describes to the network how to process the media stream (including quality of service requirements), and it can be used by another end-terminal to understand the type of the media stream.

A port type must minimally describe the bandwidth and other quality of service requirement needed to provide transport. However, specifying other parameters (such as media type) is necessary for the use of other infrastructure services beyond simple transport. For instance, the port must specify the media type and encoding information if the system is to support communication between incompatible terminals by converting between different media encodings. This flexibility in the amount of information that is described by a port type allows the system to support basic transport of media streams of any type, as well as more complex processing of streams of specific supported types.

While a logical association is created when a port is created on a channel, none of the channel resources are dedicated to the port at that time. A port simply specifies typing information and provides a "handle" for the media stream that will eventually be carried; thus, a port is in some ways analogous to a UNIX^{TM} socket. The channel resources required to transport the stream are only allocated to the port later in the process when a logical endpoint is mapped to the port, making it *active*. (An active port is actually transporting a media stream.) Many ports may be active on a channel, constrained only by the total channel resources available.

---

^{TM} UNIX is a registered trademark of Unix System Laboratory, Inc.

The separation of access link resource reservation (using channels) from specification of stream type (using ports) allows flexible use of the access link resources. For instance, a channel created with enough bandwidth to transport one high-quality HDTV video stream may support a variety of ports, one for a HDTV stream and others for streams that require less bandwidth. An application may activate an arbitrary set of these ports (by mapping endpoints to them) as long as there are sufficient resources available on the channel, thus sharing the access resources in real time without having to create or destroy any of the ports.

As outlined above, our goal is to support hybrid networks, in which transport is provided by heterogeneous technologies. In particular, the Touring Machine system must continue to support our current analog video and audio while expanding to support digital media transport. To this end, the link-channel-port abstractions can be used to describe analog network access: the circuit connecting desk-top equipment to the system corresponds to an access link of type "analog NTSC video;" this access link supports one channel with bandwidth equal to the total bandwidth of the channel; multiple ports of type "analog NTSC video" can be created on the channel, but only one can be active (have an endpoint mapped to it) at any given time. A useful default would be to have a system administrator create the channel and a single port of type "analog NTSC video" on it, allowing "naive" applications that do not understand links and channels to simply use the port.

2.2 Transport Topology

So far, we have discussed the new API network access abstractions for integrated digital access to the Touring Machine network. We now turn to defining the transport within the network. While how the API describes transport needs to be changed little for digital transport, supporting the new API will require substantial enhancements to the system internals for managing network bandwidth, meeting quality of service requirements, allocation of protocol converters, etc.

In the current system, *connectors* are used to define the transport topology for a session. A connector associates the *endpoints* of a session, where endpoints logically terminate media streams within the network and can be thought of as "logical ports." Connectors describe the presentation of the streams, and, in particular, define the bridging function when multiple source endpoints are included. In the next version of the system, connectors will be replaced by more general connection graphs, but the media streams emerging from a connection graph will still terminate on endpoints.

Currently, endpoints are typed by media and direction; this type must exactly match the type of the port to which the endpoint is to be assigned. (An endpoint is first *assigned* to a port, specifying to which port the endpoint will eventually be mapped, and then mapped, *activating* the port and causing the actual transport of the media stream.) In the new design, endpoints continue to be typed, but exact matching of types of endpoints and ports will not be required. Rather, endpoints and ports must be compatible, with the endpoint type allowed to be specified as a subset of the fully specified port type. Thus, for example, an endpoint of type "video" will map to any video port, whatever the detailed specification of the port.

This flexibility, transparently supported by the infrastructure, may be useful to the application programmer, who can specify transport in a uniform manner and treat all ports equivalently, even though the underlying technologies may vary widely. This flexibility allows, for instance, implicit allocation of media-format converters when the port types to which different endpoints in the same connector are mapped are compatible but not identical. In addition, by more completely specifying an endpoint type, the application programmer can restrict the ports to which the endpoint can be mapped, thus specifying the quality of service required for the media stream.

3   Related Work

Our work is related to various broadband signaling proposals which support multipoint, multimedia calls. The connector-endpoint-port abstractions of the current system for describing transport of media streams was heavily influenced by the EXPANSE signaling work [2] for broadband ISDN. Our generalizations of the abstractions to support digital transport and heterogeneous networks still has similarities with this work. However, there are significant differences, for instance in our use of endpoints and ports to type and control network access and in how the logical transport topology is specified.

A second broadband signaling proposal [3] describes a protocol that provides dynamic multipoint, multiconnection communication channels over a switched ATM network. They assume that each stream is available to all of the parties in a call, and allow users to specify which streams they receive (subject to permission policies). This proposal makes visible the underlying ATM transport infrastructure, and thus appears less applicable to heterogeneous networking.

Proposals for including non-operational interfaces into ANSA/ODP have introduced *connection networks* which register stream interfaces [4]. Transport of streams between interfaces is realized by connecting to an interface reference for a remote interface. Here, the typing of the streams is set by the connection network that supports the interface, with the connection networks organized hierarchically. Lower-level networks provide simple point-to-point transport, while higher-level networks might offer specific presentation services, such as video bridging.

Many other multimedia communications systems have used digital transport of media streams. However, this work has tended to focus on creating vertically integrated applications more than on developing a general set of abstractions useful for a broad range of multimedia communications applications and underlying network technologies.

4 Conclusion

In this paper we have concentrated on proposed changes to the Touring Machine API to support integrated digital transport of the media stream. As part of the next version of the Touring Machine system, efforts are currently underway to realize a new API that incorporates these changes, and to make its features available to the application programmer.

Other issues must be resolved before we will have a fully operational digital system. For instance, we are investigating software architectures for controlling the local terminal, including workstation-based video hardware. The proposed API allows flexible management of media streams: combined with software to control the workstation, the system will support multiple video streams from multiple communications sessions terminating on the workstation, each displayed in its own window, and each separately controllable; in addition, the port and endpoint typing mechanism offers a mechanism for automatic network-based conversion of media stream format.

We are currently working towards an initial workstation-based implementation using IP network technology. This introduces the special challenge of fitting IP-network end-to-end control of data streams into the Touring Machine control model, which up to now has tended to be more network-centric than end-to-end. More generally, we are unaware of any currently available networking technology that completely supports the abstractions we have proposed in this paper (future broadband ATM networks appear to hold the most promise). Supporting the abstractions we have proposed places requirements on network protocols; by demonstrating the usefulness to application programmers of these abstractions, we hope to influence new network protocols as they are developed.

References

1. Arango et al.: Touring Machine: a Software Platform for Distributed Multimedia Applications. Proceedings of 1993 IFIP Int'l. Conf. on Upper Layer Protocols, Architectures and Applications, Vancouver, CA, (1992); a revised version of this paper is scheduled to appear in CACM, January 1992
2. Minzer, S.: A Signaling Protocol for Complex Multimedia Services. IEEE J. Selected Areas in Comm. 9 (1991) 1383-1394
3. Gaddis, M., Bubenik, R., DeHart, J.: A Model for Multipoint Communications in Switched Networks. Proceedings of Supercomm-ICC (1992)
4. Corbett, R.: U S WEST Position Paper on Stream Interfaces. ANSA/ODP Workshop on Stream Interfaces, Boulder (Nov. 11-12, 1992)

Session 5: Network and Operating System Support for Multimedia

Chair: Ralf Guido Herrtwich, IBM European Networking Center

The second day of the workshop started with a session comprising three papers on communication support for multimedia. First, Hui Zhang presented experimental results from the real-time protocol suite they have implemented as part of the Berkeley Tenet project. Jo Pasquale then introduced his multimedia multicast channel abstraction. The session concluded with a presentation from Liming Wei on a resequencer model for multicast over ATM networks.

The first paper by Hui Zhang and Tom Fisher of the University of California at Berkeley and the International Computer Science Institute was entitled "Preliminary Measurement of the RMTP/RTIP". It dealt with performance studies applied to two protocol of the Tenet project protocol suite for multimedia data traffic. RMTP is the Real-Time Message Transport Protocol and RTIP stands for Real-Time Internetwork Protocol. Both protocols focus on guaranteed-performance communication by means of underlying resource reservation mechanisms.

The authors have implemented their protocols in Ultrix on DecStation 5000 workstations and in HP/UX on HP 9000/7000 workstations. Some of the results of the preliminary measurements of their prototype implementation are: The throughput obtained by using RMTP/RTIP is comparable to that obtained by using raw IP. The rate control mechanism in RMTP/RTIP effectively enforces the traffic specification of communication clients. The scheduling mechanism in RTIP protects the real-time channel so that the performance of the real-time channel is not affected by the presence of IP traffic or other real-time channels in the network.

The second paper co-authored by Joseph Pasquale, George Polyzos, Eric Anderson and Vachaspathi Kompella from the University of California at San Diego was entitled "The Multimedia Multicast Channel". It dealt with a dissemination-oriented communication abstraction providing a service analogous to that of a cable television broadcast channel. A source transmits multimedia information such as video and audio onto a channel. Any number of receivers can then tune in to the channel and receive a selected set of the channel streams.

A receiver may apply certain filters to the streams selected. A typical application of a filter is the hierarchical decoding of video messages. As filters may be propagated back into the network they may save some resource capacity, e.g., if a sink can only accept a low-quality stream, a filter propagated back to the source may already reduce the amount of data being sent.

The third paper was the result of a collaboration between the Computer Science Department at the University of Southern California and Sun Microsystems. Liming Wei, Fong-Ching Liaw, Deborah Estrin, Allyn Romanow, and Tom Lyon co-authored the paper on "Analysis of a Resequencer Model for Multicast over ATM Networks". They started of by stressing the point that multicast delivery saves bandwidth and offers logical addressing capabilities to applications. The receivers of a multicast group, however, usually need to differentiate cells sent by different sources. This demultiplexing requirement can be satisfied in an ATM environment by using multiple dedicated point-to-multipoint virtual channel connections (VCs), but only with certain shortcomings.

They introduce an alternative resequencing model to solve this problem. Cells from different sources are multiplexed onto a single VC. A designated source in the group is elected as a resequencer. All other sources send their multicast ATM cells to the designated resequencer which buffers incoming cells from each source before all cells of a PDU are received. The resequencer forwards all cells of a PDU on an outgoing VC, allowing no intermediate cell from other sources. To make this solution scale for larger networks, the authors develop a hierarchical model of resequencers, arriving at a group multicast tree model. Their strategy is useful for applications spanning large regions where it is desirable to mix streams of cells from different bursty sources onto the same virtual channel.

Preliminary Measurement of the RMTP/RTIP*

Hui Zhang and Tom Fisher
hzhang, fisher@tenet.Berkeley.EDU

The Tenet Group
Computer Science Division
Department of Electrical Engineering and Computer Sciences
University of California
Berkeley, California 94720
and
International Computer Science Institute
1947 Center Street
Berkeley, California 94704

Abstract. The Real-time Message Transport Protocol (RMTP) and the Real-Time Internetwork Protocol (RTIP) are the transport and network layer data delivery protocols in the Tenet Protocol Suite. We implemented the protocols in Ultrix on the DECstation 5000 workstations and in HP/UX on the HP 9000/7000 workstations. A preliminary measurement study has been conducted to evaluate the performance of the prototype implementation. Some of the results are: the throughput obtained by using RMTP/RTIP is comparable to that obtained by using raw IP; the rate control mechanism in RMTP/RTIP effectively enforces the traffic specification of communication clients; the scheduling mechanism in RTIP protects the real-time channel so that the performance of a real-time channel is not affected by the presence of IP traffic or other real-time channels in the network.

1 Introduction

There is an increasing demand to support real-time applications such as video conferencing, scientific visualization and medical imaging in an internetworking environment. These applications have stringent performance requirements in terms of delay, delay jitter, bandwidth and loss rate [Fer90]. The best-effort

* This research was supported by the National Science Foundation and the Defense Advanced Research Projects Agency (DARPA) under Cooperative Agreement NCR-8919038 with the Corporation for National Research Initiatives, by AT&T Bell Laboratories, Digital Equipment Corporation, Hitachi, Ltd., Hitachi America, Ltd., Pacific Bell, the University of California under a MICRO grant, and the International Computer Science Institute. The views and conclusions contained in this document are those of the authors, and should not be interpreted as representing official policies, either expressed or implied, of the U.S. Government or any of the sponsoring organizations.

service provided by the Internet Protocol [Pos81] is not adequate to support these applications.

The Tenet Group of University of California at Berkeley and International Computer Science Institute has proposed a new type of service called *guaranteed performance service*. By a guaranteed performance service, we mean that, before the communication starts, the client specifies its traffic characteristics and desired performance requirements; when the network accepts the client's request, the network guarantees that the specified performance requirements will be met provided that the client obeys the restrictions implied in its traffic description.

We believe that a reservation-oriented network architecture is needed to provide such a guaranteed performance service. The current Internet Protocol, which is connectionless and reservationless, cannot be used to support the service.

The Tenet Group has also designed and implemented a new protocol suite that provides guaranteed performance service [Fer92, FBZ92]. The protocol suite consists of five protocols: three data delivery protocols and two control protocols. The three data delivery protocols are: the Real-Time Internet Protocol (RTIP), the Real-time Message Transport Protocol (RMTP) [VZ91, ZVF92] and the Continuous Media Transport Protocol (CMTP) [WM91, MGWF92]. RTIP is the network layer protocol, while RMTP and CMTP are two transport layer protocols that provide message-oriented and stream-oriented transport services, respectively, on top of RTIP. The two control protocols are: the Real-Time Channel Administration Protocol (RCAP) [BM91], and the Real-Time Control Message Protocol (RTCMP). RCAP is responsible for establishment, tear-down and modification of the real-time channels, while RTCMP is responsible for control and management during data transfers.

In this paper, we will focus on the two data delivery protocols, RMTP and RTIP. Both protocols have been implemented on DECstation 5000 and HP9000 workstations. A preliminary measurement study has been conducted to evaluate the performance of the prototype implementation. Some of the results are: the throughput obtained by using RMTP/RTIP is comparable to that obtained by using raw IP; the rate control mechanism in RMTP/RTIP effectively enforces the traffic specification of communication clients; the scheduling mechanism in RTIP protects real-time connections so that the performance of a real-time connection is not affected by the presence of IP traffic or other real-time connections in the network.

The paper is organized as follows: Section 2 brief reviews the real-time channel scheme, on which the Tenet Protocol Suite are based; Section 3 describes the services, functions, software structure and programming interface for RTIP and RMTP; Section 4, which is the core of this paper, presents the simulation results; Section 5 briefly describes related work; Section 6 gives the conclusion and provides future work.

2 Background

The Tenet protocol suite implements the real-time channel scheme, which is a new communication abstraction proposed to support guaranteed performance service in a general packet-switched internetwork environment [FV90].

A real-time channel is a simplex unicast end-to-end connection with performance guarantees and traffic restrictions. Once established, it guarantees that the performance bounds requested by the communication client are satisfied so long as the client does not violate the traffic restrictions. The performance parameters a client can request are:

- delay bound D
- delay violation probability bound Z
- buffer overflow probability bound W
- delay jitter bound J

The traffic parameters a client needs to specify are:

- minimum packet inter-arrival time x_min
- average packet inter-arrival time x_ave
- averaging interval I
- maximum packet size s_max

A channel with both Z and W being 1 is called a *deterministic* channel; a channel with either Z or W being less than 1 is called a *statistical* channel [Fer90].

The following paradigm is proposed in [FV90] to provide guaranteed services to clients in a packet-switching network: before communication starts, the client specifies its traffic characteristics and performance requirements to the network; the client's traffic and performance parameters are translated into local parameters, and a set of connection admission control conditions are tested at each switch or gateway; the new channel is accepted only if its admission would not cause the performance guarantees made to other channels to be violated; during data transfers, each switch or gateway will service packets from different channels according to a service discipline; by ensuring that the local performance requirements are met at each switch or gateway, the end-to-end performance requirements can be satisfied.

Notice that there are two levels of control in this paradigm: at the connection level, the admission control policy allocates and reserves resources for each channel; at the packet level, the service discipline allocates resources according to the reservations made. In the Tenet protocol suite, RCAP is responsible for the connection level control, and RTIP is responsible for the packet level control.

3 RMTP and RTIP

RMTP and RTIP are the data delivery protocols in the Tenet protocol suite. Together with the Real-Time Channel Administration Protocol, or RCAP [BM91], they provide guaranteed performance communication services in an internetworking environment.

3.1 RMTP Services and Functions

The Real-time Message Transport Protocol, or RMTP, is the transport layer message-oriented data transfer protocol in the Tenet protocol suite. It provides a simplex, end-to-end, unreliable, in-order, and guaranteed performance message service. RMTP is responsible for such end-to-end functions as message fragmentation and reassembly, regulation of traffic according to traffic specifications and optional checksumming.

3.2 RTIP Services and Functions

The Real-Time Internet Protocol, or RTIP, is the network layer data transfer protocol in the Tenet protocol suite. Operating at each host and gateway participating in the real-time communication, RTIP performs rate control, jitter control, packet scheduling, and data transfer functions. It provides a host-to-host simplex, sequenced, unreliable, and guaranteed-performance packet service. All the data are transferred in packets on a simplex channel from the sending client to the receiving client. A packet is not guaranteed to be delivered; it may be dropped for two reasons: the packet may be corrupted during transmission, or there might not be enough buffer space in the case of a statistical channel. Packets that are not dropped are delivered to the receiving client in the same order as they were sent by the sending client; the relative positions of the dropped packets are indicated. The client data are not checksummed; they may get corrupted due to transmission errors. If the sending client sends packets neither larger than the maximum packet size negotiated during channel establishment time, nor faster than the specified maximum rate, all the packets delivered are guaranteed to meet the delay and/or delay jitter requirements with certain probability (the probability is 1 for a deterministic channel).

3.3 Relationship Between RMTP/RTIP and RCAP

All real-time data is carried on real-time channels. Channel state information is kept in each node along the channel's path. The RMTP state information is maintained at the end-points of the channel, while the RTIP state information is maintained at every node along the channel's path (including the end-points). RCAP is responsible for initializing the state information of a newly created channel and for destroying the state information of a channel when it is torn down.

3.4 Software Structure

RMTP and RTIP have been implemented in Ultrix on DECstation 5000 and in HP/UX on HP9000/7000 workstations. Both Ultrix and HP/UX are Unix-like operating systems; the networking software of both was derived from BSD Unix [LMKQ89].

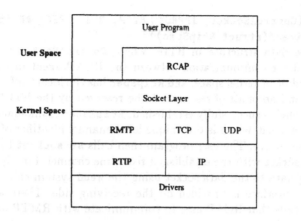

Fig.1. Software structure of RMTP/RTIP

The software structure of the RMTP and RTIP implementation is shown in figure 1. As can be seen in the figure, while RCAP is implemented in user space, RMTP/RTIP are implemented in the kernel, and co-exist there with TCP/IP. Through the socket layer software, RMTP and RTIP export interfaces to RCAP for control purposes and to user programs for data transfer purposes.

3.5 Programming Interface

RMTP and RTIP export programming interfaces to both user programs and RCAP agents. The following example illustrates how a user program sends data using the Tenet protocol suit.

```
DataSock = socket(AF_INET, SOCK_DGRAM, IPPROTO_RMTP)
RcapEstablishRequest(&ParametersBlock, &lcid, &Destination)
setsocketopt(DataSock, IPPROTO_RTIP, RTIP_ASSOC, &lcid, sizeof(u_short))
send(DataSock, msg, len, flags)
```

Fig.2. Programming interface for a user program

The user program first invokes the **socket** system call to open an RMTP socket; it then calls **RcapEstablishRequest** to establish a real-time channel, while **ParametersBlock** specifies the end-to-end traffic and performance parameters, **lcid** is used to hold a return value, and **Destination** specifies the destination IP address. **RcapEstablishRequest** is an RCAP library function; when invoked, it will contact RCAP agents at this node and other nodes on the path to reserve resources for the new request and establish a real-time channel. The agent on each node of the real-time channel, which resides in user space, will call

```
setsockopt(ControlSocket, IPPROTO_RTIP, RTIP_SPEC, &RtipSpec,
          sizeof(struct RtipSpec))
```
to set up the data structure in RTIP, where `ControlSocket` is a permanent socket opened for communication between the RCAP agent in user space and the RTIP module in kernel space, and `RtipSpec` has the local traffic, performance parameters, and amounts of resources to be reserved by the RTIP module.

If the channel is successfully established, `RcapEstablishRequest` will return a small integer `lcid`, which is the unique local channel identifier of the real-time channel at the source. The user program then calls `setsockopt` to associate the opened data socket with the established real-time channel. Finally, the user can start sending data to the data socket using the `send` system call.

A similar interface is provided on the receiving side. There are also other `setsockopt` calls that RCAP uses to communicate with RMTP and RTIP.

4   Measurement Experiments

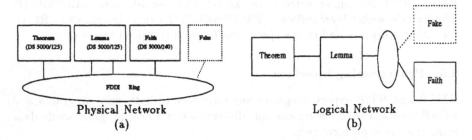

Fig.3. Testbed configuration

In order to evaluate the performance of the RMTP/RTIP implementation, we set up a local testbed and performed a set of experiments on the testbed.

The physical configuration of the testbed is shown in Figure 3 (a). Both *theorem* and *lemma* are DECstation 5000/125's, while *faith* is a DECstation 5000/240. *Fake* is a non-existent machine; it is used to be the receiver of network load. All three workstations are connected to one FDDI ring. While *theorem* and *faith* each have one FDDI interface attached, *lemma* has two FDDI interfaces. The routing tables on the three machines are set up to form a logical network as shown in Figure 3 (b).

4.1   Throughput and Processing Time

Our first experiment compares the throughput of RMTP/RTIP with those of UDP/IP and raw IP. The experiment was performed with one process on *theorem* sending 10,000 packets of same size in a tight loop to a receiving process on *faith*. No load was applied to either the gateway or the hosts.

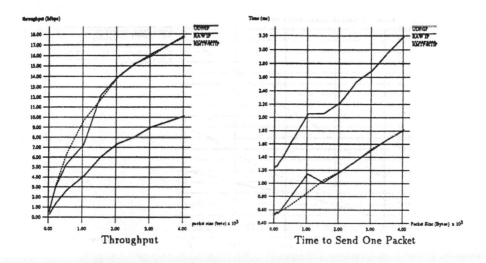

Throughput   Time to Send One Packet

Fig.4. Performance of UDP/IP, raw IP and RMTP/RTIP

Tests were performed with different packet sizes and underlying protocols. For RMTP/RTIP, the rate parameters (Xmin, Xave, I) were set so that the "reserved rate" was higher than the achievable rate. This was just for experimentation purpose, so that packets could be sent back to back without being held by the rate control mechanism. Figure 4 shows the results of the experiment. The diagram on the left shows the throughput, and the diagram on the right shows the amount of time to send a packet during a tight sending loop on a DECstation 5000/125. It can be seen from the figure that the throughput achieved using RMTP/RTIP is almost the same as that using raw IP. This result is not surprising. It has been observed in [CJRS89] that, for bulk data transfers, the cost of operations on data such as copying or checksumming dominates protocol processing. Since neither RMTP/RTIP nor raw IP do checksumming, the operations on data are the same for both of them, thus the time for sending one packet and the throughput are approximately the same in both cases. The UDP throughput was measured with UDP checksumming turned on, which explains the discrepancy between its curves and the curves of raw IP and RMTP/RTIP.

Figure 5 shows the breakdown of the processing time of a RMTP/RTIP packet on a DS 5000/125. As can be seen, the RMTP/RTIP protocol processing time is about 68 μs per packet, and the driver software processing time is about 114 μs per packet. Socket level processing includes copying the data from user space into kernel space; transmission includes copying the packet from the kernel memory into the interface adaptor and transmitting the packet onto the FDDI ring. Both types of processing involve data movement and account for a larger fraction of the total processing time.

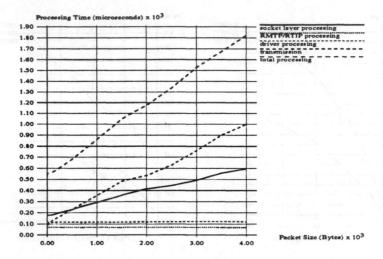

Fig.5. Breakdown of Processing Times In the Kernel

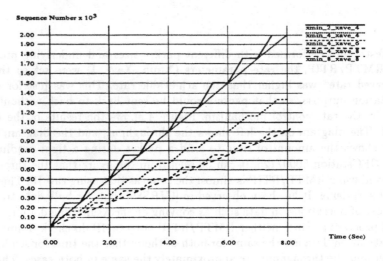

Fig.6. Effects of Rate-Control at the Source in RMTP/RTIP

4.2 Effects of Rate Control

One of the functions of RMTP is to perform rate control at the source so that the packets are injected into the network at a rate not higher than declared when the channel was established. This experiment was designed to check the effectiveness of rate control.

In the experiment, a process on *theorem* sent packets continuously to a process on *faith* through a real-time channel in a tight loop. The receiver program takes a timestamp upon receiving of each packet.

Figure 6 displays the relationship between packet sequences and timestamps. Each of the experiments whose results are reported in the figure was performed independently with different values of $xmin$ and $xave$. $Xmin$ and $xave$ are expressed in milliseconds. The same value of 1 second is used for I in all experiments.

Although the sender program sends packets in a tight loop, the rate control mechanism in the RMTP/RTIP protocol ensures that the source never sends packets faster than what is specified by traffic parameters. As can been seen in the figure, the $(xmin, xave, I)$ traffic restrictions have been indeed satisfied by the rate control mechanism.

4.3   Load on Gateway

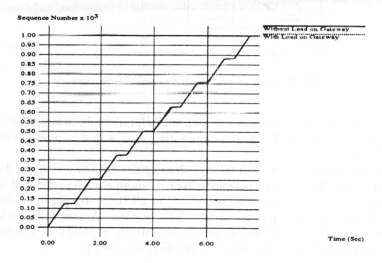

Fig.6. Effect of gateway load on real-time channel

This experiment was designed to examine the effect of gateway load on real-time channels. In the experiment, a process on *theorem* sent packets to a process on *faith* via a real-time channel. We timestamped each packet and recorded its sequence number at the receiving end. Two tests were performed: gateway unloaded and gateway loaded. To load the gateway, a couple of processes were created on the gateway machine *lemma*, and each of them sent raw IP packets in a tight loop to the non-existing machine *fake*. Figure 8 shows both the loaded and the unloaded case. We can see that the load on the gateway did not affect the performance of the real-time channel.

A similar experiment was performed for UDP. Under load, a significant fraction of UDP packets were dropped at the gateway.

4.4 Co-existence of Real-Time Channels

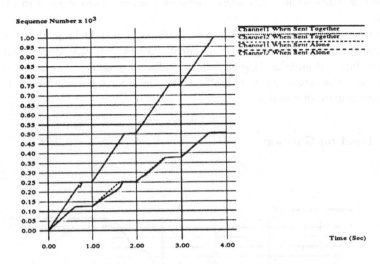

Fig.7. Interference between two real-time channels

The previous experiment shows that RMTP/RTIP traffic was not affected by the presence of IP traffic. A more interesting case is that of multiple real-time channels.

We did three tests: channel 1 active alone, channel 2 active alone, and channel 1 and channel 2 active simultaneously. We timestamped each packet and recorded its sequence number at the receiving end in all the tests. Figure 7 shows the sequence number vs. time graph for all three cases. As can been seen in the figure, for both channel 1 and channel 2, the patterns of packet arrival are almost the same regardless of whether the other channel is active.

5 Related Work

Several other algorithms or protocolshave been proposed to provide Quality of Service in an internetworking environment. Among them are: the Stream Protocol Version II (ST-II) [Top90], the Multipoint Congram-oriend High-performance Internet Protocol (MCHIP) [Par90], the Flow Protocol [Zha89], a proposal for predicted service [CSZ92], and a proposal for resource management in datagram networks [Flo92].

There have been two implementations of ST-II [Seo91, PP91], but, to the best of our knowledge, no performance data have been reported. Also, to the best of our knowledge, none of the other proposed algorithms or protocols has been fully implemented yet.

6 Conclusion and Future Work

We have conducted a preliminary measurement study to evaluate our prototype implementation of RMTP and RTIP. The results obtained are encouraging: the throughput obtained by using RMTP/RTIP is comparable to that obtained by using raw IP; the rate control mechanism in RMTP/RTIP effectively enforces the traffic specification of communication clients; the scheduling mechanism in RTIP protects the real-time channel so that the performance of a real-time channel is not affected by the presence of IP traffic or other real-time channels in the network.

There are certain limitations to this research, which we would like to address in future works:

- The current testbed is a local area network. We would like to perform similar measurements in an internetwork such as the Xunet-2 [FKK+92] network and the SequoiaNet [FPP].
- All the experiments were performed with synthetic loads. In the future, we would like to experiment with real applications such as video conferencing or scientific visualization.
- The testbed consists of only DECstations. A heterogenous environment would be more interesting. We are in the process of moving the implementation of RMTP/RTIP to the Sun 4/280 and SGI IRIS-4D workstations.

7 Acknowledgement

Domenico Ferrari guided and supervised the entire project. John Limb of Hewlett Packard Co. made it possible for Hui Zhang to implement the first prototype of RMTP/RTIP on a HP9000/700 workstation. Jim Hughes of Hughes Co. loaned us a 80 ns resolution timer board. Fred Templin shared with us his expertise in Ultrix. Bruce Mah helped us to set up the testbed and provided system administrative support.

References

[BM91]   Anindo Banerjea and Bruce Mah. The real-time channel administration protocol. In *Proceedings of the Second International Workshop on Network and Operating System Support for Digital Audio and Video*, pages 160–170, Heidelberg, Germany, November 1991. Springer-Verlag.

[CJRS89] David D. Clark, Van Jacobson, John Romkey, and Howard Salwen. An analysis of TCP processing overhead. *IEEE Communications Magazine*, June 1989.

[CSZ92]   David Clark, Scott Shenker, and Lixia Zhang. Supporting real-time applications in an integrated services packet network: Architecture and mechanism. In *Proceedings of SIGCOMM'92*, Baltimore, Maryland, August 1992.

[FBZ92]   Domenico Ferrari, Anindo Banerjea, and Hui Zhang. Network support for multimedia: a discussion of the Tenet approach. Technical Report TR-92-072, International Computer Science Institute, Berkeley, California, October 1992.

[Fer90]   Domenico Ferrari. Client requirements for real-time communication services. *IEEE Communications Magazine*, 28(11):65–72, November 1990.

[Fer92]   Domenico Ferrari. Real-time communication in an internetwork. Technical Report TR-91-001, International Computer Science Institute, Berkeley, California, January 1992.

[FKK+92]   Alexander G. Fraser, Chuck R. Kalmanek, A.E. Kaplan, William T. Marshall, and R.C. Restrick. Xunet2: A nationwide testbed in high-speed networking. In *Proceedings of INFOCOM*, Firense, Italy, May 1992.

[Flo92]   Sally Floyd. Issues in flexible resource management for datagram networks. In *Proceedings of the 3rd Workshop on Very High Speed Networks*, Maryland, March 1992.

[FPP]   Domenico Ferrari, Joe Pasquale, and Gorge Polyzos. Network issues for Sequoia 2000. In *Proceedings of COMPCOM 92*, pages 401–406.

[FV90]   Domenico Ferrari and Dinesh Verma. A scheme for real-time channel establishment in wide-area networks. *IEEE Journal on Selected Areas in Communications*, 8(3):368–379, April 1990.

[LMKQ89]   Samuel J. Leffler, Marshall Kirk Mckusick, Michael J. Karels, and John S. Quarterman. *The Design and Implementation of the 4.3 BSD UNIX Operating System*. Addison-Wesley Publishing Company, 1989.

[MGWF92]   Mark Moran, Amit Gupta, Bernd Wolfinger, and Domenico Ferrari. A continuous media communication service and its implementation, December 1992. to appear in GLOBECOM '92.

[Par90]   Guru Parulkar. The next generation of internetworking. *ACM SIGCOMM Computer Communication Review*, January 1990.

[Pos81]   Jon Postel. Internet protocol, September 1981. RFC 791.

[PP91]   Craig Partridge and Stephen Pink. An implementation of the revised internet stream protocol (ST-2). In *Proceedings of the Second International Workshop on Network and Operating System Support for Digital Audio and Video*, Heidelberg, Germany, November 1991. Springer-Verlag.

[Seo91]   Karen Seo. ST-II – new release, November 1991. Connection IP (cip) mailing list.

[Top90]   Claudio Topolcic. Experimental internet stream protocol, version 2 (ST-II), October 1990. RFC 1190.

[VZ91]   Dinesh Verma and Hui Zhang. Design documents for RMTP/RTIP, May 1991. unpublished internal technical report.

[WM91]   Bernd Wolfinger and Mark Moran. A continuous media data transport service and protocol for real-time communication in high speed networks. In *Proceedings of the Second International Workshop on Network and Operating System Support for Digital Audio and Video*, pages 171–182, Heidelberg, Germany, November 1991. Springer-Verlag.

[Zha89]   Lixia Zhang. *A New Architecture for Packet Switched Network Protocols*. PhD dissertation, Massachusetts Institute of Technology, July 1989.

[ZVF92]   Hui Zhang, Dinesh Verma, and Domenico Ferrari. Design and implementation of the real-time internet protocol. In *Proceedings of the IEEE Workshop on the Architecture and Implementation of High Performance Communication Subsystems*, Tucson, Arizona, February 1992.

The Multimedia Multicast Channel *

Joseph C. Pasquale
George C. Polyzos
Eric W. Anderson
Vachaspathi P. Kompella

Computer Systems Laboratory
Department of Computer Science and Engineering
University of California, San Diego
La Jolla, CA 92093-0114.

{pasquale, polyzos, ewa, kompella}@cs.ucsd.edu

Abstract. The Multimedia Multicast Channel is a dissemination-oriented communication abstraction providing a service analogous to that of
a cable television broadcast channel. A source transmits multimedia information such as video and audio streams onto a channel, and a varying number of receivers "tune in" to the channel to receive a selected
set of the streams. To support heterogeneity, each receiver may tailor
the selected streams to meet individual needs through the use of filters.
The design encourages a very loose coupling between the source and
the receivers, promoting open-loop control for the underlying network
protocols.

1 Introduction

The Multimedia Multicast Channel (MMC) is a programming abstraction which
supports dissemination-oriented communication [4]. The abstraction is analogous
to that of a cable television channel: a source transmits onto a specific channel,
and receivers which have subscribed to that channel receive media streams (e.g.
audio and video) without explicit interactions with the source.

The MMC communication paradigm is a major departure from more traditional ones, in that a source and a set of receivers are very loosely coupled in
their control and data exchange interactions. In general, the source's main concern is to push various media streams onto a channel, without emphasis on where
they end up (i.e. who the actual receivers are), and how they are used (i.e. what
specific receivers will extract from any or all of the streams). A receiver's main
concern is what to extract from a channel, which is viewed as offering multiple
media streams, some or all of which are of interest.

Perhaps the most unique feature of this communication paradigm is in how it
addresses heterogeneity, in particular that it is not unusual for the forms of the

* This research is supported in part by grants from DEC, IBM, NCR, NSF, TRW,
and UC MICRO. The views expressed are those of the authors, and not necessarily
those of the supporters.

media streams, generated by the source and required by the different receivers, to be quite different. For example, the source may generate HDTV-quality video and CD-quality audio, whereas some receivers can only use NTSC video and its associated audio, while other receivers can use only audio and no video. Indeed, every receiver may have very different and independent requirements. Therefore, it is expected that the receivers will individually tailor, to a high degree, what streams are actually received. These streams may be a subset of the source streams, or new ones computed from the source streams. The use of component coding at the source to facilitate tailoring is encouraged.

We believe this communication paradigm is highly appropriate for multi-media distribution services required by a large class of multimedia multipoint applications. A prime example is "video distribution" [20, 21, 14], as in cable television systems where a single source generates video (and associated audio) distributed to a large set of receivers who generally have little or no interaction with the source. Another application is video conferencing [3, 10, 24], where each member is both a source and receiver. This application would require a separate MMC per source to support base-level audio-video distribution. However, video conferencing also requires other control mechanisms which are outside the scope of what is provided by the MMC.

More generally, one can distinguish between higher-level *application-specific* control mechanisms such as floor-control and voting, and lower-level *media-oriented* control mechanisms such as modifying the resolution, granularity, or intensity of a media stream, and mixing or synchronizing multiple media streams. Media-oriented control does not imply (explicit) control between source and receivers, and is generally useful in most multimedia applications. Consequently, the MMC supports media-oriented control through its filter mechanism. All other required control, particularly application-specific control between source and receivers, is expected to be provided by the application itself (e.g. through the use of libraries/toolkits).

In this paper, we describe the architecture of the MMC. The paper is organized as follows. Section 2 contains a description of MMC concepts. Section 3 describes the programming interface, and Section 4 contains an example of its usage. Section 5 contains a discussion on underlying network support issues, such as how to take advantage of hierarchical coding, and why open-loop control is important for efficiency. In Section 6, we review related work. Finally, in Section 7 we present conclusions.

2 Concepts

Four goals motivate the design of the MMC:

1. *dissemination-oriented communication*: that which a source sends is transported to multiple receivers.
2. *loose-coupling between source and receivers*: a channel's source and receivers need not know each others' identities. The channel is the only common object

that both source and receivers are aware of, reducing the complexity of connection establishment.

3. *heterogeneity*: the receivers can have different needs, and satisfy them by individually tailoring the media streams extracted from a channel.

4. *efficiency*: the first three goals lend themselves to the following support mechanisms, respectively: (1) multicasting, which minimizes bandwidth usage; (2) open-loop control, which minimizes source-receiver interactions; (3) filtering for transforming a shared set of streams, and filter propagation, which further reduces bandwidth usage.

We now describe the basic concepts of the MMC. A *channel* is one-to-many communication abstraction which transports typed media streams between a source and multiple receivers. A channel is created with a specification of the types of media streams it will transport and an access control list identifying potential channel members.

A *port* is an access point to a channel. A potential member can dynamically "tune in" to the channel by opening a port, and specifying whether it will act as a source or as a receiver. There can be at most one active source per channel.

A port provides access to one or more *streams* generated by the channel source. Streams travel in one direction, from source to receiver. A stream is typed, with different types corresponding to different media or media components. For example, a source may generate the following streams:

audio left, audio right, video luminance, video chrominance

The first two streams are both of type "audio." The video luminance and video chrominance streams correspond to two different types of streams.

Each stream within a channel has a number of parameters describing priority, translation, and scaling. The priority describes the stream's importance/precedence relative to other streams on the same port. For example, video luminance and video chrominance are hierarchically related, with the former taking precedence over the latter. When provided to the network, this priority information may be used advantageously for congestion control and packet scheduling policies. The translation and scaling parameters are used to "align" multiple streams, described next.

A stream is actually a sequence of *segments*. A segment is the basic unit of information sent or received through the MMC programming interface, and is also the basic unit of information operated on by filters. Associated with each segment are its *stream position* and *stream length*, or simply its position and length. The position locates the segment relative to the beginning of the stream (which is always position 0). The length describes the distance covered by the segment in the same scale. A segment's position plus its length is equal to the position of the end of the segment in the stream. A useful (but not necessary) interpretation of the position and length is with respect to time: the position is the (logical) time of the segment since the beginning of the stream; the length is the duration in (logical) time of the segment within the stream.

One purpose for the segment's position (and length) is to logically identify and distinguish the segment from other segments in the same stream. Another purpose of the segment's position and length is to allow *alignment* of multiple streams (often referred to as "media synchronization" for a time-based interpretation), achieved by identifying segments with the same *channel positions*. A segment's channel position is calculated as follows: the segment's (stream) position number P is scaled and translated by the stream's scaling and translation parameters, S and T, resulting in a channel position $P' = T + S \times P$. A segment's channel length L' is calculated from the its (stream) length L as follows: $L' = S \times L$. Segments from different streams are said to be "co-located" if their channel position numbers are the same. Multiple-stream alignment makes sophisticated filters possible by allowing specification of computations on co-located segments.

A *filter* is an executable module which may be placed on a port, and implements a function which takes a specified set of streams associated with that port and produces a new stream (which then also belongs to that port). When a segment encounters a filter, it may be queued, passed through, or both. After collecting enough segments (governed by filter code) or after a programmable timeout interval, the filter operates on the collected segments to produce a new segment. The filter uses the channel position numbers and channel lengths for each segment to determine which are the corresponding segments on which to operate. Some standard filters include AVG (take the average of streams), and μ-LAW-ADD (mix μ-law formatted audio streams).

A network implementation supporting the MMC may allow *filter propagation*. After a filter is placed on a port, it may propagate into the network on connections which are part of the channel and execute as close to the source as possible if it is efficient to do so. For instance, if a receiver is only interested in the average of two streams, but not the two streams themselves, then it would be inefficient to transmit the two streams to the receiver only to discard them after they are averaged. Filters may also combine, such as when two receivers on the same channel install identical filters on their ports. If channels are built using source-rooted multicast tree connections [16], the filters can independently propagate upstream and encounter each other at some intermediate node. There they would be combined to form a single filter which may then continue propagation upstream.

While there are benefits to filter propagation, we note that there may be serious difficulties in its support, such as a switch's inability to execute filters, or the overhead in their execution on intermediate nodes and the corresponding adverse effect on the transport of other streams. We only mention filter propagation as a novelty with potential performance benefits which may outweigh the drawbacks for some networks. We expect to report on these issues as we gain more experience.

3 Programming Interface

```
channel = CreateChannel (access_control, ac_count,
                           stream_spec, ss_count)
DestroyChannel(channel)
```

These calls create and destroy channels. Once created, channels exist until destroyed, even across reboots. The channel identifier returned by `CreateChannel` is unique within an internetwork domain where channels are used as communication objects. Parameters include arrays (and their sizes) describing access control and the nature of the streams in the channel. These characteristics are static for the lifetime of the channel.

For each stream, the specification includes several parameters. The stream is typed and subtyped. The type specifies a general media category such as `Audio`, `Video`, `Text`, or `Data`. The subtype identifies a specific format of that category, such as `NTSC`, `JPEG`, or `MPEG`. More parameters describing finer distinctions, such as for a specific format, may be encoded in the data sent on the stream (e.g. by using libraries for the formatting of individual stream segments). In this way, specific details such as resolution may be chosen dynamically. Stream parameters also include priority, scaling, and translation as described in Section 2.

```
port = OpenPort (channel, stream_spec, ss_count, mode)
ClosePort (port)
```

These calls open and close ports on an existing channel. When a port is opened, a copy of the stream specification for the channel is returned. Ports are opened in either `SOURCE` or `RECEIVE` mode. Only one port may be opened in `SOURCE` mode on a particular channel at any one time. The opening of ports is subject to the access control specified at channel creation, if any.

```
EnableStream (port, stream, buf, buf_size)
DisableStream (port, stream)
```

These calls enable and disable the data flow of a single stream for a given port. A stream must be enabled before data can be sourced or received. Streams are referenced by their index in the stream specification array returned by `OpenPort`. When enabling a stream, a buffer area is specified which will be used by the `GetSegment` and `ReleaseSegment` calls. By not enabling (or disabling) streams which are not needed, it becomes possible for the underlying network to avoid transporting data which would not be used by the calling process. All streams are initially disabled.

```
GetSegment (port, stream, buf, buf_size, position, length, blocking)
ReleaseSegment (port, stream, buf, buf_size)
```

These two calls form a shared memory I/O interface to a stream. Buffers containing sourced or received segments are all located within a shared memory area whose size is set in the `EnableStream` call.

To receive a segment, the user calls `GetSegment`. When the user has finished using the segment, `ReleaseSegment` is called to free the segment's buffer area (and allow it to eventually be used for a new segment).

When sourcing a stream, the user calls `GetSegment` to obtain some memory into which the segment is placed. `ReleaseSegment` is then called to actually send the segment.

The `GetSegment` call can be performed with or without blocking. When blocking is specified, the call will block until the request can be satisfied.

```
stream = InstallFilter (type, port, streams, stream_count)
RemoveFilter(port, stream)
```

`InstallFilter` places a filter onto a port, and returns the index of a new stream. The `type` parameter identifies a filter from a library of predefined filters. These filters include HDTV to NTSC reducers, stereo to mono mixers, bi-level video coders, etc. A filter receives input from the streams (which need not necessarily be previously enabled) whose indices are given in the `streams` array. Like any other stream, a filtered stream is initially disabled, and must be enabled for use. After disabling the filtered stream, the filter may be removed with `RemoveFilter`.

4 An Example

An application which distributes audio and video from one location to several observers can be constructed using the MMC. A channel is constructed beforehand, and its identifier is given to all participants. Two streams are declared. One is of type `Audio`, subtype μ-`law`, with a sampling rate of 8K samples/second. The other stream is of type `Video`, subtype `NTSC`, with 640x480 8-bit pixels.

The server process performs simple setup operations:

```
port = OpenPort(channel, stream_spec, &ss_count, SOURCE);
EnableStream(port, 0, audio_buf, 10000);
EnableStream(port, 1, video_buf, 1000000);
```

The buffer sizes were chosen to approximate those which might be used for uncompressed audio and video. The video buffer can hold three complete frames of video at one time. The main loop in the server includes the following operations to send audio. First, the number of bytes waiting in the codec buffer is determined. Then, a suitable segment is obtained, filled with audio, and returned to the kernel. For this audio stream, the stream position and length of segments are defined in units of bytes.

```
avail = query_codec_buffer();
GetSegment(port, 0, &audio_seg, avail, position, avail, 1);
read_codec_buffer(audio_seg, avail);
position += avail;
ReleaseSegment(port, 0, audio_seg, avail);
```

The operations for video are similar. The client program performs similar setup (specifying RECEIVE instead of SOURCE). The client's main loop is similar to that of the server:

```
GetSegment(port, 0, &audio_seg, &size, &position, &length, 1);
write_codec_buffer(audio_seg, size);
ReleaseSegment(port, 0, audio_seg, size);
```

If the client desires only one-bit dithered video, the following calls could be used to produce a suitable filtered stream:

```
streams[0] = 1;
new = InstallFilter(DITHER_1_BIT, port, streams, 1, &spec);
DisableStream(port, 1);
EnableStream(port, new);
```

Segments may now be obtained from the new stream which will contain frames of dithered video. If the underlying network supported filter propagation, it would be possible to propagate this filter towards the source until it reaches an intermediate node (possibly even the source) which must forward color video to the other receiver nodes. If the filter operates at that node, only dithered video needs to be sent to the node that installed the filter, thus reducing the consumption of network bandwidth.

5 Discussion

Our main goal is to support real-time dissemination of continuous media from a source to multiple destinations. With multiple receivers, end-to-end closed-loop control is difficult and cumbersome. For example, flow control of a connection to multiple receivers typically results in the slowest receiver impeding the progress of the others. Similarly, error-control based on multipoint ARQ (Automatic Repeat reQuest) protocols is complex and can slow down the source and the other receivers, even when only one of the receivers has a poor quality network path.

We believe that tight, closed-loop, end-to-end control is inappropriate for applications that expect a large number of receivers having different capabilities, that are distributed across a wide geographical area, and are interconnected through networks providing different Quality-of-Service (QoS). Instead, we have adopted an alternative approach that relies on very loose coupling between the source and the receivers, i.e. an open-loop approach, which is better suited to real-time continuous media and broadband networks.

The approach we propose relies on preventive actions to minimize reliance on feedback [18]. For error control, this means using a combination of forward error correction, to anticipate errors and provide information redundancy, and careful coding to aid in error localization and concealment, allowing the receivers to maintain continuity without requesting the sender to retransmit. For flow and congestion control, this means using one or some combination of the following techniques: (i) reserving resources so that receivers and intermediate switching nodes are always able to support the flow rate dictated by the sender; (ii) using admission control to limit the amount of traffic in the network; (iii) dropping the excessive flow at the point of congestion. For the latter technique to be effective, it should be combined with hierarchical coding (see below) and priority-based discarding of excess load by the network. Various congestion control schemes based on these ideas are under consideration for Broadband ISDN using the Asynchronous Transfer Mode (ATM) [9].

Hierarchical coding techniques (also referred to as component, layered, or sub-band coding) split continuous media signals into components of varying importance [15, 13]. The aggregation of these components reconstructs the original data, but subsets of them can also provide various degrees of approximation to the original signal. Many compression standards support various forms of hierarchical coding [26].

The application of hierarchical coding to continuous media gives the system software at the receiver the capability of allocating resources based on local (i.e. the receiver's) specifications and priorities. This might entail deciding to gracefully and dynamically degrade the quality of the received (and played-back) signal when resources are limited. Also, it facilitates local processing and presentation of the signal in ways not intended by the sender. In the case of multipoint communication, hierarchical coding enables receivers to adjust the quality of the signal they receive, independently of one another and without the source actually being aware of this adjustment (of course, the adjustment is only possible towards lower quality). This is a useful property, considering the feedback control problems of multipoint communication, and can also be used to effectively address many compatibility problems.

Hierarchical coding fits well with an open-loop approach to congestion control of high-speed networks. With layered coding of continuous media, when network congestion arises, it is possible to drop the less important signal components without causing service interruption, and without the need for retransmissions. This should only lead to a reduction in service quality (which should be engineered to be tolerable by the users). Since continuous media will constitute the bulk of the network traffic, this technique can be very effective as a last resort for congestion control. Furthermore, routing algorithms may take into account the different needs and capabilities of the destinations by, for example, forwarding only usable components to select destinations.

The design of the MMC was influenced by these ideas, particularly the open-loop approach. Even though many of these ingredients are not prescribed (and even opposing views could be supported), the overall design attempts to facilitate the use of these techniques.

For example, a segment, with its size defined by the application, is a relatively independent piece of information, similar to the "Application Data Unit" concept described in [6]. We expect that for many multimedia applications, guarantees of reliable delivery will not be necessary for various media component types, and some segments could be dropped at times of heavy congestion. In addition, some of these applications may actually be quite tolerant to variability of delays, as described in [5]. Particularly for lower priority components, applications would be expected to recover gracefully from loss of segments, or adapt to changes in the delays of their arrivals.

To enable such adaptability and recovery, segment information (e.g. priority, position, and length) is made available to the receiver. In particular, the segment position and length allow the receiver to recognize the segment's identity and relationship to other segments in the stream. This allows, for example, out-of-order operation on segments which improves performance [6] as well as fault tolerance.

With regard to underlying system software support, it is advantageous to treat the segment as the data unit of manipulation [6]. Cooperation between the network system software and the operating system dictates that a segment be recognized as a basic unit through the protocol stack and the network. Even though this is not a requirement, the approach taken here would suggest that if any part of a segment is damaged or lost, the whole segment should be discarded.

Stream performance and semantics depend on the QoS provided by the underlying system. Each stream data type dictates what QoS is expected from the network and I/O system. The MMC philosophy expects that applications will specify the least stringent QoS necessary for each stream. Admission control blocking probability and network pricing policies will encourage this practice (as well as the definition of standard media component types). How the underlying system makes provisions to implement QoS demands is outside the scope of this paper, but is an important issue – see [12, 27] for various schemes.

The filter mechanism, operating on multiple streams, is well matched to the approach of the source transmitting multiple hierarchically coded media components from which the receivers pick and compose the presentation signals according to their individual specifications and capabilities. Filter propagation through the network towards the source will permit conservation of network resources (and potentially even local resources by off-loading standard operations to specialized network servers).

The value of this approach is that there is considerably less complexity due to the absence of (many) feedback control mechanisms, which are typically useless for real-time continuous media. [2] In addition, considerably more functionality can be provided (at low cost) because of the support of independent media components, which allows receivers to easily and efficiently tailor presentation to individual requirements. The cost is the additional complexity due to these anticipatory/preventive action schemes. We believe that this solution is the most appropriate for multicasting of continuous media.

---

[2] Note that resource reservation schemes will generally employ a protocol which uses feedback to determine resource allocations. However, since resource reservation takes place once, before the start of actual information flow (rather than multiple times throughout the information flow, as do the dynamic feedback control mechanisms described above), its cost may be acceptable.

6 Past Work

Multicasting has received considerable attention lately because of the interest in collaboration technologies. However, little attention has been paid to appropriate system support for this mode of communication. Most of the work in this area revolves around low-level either communication protocols or groupware applications, which typically simulate multicast through a series of unicasts. However, the support for dissemination-oriented communication paradigm is beginning to receive attention.

Pioneering work on IP multicasting by Deering *et al.* [8] is fueling some of the current protocol work, which is based on a group communication protocol, the Internet Group Management Protocol (IGMP) [7]. This protocol specifies how multicast groups can be formed and managed with little change to the basic infrastructure of IP. Multicast groups are created by individual hosts explicitly joining the groups, i.e., using IGMP to include them in the address list corresponding to a multicast address. The actual multicast routing is done by routers that learn the shortest path routes to the destinations [8].

Recent experiments using IP multicasting include the telecasting of the recent IETF meetings [2]. Live audio and slow scan black-and-white video from the most recent meeting was distributed to hosts that wanted to be included in the multicast group. Tunneling [25] was used to send data over segments of the Internet that did not support multicast routing.

An alternative network layer protocol in the Internet suite, specially designed for continuous media transmission, is ST-II [23]. This experimental protocol supports multicasting and resource negotiation appropriate for continuous media, but the protocol itself does not include specific implementation mechanisms, e.g., for multicast route set-up or network resource reservation (instead, such mechanisms must be provided outside of the protocol).

The multicast routing algorithm described in [16] optimizes network performance for continuous media, making it a strong choice for MMC support. It constructs a source-rooted multicast tree which efficiently uses bandwidth while bounding delay between the source and all destinations. Other work on optimization of multicast tree set up, some of it in the context of multimedia communications, can be found in the references in [16].

Hierarchical coding has been studied extensively in the area of signal processing. It has recently been recommended as a potentially effective mechanism for congestion control for high-speed networks carrying digital continuous media [15, 13, 17, 11, 19]. However, we are not aware of any other operating system or network system software level support for layered coding, in particular, in relation to multicasting.

Forward error correction (FEC) has been suggested as a method for avoiding feedback for error notification and correction. FEC consists of sending enough redundant information so that occasional dropped packets can be reconstructed from the packets that were correctly received. More on FEC can be found in [22, 1].

7 Conclusions

The dissemination-oriented service provided by the Multimedia Multicast Channel supports group multimedia applications. Application writers are freed from the task of maintaining network state and performing connection setup and teardown. The support for specific media types and the selective reception of original or filtered media makes possible optimization in the network layer. For loss-tolerant media, the open-loop approach simplifies the internal maintenance of network state and prevents problems at one receiver from causing service degradation for any other participants.

8 Acknowledgments

The design of the MMC is the result of countless discussions taking place over the last few years at the UCSD Computer Systems Lab. One could not have been both a member of the lab and ignorant of "The Channel," the MMC's insider name and now part of the local vernacular. In addition to the authors, the participants of those discussions included Mark Bubien, Kevin Fall, Jonathan Kay, Scott McMullan, Keith Muller, Dipti Ranganathan, Robert Terek, and many others, who all contributed to the design through their recommendations, objections (which were often vociferous), and trial implementations. Of course, there is still controversy between us regarding this design, and we look forward to more arguments. To all the contributors, we are indebted to you, and we realize that we still probably have not "gotten it right" once and for all, but we think we are close.

References

1. E. W. Biersack: Performance Evaluation of Forward Error Correction in ATM Networks. Proc. ACM SIGCOMM '92, Baltimore, MD (Aug. 1992) 248–257.
2. S. Casner, S. Deering: First IETF Internet Audiocast. ACM SIGCOMM Computer Communications Review, 22 (July 1992) 92–97.
3. S. Casner, K. Seo, W. Edmond, C. Topolcic: N-Way Conferencing with Packet Video. Proc. 3rd International Workshop on Packet Video, Morristown, NJ (Mar. 1990).
4. D. Cheriton: Dissemination-oriented Communication Systems. Distinguished Lecturer Series, Department of Computer Science and Engineering, University of California, San Diego. (Apr. 1992)
5. D. D. Clark, S. Shenker, and L. Zhang: Supporting Real-Time Applications in an Integrated Services Packet Network: Architecture and Mechanism. Proc. ACM SIGCOMM '92, Balitimore, MD (Aug. 1992) 14–26.
6. D. D. Clark and D. L. Tennenhouse: Architectural Considerations for a New Generation of Protocols. Proc. ACM SIGCOMM '90 (Sep. 1990) 200–208.
7. S. Deering: Host Extensions for IP Multicasting. ISI, RFC 1112, Network Information Center, SRI International, Menlo Park, CA (Aug. 1989).

8. S. Deering and D. Cheriton: Multicast Routing in Internetworks and Extended LANs, ACM Transactions on Computer Systems **8** (May 1990) 85–110.
9. A. E. Eckberg: B-ISDN/ATM Traffic and Congestion Control. IEEE Network **6** (Sep. 1992) 28–37.
10. J. R. Ensor, S. R. Ahuja, D. N. Horn, and S. E. Lucco: The Rapport Multimedia Conferencing System - A System Overview. IEEE Computer (Apr. 1988) 52–58.
11. D. Ferrari, J. C. Pasquale, and G. C. Polyzos: Network Issues for Sequoia 2000. Proc. IEEE Compcon Spring '92, San Francisco, CA (Feb. 1992) 24–28.
12. D. Ferrari and D. Verma: A Scheme for Real-Time Channel Establishment in Wide-Area Networks. IEEE J. on Selected Areas in Comm. **8** (Apr. 1990) 368–379.
13. M. Ghanbari: Two-Layer Coding of Video Signals for VBR Networks. IEEE J. on Selected Areas in Comm. **7** (Jun. 1989) 771–781.
14. R. Gusella and M. Maresca: Design Considerations for a Multimedia Network Distribution Center: Proc. 2nd International Workshop on Network and Operating System Support for Digital Audio and Video, Heidelberg, Germany (Nov. 1991).
15. G. Karlsson and M. Vetterli, Packet Video and its Integration into the Network Architecture, IEEE J. on Selected Areas in Comm. **7** (Jun. 1989) 739–751.
16. V. Kompella, J. Pasquale, and G. Polyzos: Multicasting for Multimedia Applications. Proc. IEEE INFOCOM '92, Florence, Italy (May 1992) 2078–2085.
17. J. Pasquale, G. Polyzos, E. Anderson, K. Fall, J. Kay, V. Kompella, D. Ranganathan, S. McMullan: Network and Operating Systems Support for Multimedia Applications. Tech. Report CS91-186, Univ. of Calif., San Diego (Mar. 1991).
18. J. Pasquale, G. Polyzos, and V. Kompella: The Multimedia Multicasting Problem. Submitted for publication.
19. N. Shacham: Multipoint Communication by Hierarchically Encoded Data. Proc. IEEE INFOCOM '92, Florence, Italy (May 1992) 2107–2114.
20. W. D. Sincoskie: Video On Demand - Is It Feasible? Proc. IEEE Globecom '90, San Diego (1990).
21. W. D. Sincoskie: System Architecture for a Large Scale Video On Demand Service. Computer Networks and ISDN Systems, North-Holland. **22** (1991) 155–162.
22. N. Shacham and P. McKenney: Packet Recovery in High-Speed Networks Using Coding and Buffer Management. Proc. IEEE INFOCOM '90, San Francisco, CA (Jun. 1990) 124–131.
23. C. Topolcic, Ed.: Experimental Internet Stream Protocol, Version 2 (ST-II). RFC1190, NIC, SRI International, Menlo Park, CA (Oct. 1990).
24. H. M. Vin, P. T. Zellweger, D. C. Swinehart, and P. V. Rangan: Multimedia conferencing in the Etherphone environment. IEEE Computer **24** (Oct. 1991) 69–79.
25. D. Waitzman, C. Partridge, and S. Deering: Distance Vector Multicast Routing Protocol. RFC 1075, NIC, SRI International, Menlo Park, CA (Nov. 1988).
26. G. K. Wallace: The JPEG Still Picture Compression Standard. Communications of the ACM **34** (Apr. 1991) 30–44.
27. L. Zhang: Virtual Clock: A New Traffic Control Algorithm for Packet Switching Networks. ACM Trans. on Computers, **9** (May 1991) 101–124.

Analysis of a Resequencer Model for Multicast over ATM Networks

Liming Wei[1] [*], FongChing Liaw[2], Deborah Estrin[1], Allyn Romanow[2], Tom Lyon[2]

[1] Computer Science Department, University of Southern California, Los Angeles, CA 90089-0782
[2] Sun Microsystems Computer Corporation, Mountain View, CA 94043-3156

Abstract. Multicast delivery saves bandwidth and offers logical addressing capabilities to the applications. The receivers of a multicast group need to differentiate cells sent by different sources. This demultiplexing requirement can be satisfied in an ATM environment using multiple dedicated point-to-multipoint virtual channel connections (VCs), but with certain shortcomings. This paper discusses an alternative resequencing model to solve this problem. It scales well in large networks. Three resequencing methods are are developed and simulation results reported. The strategy is useful for applications spanning large regions where it is desirable to mix streams of cells from different bursty sources onto the same virtual channel. [3]

1 Introduction

It is important for future ATM networks to have multicast capability, as they will support such applications as teleconferencing and information distribution services. Applications based on multicast have two major advantages over unicast: logical addressing and bandwidth savings [Deering]. With logical addressing, an application uses a single multicast group address to send and receive data. It is not necessary for senders and receivers to know the number or location of group members. Multicast provides significant savings in bandwidth because the sender transfers only a single copy of the data. Data cells are not replicated until they reach a branching point in the multicast delivery tree, at a location closer to the destinations.

The results of research and experimental implementations of multicast in IP networks have demonstrated the benefits of supporting multicast in public data networks [RFC1112] [MOSPF]. Although it would be desirable to adopt the implementation approaches successfully used in IP networks, the differences between an ATM network and a conventional packet switched network make it difficult to directly map multicast models for IP into ATM networks.

---

[*] Work supported by SUN summer intership 92 at SMCC, Mountain View, California.
[3] We would like to thank Steve Deering, Bryan Lyles, Lixia Zhang and the referees for helpful discussions and comments.

In IP networks, for a multicast group having N senders, N multicast delivery trees are established, each tree delivering packets from a source to all destinations in one-to-many fashion. Each IP packet carries a source address and a destination (i.e. multicast group) address in its header [RFC1112]. To route multicast packets both destination address and source addresses are used for lookup in the routing table. A receiver uses the source address field of the IP header to distinguish packets from different senders. Thus multicast packets can be arbitrarily intermixed with each other on the forwarding path.

ATM, however, is a connection oriented technology, in which connections must be explicitly setup before data can be transferred. An ATM connection can be viewed as a cached route [LLR], in the sense that each cell carries only a small routing tag — the virtual path identifier and virtual channel identifier (VPI/VCI). Unfortunately, if cells from different sources are multiplexed onto the same virtual channel, they will carry the same routing tag, or VPI/VCI, upon arrival at a receiver. There is not a straightforward way to distinguish cells sent from different sources. This is often referred to as the *cell demultiplexing* problem. Although several solutions have been proposed, they have shortcomings and, most importantly, they do not scale well.

This work proposes a solution that both solves the cell demultiplexing problem and scales well. The following section describes currently proposed solutions to ATM multicasting and discusses their shortcomings. The third section develops the proposed resequencer model. In section four, simulation results for the performance of several different methods of resequencing are presented.

2 Strategies for Implementing Multicast in ATM

The cell demultiplexing problem can be avoided if one VPI/VCI is used for each source. In this case, each multicast VPI/VCI is a one-to-many connection representing one multicast tree and is independent from other VPI/VCI's. Each tree can be optimized according to some criterion such as least delay or least cost.

The VPI is an 8-bit field in the cell header. If it is used to identify a multicast group, it restricts the number of multicast connections to even less than the number available with VCI's. In addition, VP switching may be used to bundle large number of VC connections, or to separate VCs of differing quality of service classes. Therefore, we assume that the VP will not be available for use as a multicast tree identifier. For the remainder of this discussion we refer to the use of VCI's. Figure 2 shows the *one-vc-per-source* multicast model for groups with multiple senders.

The one-vc-per-source strategy is compatible with the current IP multicast models [RFC1112], provided that the switch signaling mechanisms have access to the installed IP multicast routing tables [LLR]. The scheme is useful for (a) applications with small numbers of sources; (b) local multicast groups where there is not a shortage of VCs relative to the demand; and (c) applications where all sources continuously transmit (e.g., video).

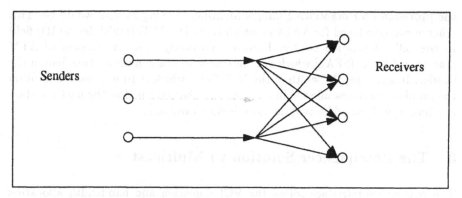

Fig. 1. One-vc-per-source multicast model

However this scheme uses a large number of VCIs and therefore does not scale well to networks with many nodes and large numbers of long-lived multicast groups.

In particular, the scheme has a number of limitations. The number of virtual channels is too limited to accomodate a sizable number of large, long-lived groups. Furthermore, if bandwidth is reserved or allocated on a per VC basis, the large number of VCs resulting from a multcast group will quickly use up the available bandwidth. Similarly, In the public network, if the cost of cell switching depends on the number of virtual channels, the additional VCs used in per-source-VC will be a costly disadvantage. In addition, since each sender defines a tree, when a new receiver joins a group, it must be added to all existing trees.

It is important for ATM multicast to scale to large size. Applications involving hundreds, or even thousands, of end users globally are not too far distant. Thus it is desirable to explore alternatives in which the number of available VCs in a switch or host interface is not an upper bound on the number of sources an application can support.

Recently several models have been proposed for ATM multicast, aimed at solving the VC depletion problem, and achieving high throughput and scalability [BGD] [GRELYL] [SBO] [KPP]. One proposal [GRELYL] is for all sources in a multicast group to share one virtual channel, and to use collision detection techniques, such as Aloha, to avoid intermixing cells from different sources. The CRC will detect when a cell is out of order, i.e., not from the same sender, and the sender retransmits. In this scheme, the common virtual channel is treated as a shared media, similar to an ethernet cable, with some of the same advantages and disadvantages ethernet has. The scheme is useable for local, low bandwidth and non real-time applications.

In another scheme [BGD], an identifier in the ATM payload field is used to distinguish cells from different sources. It is not clear how simple or complex the procedures for negotiating unique identifiers among sources would be. This scheme was developed for AAL4 in which there is a MID (multiplexing ID) field in the cell information header. However, currently many developers of ATM plan to use AAL5 [SEAL] which does not have such a demultiplexing field in the header. In any case, the width of the MID field, which is 10 bits, puts a limit on the number of sources allowed in a group. Another cost is that the host interface is required to have additional demultiplexing hardware.

3   The Resequencer Solution to Multicast

The scheme we introduce solves the VCI depletion and bandwidth allocation problems without requiring specialized bandwidth allocation schemes for certain common cases of multicast. The traffic classes for which this approach is well suited are: (a) transmission per source is not continuous, (b) small to modest size PDUs, (c) large multicast groups. First we describe the basic resequencing strategy, followed by several extensions and refinements to the basic idea. Next we present a group tree model which extends the scheme to a larger network.

3.1   Resequencers

To achieve better bandwidth sharing without complicating the VC-based resource management mechanisms, and to avoid the VC availability dilemma, cells from different sources are multiplexed onto a single VC. A designated source in the group is elected as a *resequencer*. All other sources send their multicast ATM cells to the designated resequencer, which buffers incoming cells from each source before all cells of a PDU are received. After the last cell of a PDU has been received, the resequencer forwards all cells of that PDU onto a single outgoing VC, without interleaving it with cells from other sources.

Note that the outgoing VC of the resequencer is a point-to-multipoint connection. Cells are duplicated at the branching point(s) and the order of cells is preserved across all links in this one-to-many connection. If a sender is also a receiver, there is a branch of the one-to-many VC leading back to the sender.

Although this single resequencer scheme provides the benefits of multicast, it has several problems. The processor and link speed of the source designated as a resequencer could become a bottleneck. Choosing the resequencer is complex, and there may be situations where no ideal selection of designated resequencer exists. This scheme is best when all senders are close to the resequencer.

To extend the scheme to accommodate the case where members are spread over different regions, and each region has many members, members in each region elect a source to act as a resequencer for local senders. A one-to-many VC is created for each resequencer which delivers cells to all other resequencers of the group, and cells are therefore delivered to all receivers. The number of VCs required is now reduced to the number of participating regions, instead of the number of sources (N one-to-many VCs for N resequencers).

To make the scheme more efficient, resequencing can be done in switches instead of in hosts. Normally switches sustain higher aggregate cell rates. To participate in multicast, the host only needs to know the address of its nearest multicast resequencer. In a local area ATM, the location of a multicast resequencer can be either statically assigned, or dynamicaly discovered, so that the native nodes are kept updated about the available multicast servers. To avoid long call setup latency, the resequencer information can be cached in the host. This approach eliminates the necessity of going through a decision process to choose a host resequencer.

3.2 Multicast Group Trees

Our final refinement of this approach is to further reduce the number of VCs maintained for a group. We do this by building a group tree instead of per-source (or per-source-region) rooted trees. Connections among the resequencers are bidirectional. A resequencer serializes all cells destined for a common group and duplicates them onto the appropriate outgoing ports with correct VPI/VCIs — some lead to other resequencers, some go to local members. Figure 2 depicts a group tree of resequencers and receivers. The group tree is a delivery path in common for cells from all sources. As Figure 2 shows, each source establishes a point-to-point connection to its resequencer and sends cells towards a multicast group along that connection, as if it were a unicast destination.

Either point-to-point or point-to-multipoint connections can be used to distribute cells from a resequencer to its local receiving members, depending on support from the local network. If a member is both a sender and a receiver, a bidirectional point-to-point connection is used. The connections between s2, s3 and Resequencer M2 in fig 2 are such examples. To reduce the bandwidth used in local multicast cell delivery, local switches must support point-to-multipoint connections. The connection between resequencer M3 and r0, r1 and r2 in fig 2 is such a point-to-multipoint connection, where r0, r1 and r2 are all receivers [4]. Note also that there may be many "pure" ATM switches between any two resequencers, and between any resequencer and its local members. The paths drawn in figure 2 are logical channels.

3.3 Discussion

The resequencer approach requires algorithms different from those used for IP multicast to compute the group multicast tree. The tree will not be optimal in terms of shortest path or least delay. However, suboptimal solutions should suffice for a number of applications [KPP] [CBT].

Since this resequencer and group tree model can be built upon one-to-one and one-to-many virtual channel connections, it can coexist with the one-vc-per-source model. An application can choose one of the two models. For example,

---

[4] Note that it is not suggested to replace the reverse channels of s2 s3 and s4 with a single point-to-multipoint connection, unless the senders don't care if they receive duplicated cells looped back by resequencer M2.

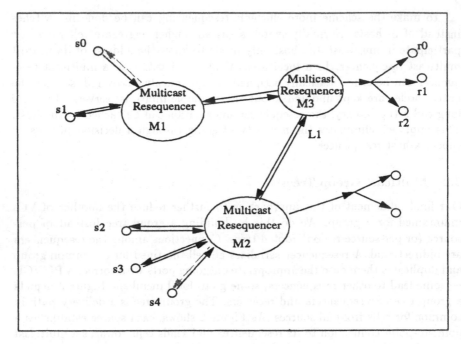

Fig. 2. The *resequencer and group multicast tree* model

video applications where sources transmit continuously may choose to use the one-vc-per-source model, whereas video conferencing with compressed streams and multiple changing sources may choose the resequencer approach. The resequencer and group tree model is also flexible in that it works both with and without point-to-multipoint virtual channels. The difference is in the bandwidth savings of the local cell delivery paths. This approach turns out to be especially well suited for voice conference applications, where one bandwidth allocation for the whole multicast group may be sufficient and is independent of the number of sources. In an audio conference, it is unusual to have multiple simultaneous transmitters. Usually only one or two persons speak at a time, as is consistent with higher level human protocols.

In order to setup and maintain the group tree, a higher layer protocol — either a network layer protocol or a call control level protocol, is needed. Different protocols may be appropriate for this purpose. One possibility is to adapt Deering's host membership protocol and multicast tree setup protocols [Deering].

The resequencer and group tree multicast model has a nice scaling property. Only one virtual channel is needed for each receiver to receive cells from all sources. When a new member (be it a sender or a receiver or both) joins, it only needs to do the setup necessary to reach a resequencer already in the group. The operations of join and leave are inherently local, growing a branch or chopping off a branch does not have to involve a distant part of the tree.

Two performance parameters, maximum throughput and delay, will be used to assess the resequencer and group tree model, and will determine the range of applications of this model. Different implementation methods result in different maximum throughputs ranging from tens of megabits per second to that approaching the maximum unicast rate. We will show in the next section that as the needs for higher speed multicast service appear, there are ways to increase the maximum throughput without changing the end system.

4   Three Approaches to Resequencing

In the resequencer and group tree model, resequencers are the crucial components that determine performance. We describe three different schemes for resequencing cell streams from different sources: in software at the network layer; in hardware; and an optimized hardware scheme. The performance of these approaches is evaluated with respect to throughput, delay and delay jitter.

Software Resequencing Multicast cells from the same source can be reassembled at the resequencer in the same way that signaling messages are reassembled. An ATM switch with signaling capability has associated CPU and memory to process and forward signaling messages [Q.93B]. Similarly for resequencing of multicast cells, the reassembled PDUs are passed to the control processor, which makes routing decisions by inspecting the address field in the header of the PDU. The forwarding process is shown in figure 4(a). The route lookup and PDU forwarding are done in software, similar to a datagram router.

To examine performance, as shown in figure 4(b), the cell forwarding delay can be decomposed into three parts: (1) the delay to collect cells of the PDU; (2) the delay due to the route lookup algorithm; and (3) transmission delay. [5] We observe that the cell collection time and transmission time of consecutive PDUs can be overlapped. Therefore, route lookup is the dominant factor affecting throughput.

With certain optimizations in the implementation, the software forwarding scheme can achieve high throughput. Consider a switch equipped with a 50 MIPS processor. Assume its multicast routing table has 1000 entries and it takes about 1000 instructions to forward one PDU[6]. The per PDU route lookup and forwarding time will be $\sim$20 μs. For 256-byte PDU, this corresponds to a maximum throughput of 100 Mbps, if the cell collection times and transmission times of consecutive PDUs are completely overlapped.

The delay experienced by each individual PDU depends on a number of factors. In the one-vc-per-source model, a PDU is not reassembled until reaching an end host. Using the software resequencing method, a PDU has to be reassembled

---

[5] PDU queuing delay needs not be considered here, if we assume no congestion.

[6] This is a rough estimation. If the operating system in ATM switches are better designed for communication protocol processing, the number of instructions would be smaller.

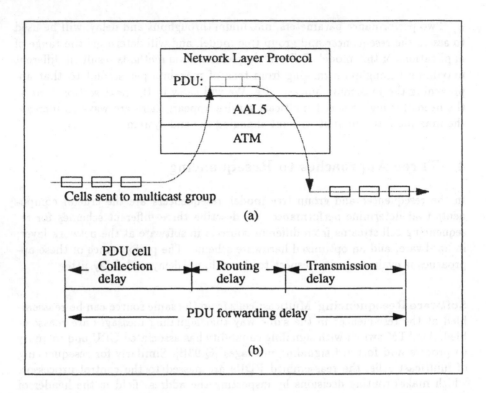

Fig. 3. Software PDU forwarding

in every resequencer it passes. The additional delay is the sum of cell collection delays and routing delays the PDU experienced in all resequencers along its path. The PDU cell collection delay depends on the PDU size and link speed.

Figure 4 shows the PDU cell collection delay under different link speeds and PDU sizes. Multimedia applications with delay jitter or other real-time requirements tend to use small packet sizes, between 128 to 256 bytes. Even with larger PDU sizes (1000 Bytes) and slower link speed (155Mbps), the worst case aggregate reassembly and routing delay in the above example can still be less than ~7.7ms, if less than 100 resequencers are involved in the longest path.

Thus the performance of the software resequencing method, in terms of speed and delay, is good for applications requiring less than 100Mbps and involving less than thousands of regions — provided the longest path of the multicast group tree has less than a few hundred resequencers.

Hardware Resequencing When higher throughput is required, we propose a method using hardware support in the switch. Instead of a software route lookup, multicast cells can be forwarded directly by hardware switch fabrics using mechanisms similar to that used for forwarding point-to-point connection

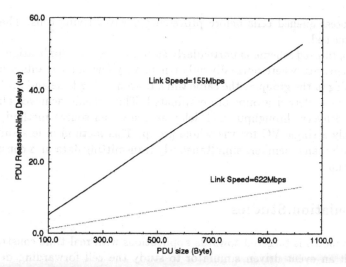

Fig. 4. PDU reassembly delay

cells. Incoming VCs carrying multicast cells are treated separately from those for normal point-to-point connections; they are queued in buffers, neither forwarded nor reassembled. The switch monitors the End of Packet (EoP) cell. When the EoP cell arrives, the hardware dumps all cells in that PDU back-to-back onto outgoing links.

Hardware resequencing requires extra buffers and buffer management mechanisms in the resequencing switch hardware. The amount of buffer required depends on a variety of factors, including the PDU size, traffic characteristics, congestion control algorithms, etc. We can estimate the number of buffers required by a resequencer in the restricted case assuming no congestion and that sources obey negotiated resource allocations. In this case, a switch has sufficient bandwidth to carry the incoming traffic, and the queue size for each incoming VC should not grow beyond 2 PDUs. More buffer space is needed only when there is congestion. If the maximum number of active multicast groups is N, the maximum number of active source streams coming in to a resequencer for each active group is S, and the maximum PDU size used by all multicast sources is P, the maximum amount of buffer required in a resequencer is $2 \times N \times S \times P$.

Optimized Hardware Cell Forwarding The third method is an optimization of the second. When the resequencer receives the first cell of a PDU from a source, if there are no cells from other sources queued or being forwarded, it directly forwards all cells on the outgoing links without queuing them first, until it forwards the End of Packet cell. During the forwarding action, if cells from other sources arrive (i.e., on other incoming links), they will be queued and processed according to either the software or hardware resequencing method.

To reduce delay variance, after an EOP cell has been sent, the switch should try to process queued cells before processing newly-arriving cells. These latter cells are queued.

This optimized scheme is particularly suited for certain applications, such as voice conferences, where normally only one or very few participants are actively transmitting to the group at the same time. Both queuing time and route lookup time for the preferred groups are eliminated. This scheme achieves almost the same cell delivery throughput and delay as one-vc-per-source method, however it uses only a single VC for the whole group. This method is less suitable for groups with many members simultaneously transmitting data at a constant rate to the group.

5   Simulation Studies

Since this model is targeted towards applications with real-time constraints, we have built an event driven simulator to study the cell forwarding delays and delay jitters. End-to-end delay and variance are dependant on per switch delays and other parameters. We simulate and measure the switches under different traffic conditions and with each of the resequencing methods described.

The simulator uses a queuing model to characterize the behaviors of different buffers in a switch. The input queues buffer incoming cells; the PDU queue is used by the software resequencer to save reassembled PDU's before route-lookup; the output queues store outgoing cells when the cell arrival rate is larger than the link speed. For simplicity, a FIFO algorithm is used to schedule the resources whenever competition occurs. The link speed, link delay, network connectivity, software PDU forwarding speed, end-to-end route, and multicast group information etc. are entered by the user and stored statically.

A simple traffic model has been used to approximate Audio/Video style traffic sources in the simulations. The traffic model assumes that each source generates data at a constant rate, and every t seconds, a PDU is ready to send. Then the PDU is chopped into ATM cells and sent onto the network at the link speed, while at the same time the application (slower than the network) is generating data for the next PDU. The source stops generating data after sending several PDU's and stays silent for a randomly selected period of time. Then it starts sending again.

A link during the interval between delivery of two consecutive PDUs is like a vacuum. This partially explains why in the following simulation, even when the total active sending time of all the relatively slow members of the group exceeds 100%, the delay caused by resequencing different PDU's in collision is not very significant. Of course, there is a chance that the number of collisions can be larger if the slow sources together with the link delay make different PDU's frequently arrive at a resequencer at the same time.

A fully connected 17-node network (including 11 hosts and 6 switches) was constructed. A total of 10 multicast groups were created. The resequencer switch located at the network bottleneck location was setup for measurements. Figure 5 shows the distributions of the cell delays across that switch. The same traffic sources are used for the three runs with different resequencing methods. The measured path has four sources sending to the same group and whose percentage of time actively sending are 80%, 50%, 50%, 30%. Although the source rates are below the saturation point, they do collide inside the resequencer and get queued and reordered on exits. The tails of the curves are caused by the queuing delays. The horizontal distance between the Hardware resequencing curve and the software resequencing curve reflects the route look up delay (20 μs). The horizontal distance between the optimized and the hardware resequencing curves reflects the cell collection time (related to link speed and PDU size). The differences in heights of the three curves depict the fact that the more time a PDU spends in a resequencer, the more chance it may collide with another PDU. The widths of the delay distribution curves shows the jitter the cell experienced.

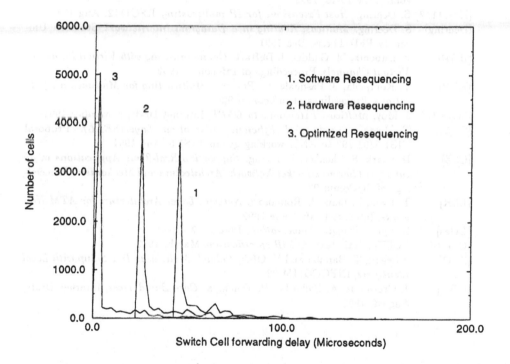

Fig. 5. Distribution of cell forwarding delays

Most of the sources simulated use 20-cell PDUs, except for one that uses smaller 9-cell PDU, and one that uses larger 30-cell PDUs. With larger PDU sizes, the curves in the picture will move horizontally. Curves 1 and 2 will move towards the right, indicating larger cell collection delay, and more chances of collision would occur. The lines at the measured resequencer are heavily utilized at 50% to 90% of available bandwidth.

6 Conclusion

We have introduced a resequencer and group tree multicast model for multicast over ATM networks and the traffic classes for which this approach is well suited.

References

[GRELYL] D. H. Greene and J. Bryan Lyles. *Reliability of Adaptation Layers*, 3rd IFIP WG6.1/6.4 workshop on protocols for high-speed networks, Stockholm, May 13-15, 1992

[RFC1112] S. Deering. *Host Extensions for IP multicasting*, RFC1112, Aug 1989

[Deering] S. Deering. *Multicast Routing in a Datagram Internetwork*, Stanford University PhD Thesis, Dec 1991

[BGD] R. Bubenik, M. Gaddis, J. DeHart. *Communicating with Virtual Paths and Virtual Channels*, Proceedings of Infocomm 1992

[KPP] V. Kompella, J. Pasquale, G. Polyzos. *Multicasting for Multimedia Applications*, Proceedings of Infocomm 92

[MOSPF] J. Moy. *Multicast Extensions to OSPF*, Internet Draft, November 1991

[SEAL] T. Lyon, *Simple and Efficient Adaptation Layer(SEAL)*. Proposal T1S1.5/91-292 to ANSI working group T1S1.5, Aug 1991

[CSZ] D. Clark, S. Shenker, L. Zhang. *Supporting Real-Time Applications in an Integrated Services Packet Network: Architecture and Mechanism*, Proceedings of SigComm 92.

[LLR] T. Lyon, F. Liaw, A. Romanow. *Network Layer Architecture for ATM Networks*, Internet Draft, June 1992

[Lyles] B. Lyles. *Private Conversation*, June 1992

[Q.93B] CCITT, draft text, *Q.93B specification*, March, 1992

[SBO] A Segall, T. Barzilai and Y. Ofek, *Reliable Multi-user Tree Setup with Local Identifiers*, INFOCOMM 92

[CBT] J. Crowcroft, A. Ballardie, P. Tsuchiya. *Core Based Trees*, Internet Draft, August, 1992

Session 6: Multimedia Models, Frameworks, and Document Architectures

Chair: Rita Brennan, Apple Computer

This session consisted of two long papers, one which targeted multimedia model integration and object management, while the other focused on the scheduling of multimedia documents.

"An Integrated Platform and Computational Model for Open Distributed Multimedia Applications" by G. Blair, et. al. was presented by the gregarious Francois Horn of CNET, France. Francois opened up by explaining why the paper's researchers chose a synchronized processing graph model approach to their work. Francois explained that the primary reason for this decision was so they could study synchronization constraints closely.

Francois then discussed the object modeling of their research. There are basic objects (defined by earlier research from ANSA) and reactive objects which exhibit synchronous behavior programmed by Synchronization Managers. The programmer must be aware of reactive objects in the system. Hence the programmer needs a coherent view of the application-system relationship.

Francois provided an example of a machine to illustrate the model's concepts of integration, object encapsulation and synchronization. Francis stressed that synchronization is key to the application's success and asked that more research be done in global system synchronization.

At the end of his session Francois addressed several questions on synchronization for reactive systems.

"Scheduling Multimedia Documents Using Temporal Constraints" was presented by M. Cecelia Buchanan. Cecelia, a PhD candidate at the University of Washington, discussed her joint research with Polle Zellweger at Xerox PARC.

Cecelia described their algorithm and application, known as Firefly, which runs on Xerox's Cedar system. Their primary goals for this algorithm are to enhance synchronization of asynchronous components and dynamic media in existing synchronous documents. Another key goal of Firefly is to ease authoring tasks.

The algorithm performs four key chores: computing media item durations, finding connected components, assigning time to events and creating commands.

According to Cecelia, temporal constraints have three parts: media, operations, and synchronization constraints. To address temporal constraints, they designed a Firefly Scheduler to automatically produce a schedule for controlling the display of a Firefly document.

Firefly is a runtime architecture with three components: a Viewtime System, which manages a documents schedule; the Event Handler, which filters asynchronous events and the media items themselves which execute the operation lists and watch out for naturally occurring asynchronous events.

To illustrate examples, Cecelia briefly described their technique for composing a graph representation of documents. Asynchronous events are managed first. There are three parts to the Firefly system on Cedar: the first is an authoring environment for creating test documents, the second is the Firefly Scheduler which consists of a suite of analysis tools and the third part is the display on a run-time system. Components are connected with time events assigned to them.

Cecelia's presentation generated a large amount of questions and future discussion. Most of the questions were on synchronization and several on how to handle "dead-air" and error-recovery.

An Integrated Platform and Computational Model for Open Distributed Multimedia Applications

G. Blair, G. Coulson (Department of Computing, Lancaster University, U.K.),
P. Auzimour, L. Hazard, F. Horn, J.B. Stefani (France-Telecom/CNET, France)[1]

ABSTRACT : Distributed multimedia applications require an adequate platform support. Towards a fully integrated infrastructure, we propose an approach, that combines a distributed object-based platform, with programming multimedia abstractions, and a synchronous model to control multimedia data flows.

1. Introduction

It is recognised that distributed multimedia applications introduce new design challenges at all systems levels from network protocols and operating systems to application support platforms. Examples of the new requirements are: the need to represent continuous media, storage and transmission issues, the real-time synchronisation of continuous media streams and quality of service configurability. This paper presents a step towards an integrated distributed system platform for multimedia applications. It is a result of an ongoing cooperation between France Telecom research center (CNET) and Lancaster University, UK. It integrates the CNET computational model into the Lancaster platform structure.

The work of the two groups has been presented separately elsewhere [CBD92], [DCW91], [HHS91], [SHH92]. Compared to other works concerned with providing support for multimedia applications, e.g. [ATW90], [HDJ91], [HHSSS91], [Mei91], the work in CNET and in Lancaster is set in the ANSA object-based architecture for heterogeneous distributed systems developped by the ISA Esprit Project [ANS89]. In ANSA, a distributed application consists of a collection of interacting objects. An object encapsulates some state which can be accessed and manipulated by invoking operations on an interface. An object can have several interfaces.

The integration discussed in this paper provides extensions to ANSA for supporting multimedia applications, i.e. : (i) a set of object types (and associated encapsulated engineering mechanisms) for handling multimedia data flows and devices ; (ii) means to program the real-time behaviour of multimedia applications, including fine-grained synchronisation within or between multiple media flows.

The paper is organized as follows. Section 2 presents a graph model of distributed multimedia applications developed in CNET. Section 3 presents the Basic Service Platform (BSP) developed in Lancaster. Section 4 motivates and describes the integration between both approaches. Section 5 reviews what has been achieved and discusses directions for further research.

1 Work in CNET was partially supported by Esprit Project n° 2267 ISA

2. A basic model for distributed multimedia applications

2.1. A graph model

Media flow models appear useful to describe the handling of representation media in a distributed multimedia application (DMMA) [Her90]. The model proposed in [HHS91] describes a DMMA as a directed graph of processing modules through which media data flow. The arcs in the graph express unidirectionnal communication between the processing modules. The processing modules which may appear as nodes in the graph are : data sources (e.g. A/D converters, digital data generators, etc...), data sinks (e.g. D/A converters, storage drivers, etc...), data filters (e.g. compression/decompression modules), data transport, merging/multiplexing tasks, splitting/demultiplexing tasks.

This simple model is of interest for two reasons: firstly it can be used to decompose a DMMA into different elements; secondly it enables a precise localisation of all points where synchronisation constraints apply. Synchronisation constraints are identified as either elementary or induced: elementary constraints depend on the presentation activity and constitute absolute invariants on the correct behaviour of the application; induced constraints are derived through back-propagation from the elementary constraints.

Elementary constraints comprise:

- constraints which apply on continuous device interface, i.e. an interface that has a temporal behaviour imposed by the structure of the device (e.g. an audio card which raises an interrupt when a sample is needed);

- constraints which apply on discrete device interface, i.e. an interface that has no temporal behaviour (e.g. the interface of a video bitmap);

- concurrency presentation constraints either between several interfaces (e.g. lip-synch) or at the same interface (e.g. to insert a subtitle in a picture);

- interactivity constraints (e.g. response time between acquisition and its effect on restitution).

2.2. Programming synchronisation

Synchronisation constraints identified on a graph of a DMMA derive from the nature of the media handled by the application or originate directly from the application (e.g. to achieve certain presentation effects). In general, application programmers must deal with them and construct programs that ensure they are met. The approach in CNET is to consider that programming multimedia synchronisation is an instance of reactive programming, which is best achieved using synchronous languages [Ber89].

Synchronous languages are those which adopt the following set of semantical assumptions: a synchronous module accepts events from its outside environment and "instantaneously" responds by generating new events. Also, all events are propagated to all internal processes in zero time, and statement execution takes zero time. A representative synchronous language, Esterel [BeG88], is used throughout this paper and is briefly introduced in the annex.

Synchronous languages have attractive formal characteristics such as simple operational semantics and natural support for reasoning about real-time properties (for example, by preserving arithmetic properties on time). The synchronous assumptions can actually be met in practice as *'zero time'* effectively means *'less time than the inter-arrival gap between events from the surrounding environment'*. Synchronous programs are compiled into highly efficient finite state automata, and as long as these automata are capable of making transitions faster than state changes in the outside environment, the synchrony assumptions are effectively upheld.

Multimedia synchronisation constraints can thus be enforced using synchronous modules called reactive objects which provide real-time control. More generally, an application with multiple, distributed synchronisation constraints will comprise several interacting reactive objects controlling sets of local, non-reactive ones [SHH92]. All the objects communicate asynchronouly with each other, as in the standard ANSA architecture.

2.3. An example

We now consider an example distributed multimedia application and illustrate how it is described in terms of the graph model and synchronisation model. The example application is a remote learning scenario with a single teacher and a number of remotely sited students. The teacher controls the video display of a recorded ballet dance and may start, suspend, slow down, resume and stop the video which is being displayed on the students' workstations. On each student workstation there are two windows, one showing the ballet video and the other showing the lecturer captured live with a camera. The digital video is stored in a compressed format and is accessed from a central server.

The teachers voice and the video soundtrack are also relayed to the students. The voice is captured live and the musical track is stored in a MIDI file (in MIDI, tempo can be slowed down without affecting the tone). The voice and music are mixed on a sound-board equipped with a DSP and a FM synthesiser. Fig. 1a shows the general configuration of this system in terms of the graph model and localises the elementary synchronisation constraints of the application which must be realised.

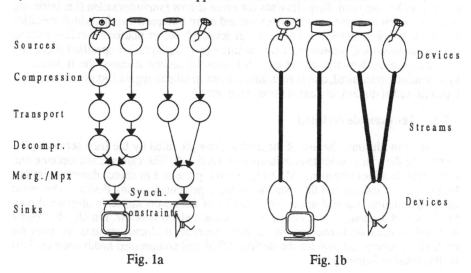

Fig. 1a   Fig. 1b

3. A Programming Platform for Distributed Multimedia Applications

3.1. The base services platform

The multimedia base services platform developed at Lancaster permits the design and implementation of distributed multimedia applications through the use of a standard set of services which have been added to the ANSA distributed systems abstractions [CBD92]. The platform consists of a set of ANSA services with location independent abstract data type interfaces. Application programmers build distributed multimedia applications by creating and configuring instances of the base service objects. The platform consists principally of four types of service: *devices, endpoints, streams and synchronisation managers.*

Devices are an abstraction of physical devices, segments of stored continuous media, or software processes. Devices have three interfaces: *endpoint* interfaces (see below), a standard interface for the control of continuous media information flow (e.g. start, stop, seek, re-wind etc.), and an optional device dependent interface which, in the case of a camera device, would contain operations such as focus, pan, tilt, etc...

Endpoints are connection points on to devices at which continuous media streams enter and leave. They are abstractions over resources allocated for continuous media production/consumption and over a quality of service associated with the interface. They are connected together by means of *stream* services.

Streams are abstractions over communication which may be either local IPC or an underlying transport protocol. The Stream interface contains operations to alter the QoS of connections, and also start and stop the flow of continuous media information. Streams support 1:1 connections, but the connectivity can be extended to arbitrary M:N connections through the use of *groups* which are a general purpose architectural facility for collecting interfaces together so that they may be invoked as a single unit.

Synchronisation Managers offer a service to which application synchronisation needs may be delegated. Requirements for *event-driven synchronisation* (i.e. initiating an action when an event occurs) are expressed as an interpreted script which specifies actions to be taken on the basis of events generated by other platform services such as devices or streams. *Continuous synchronisation* (e.g. 'lip-sync') is specified according to parameters such as the packet ratio between the related streams, the tightness of synchronisation required, and permissible actions to take to regain lost synchronisation (such as packet discard, increased throughput, etc...).

3.2. The example revisited

We can illustrate the use of the abstractions provided by the base services platform with the video ballet teleconference of section 2. The camera, microphone and file server data files (video and MIDI) are each represented as source devices. Similarly, the video windows and the soundboard are represented as sink devices. The video merging function required at the frame buffer of each workstation is abstracted over by the use of separate device abstractions for each video window. Finally, the source and sink devices are connected via streams. Separate multicast streams are used for each video source and also for the distinct MIDI and compressed audio sources. This is illustrated in figure 1b.

4. An integrated model

The ANSA based platform described in section 3 is well adapted to building distributed applications because of its flexibility, of the abstractions and the distribution transparency it provides. Programmers of distributed multimedia applications need not be aware, e.g. of the actual details of the communication mechanisms used, or of the actual locations of devices. Programming multimedia synchronisation, however, is an area which is currently not well covered by the base services platforms. Synchronisation managers provide a level of real-time control across streams but in a less general way than the use of synchronous programs. In particular, programming synchronisation managers scripts with the required level of rigour and precision in real-time control remains problematic. What is required is an integration of the base services platform with the synchronous programming of multimedia synchronisation advocated by CNET. This would result in a powerful infrastructure where the abstractions of the base platform are retained while providing a versatile and rigorous model for controlling the real-time aspects of multimedia applications.

In this section, we discuss the first steps towards such an integration. We first discuss the mapping between the graph model in section 2 and the abstractions provided by the base platform. We then consider elements of a computational model that integrates object types from the Lancaster platform with CNET's proposal for programming multimedia synchronisation. We use the teleconference example as an illustration.

4.1. Mapping the base platform abstractions and the graph model

The abstractions proposed in the Lancaster platform encapsulate the elementary processing modules identified in the processing graph of the basic model and the associated resources management. Devices abstract over a subgraph of processing modules such as merging/demultiplexing and sourcing/sinking. Streams appear as connections between arbitrary numbers of source and sink devices and thus abstract over transport and multicast/multidrop graph topologies. Stream may also encapsulate processing modules such as coding/decoding. In general, to meet a specified communication QoS, the abstract stream service can be realized using a number of engineering mechanisms, including transport connections, multiplexing, compression/decompression modules, etc.

Synchronisation constraints identified on the graph model now appear as constraints applying on the endpoint, device and stream interfaces provided by the base services platform. The level of synchronisation achievable in a given application will of course depend on the nature of the processing modules encapsulated by these interfaces. Enforcing synchronisation constraints is done by synchronisation managers in the base platform. Synchronisation manager scripts can thus be programmed using synchronous languages, following the proposal by CNET. Such scripts can be used to realize both event-driven and continuous synchronisations. This will be illustrated below with the teleconference example and is further discussed in [HoS92].

4.2. Synchronisation managers as reactive objects

The use of synchronous languages for the construction of synchronisation managers results in a hybrid computational model, containing reactive objects and non-reactive objects [SHH 92]. A reactive object behaves according to the synchronous semantical assumptions. Communication between objects (reactive or not) is based on operation invocation and is asynchronous (i.e. invoking an operation and receiving a result takes a non-zero, indeterminate amount of time), as in the standard ANSA computational model. In general, then, synchronisation management can be realized through asynchronously interacting reactive objects.

Overall real-time effects in a multimedia application are not realized only through the use of synchronisation managers. They also depend on the provision of temporal QoS guarantees for the execution of synchronisation managers and for the communication between objects (e.g. synchronisation manager to device) and within objects (e.g. end-to-end communication provided by a stream). A full computational model would contain means to express the required QoS guarantees, in particular, timing guarantees concerning the communication between reactive objects. Coupling the model with engineering support as suggested e.g. in [Kop92] would enable the construction of distributed applications with predictible real-time behaviour. This is outside the scope of this paper. Note however that the stream abstraction of the Lancaster platform indeed encapsulates regulation mechanisms and manages transport-level resources for the provision of a specified communication QoS.

Defining a synchronisation manager requires the identification of its interactions with other platform interfaces (devices, endpoints and streams). Interaction between these objects takes place of course through operation invocations. In this paper, our synchronous language of choice is Esterel. Operation invocations will thus appear in the form of emission or reception of *signals*, which is how modules communicate with their environment in Esterel (see Annex).

In general, the types of operations and the corresponding behaviour of the synchronisation manager depend on the application, on the representation media used, and on the nature of the processing modules encapsulated by streams and devices. Some of these modules may be application-dependent. For instance, if some special action must be taken on receipt of a video frame that contains a certain pattern (a face, say), then some means must be available in streams or devices to recognize that pattern.

It is however possible to specify generic Esterel modules that can be used in the construction of application-dependent synchronisation managers. These modules specify generic interactions (i.e. operation invocations or Esterel signals) between synchronisation manager and stream, endpoint, and device interfaces.

4.3. Generic modules for synchronisation management

4.3.1. The ENDPOINT module

The ENDPOINT module performs the following functions:
- (i) it controls the resources (in particular memory resources) shared between a device and a stream;
- (ii) it notifies the device synchronisation module that some data is available;
- (iii) it notifies the (application-specific part of the) application that an anomaly (or any remarkable event) has occured.

Control takes place in particular on buffers shared between the device and the stream. The application is entirely responsible for the action to be taken in case of anomaly (e.g. no free buffers may lead to either new buffer allocation, data drop-out or the end of the application).

The ENDPOINT module reacts to the following input signals :

- DATA_IN, which tells that new data or a specific event has been brought by the stream into the environment controlled by the endpoint, and

- DATA_OUT, which means that a piece of data has been consumed, i.e. has left that same environment.

Output signals are, on the one hand, DATA_AVAILABLE, to tell the device that a new data is available, and, on the other hand, INDICATION, to notify any anomaly.

Each occurence of the signals DATA_IN and DATA_OUT causes the execution of a function *flow_ctl ()*, which updates some state variables regarding the resources managed in the endpoint. If the device is free (as deduced by the execution of *flow_ctl ()* under DATA_OUT reception) and a new data is available, DATA_AVAILABLE is emitted. If a problem is detected, INDICATION is also emitted. To be generic, this module makes use of various external functions :

- *flow_ctl()*,

- *data_in()* and *data_out()* which extract from the signal values the information pertinent for the control of the endpoint and convert it into the appropriate format (input for *flow_ctl()*),

- *sig_indic()* and *sig_avail()* return boolean values for 'exception' raising and data availability,

- *indic_value()* and *data_header()* calculate the values for the signals INDICATION and DATA_AVAILABLE.

```
module ENDPOINT:
input  DATA_IN (data_header_type), DATA_OUT (data_header_type),
SET_EP_PARAM (ep_ctl_type);
output  INDICATION (indication_type),
        DATA_AVAILABLE (data_header_type);

signal RESOURCE_CTL (combine ctl_type with ctl_comb) in [
  every DATA_IN do emit RESOURCE_CTL (data_in (?DATA_IN)) end
||
  every DATA_OUT do emit RESOURCE_CTL (data_out (?DATA_OUT)) end
||
  var status: ctl_type in
    every RESOURCE_CTL do
      status:=flow_ctrl (?RESOURCE_CTL);
      if (sig_indic (status)=true)
        then emit INDICATION (indic (status)) end
      if (sig_avail (status)=true)
        then emit DATA_AVAILABLE (data_header (status)) end
    end     %every RESOURCE_CTL
  end       %var status declaration block
] end    % signal RESOURCE_CTL declaration block²
```

---

2 All pieces of code in this paper are only examples. To remain simple, they have been reduced and are not fully complete ; in particular, type and function declarations are missing.

4.3.2. The BASIC_DEVICE_SYNCHRO module

The behaviour of a device encompasses the presentation (acquisition) of data to (from) a physical interface according to a precise timing constraint, and the interaction with the stream/endpoint which delivers (consumes) the data. The timing constraint associated with a device is enforced by the BASIC_DEVICE_SYNCHRO module which is defined for a generic discrete device (bit-mapped video screens and MIDI synthesizers immediately restitute data presented at their physical interface), and also provides control functions such as starting and stopping.

The BASIC_DEVICE_SYNCHRO module builds a logical device able:

- (i) to delay the presentation of data delivered before schedule, and

- (ii) to inform the (application-specific part of the) application when the next data unit is to be presented.

It is assumed here that this logical device can only handle a single unit of data; the application must then group into a single block all data to be presented simultaneously (several MIDI notes, for example).

The principle for synchronisation is the following : the module updates two local dates. The proper date of presentation of a new available data is compared to the current date. If it is already too late, an exception signal is emitted and as soon as the current date has reached an acceptable value, the presentation is requested.

Input signals are as follows : DEV_START et DEV_STOP permit the device to be started/stopped ; DATA_AVAILABLE informs the device that data is available for presentation (its value contains the data presentation date) ; TB_TICK is the device clock signal.

Output signals are : PRESENT_DATA, which is the request to present the data to the physical device interface, and INDICATION, which is the signal warning the environment that an anomaly has been detected.

The module presented below uses some external functions: *fdate()* which extracts the presentation date from the value of the DATA_AVAILABLE signal, *ftick()* calculates a date value. *f_toolate()* and *f_not_ahead()* are boolean evaluations with the values of CURRENT_DATE and PRESENT_DATE local signals.

```
module BASIC_DEVICE_SYNCHRO :
input    TB_TICK (...), DEV_START, DEV_STOP, DATA_AVAILABLE (...);
output   PRESENT_DATA (...), INDICATION (...);

every DEV_START do
do
  signal CURRENT_DATE (...), PRESENT_DATE (...) in [
    every DATA_AVAILABLE
         do emit PRESENT_DATE (fdate (?DATA_AVAILABLE)) end
    ||
    every TB_TICK do emit CURRENT_DATE (ftick (?TB_TICK)) end
    ||
    every PRESENT_DATE do
      trap PRESENTED in [ loop
        if (f_toolate (?PRESENT_DATE, ?CURRENT_DATE)
          then emit INDICATION (late_msg (?DATA_AVAILABLE)) end
```

```
        if (f_not_ahead (?PRESENT_DATE, ?CURRENT_DATE)
          then [ emit PRESENT_DATA (?DATA_AVAILABLE);
                exit PRESENTED ] end
        await TB_TICK end % loop
     ] end % trap block
    end % every PRESENT_DATE
  ] end         % internal signals declaration
watching DEV_STOP
end     % every DEV_START
```

4.4. The example completed

We shall now study how the generic modules defined in the previous section can be used to build the synchronisations needed for our teleconference. For the sake of simplicity, we will only consider the synchronisation required by the presentation of the stored data (actually, moving video and MIDI data) and ignore the problem raised by the copresentation of the other media (e.g. merging on the screen video data coming form several sources).

The device, stream and endpoint interfaces necessary for our example have been identified in section 3. According to our integrated model, the application behaviour (and corresponding synchronisation) is obtained in each client workstation by a local reactive object that realizes:

- (i) an *intramedium* synchronisation, which is required to achieve a correct presentation of each medium,

- (ii) an *intermedia* synchronisation (between moving video and MIDI).

Both synchronisation must be maintained in case of suspend/resume, slow_down/speed up request from the teacher.

In this example, the two forms of synchronisation rely on a single datation mechanism: the reactive object constructs in each workstation a local current date, from ticks generated by an external clock, and synchronizes the data presentation activity using this date, which may be an absolute time, a duration elapsed since the beginning of presentation, etc.... This datation mechanism is the basis for the presentation of both video and MIDI data.

4.5. Programming intramedium synchronisation

The BASIC_DEVICE_SYNCHRO module controls data presentation on the physical device interface, through a current date which must be regularly updated. The TB_GENERATOR module is responsible for this function and, thus, constitutes the actual internal clock for the device.

4.5.1. The Time-Base module : TB_GENERATOR

The time interval between two updates of the internal clock is defined by the temporal granularity of the medium (e.g. 40 ms for a 25 image/sec video). The module receives one input signal CLOCK_TICK bound to a physical clock (e.g. a quartz) and emits the output signal TB_TICK, the value of which is the associated device specific date.

This module is responsible for time control and has in charge the control of operations such as slow_down/speed_up and suspend/resume. Slow_down/speed_up operations appear in the form of a single valued input signal SET_TB_PARAM, which modifies the parameters used in the generation of the signal TB_TICK. The occurence of the SUSPEND signal freezes the value of current date and inhibits TB_TICK emission until the occurence of the RESUME signal. Note that only the general structure of the module is described here.

```
module TB_GENERATOR:
input  CLOCK_TICK,SUSPEND,RESUME, SET_TB_PARAM;
output TB_TICK (...);
  every SET_TB_PARAM do
    <compute new parameters controling the generation of TB_TICK>
    end    %every SET_TB_PARAM
||
loop
  do

    every CLOCK_TICK do
      <update the value of variables and local signals
      controling the generation of a TB_TICK signal>
      if (it is time to emit TB_TICK) then emit TB_TICK (current date)
    end %every CLOCK_TICK
  upto SUSPEND;
  await RESUME
end    %loop
```

4.5.2. The DEVICE module

Intramedium synchronisation is obtained by combining the blocks defined above (i.e. ENDPOINT, BASIC_DEVICE_SYNCHRO and TB_GENERATOR). This is simply achieved by means of the Esterel parallel operator. The signals DATA_AVAILABLE et TB_TICK become local to the synchronisation manager and thus an instantaneous communication is provided. The resulting module is defined as follows :

```
module DEVICE:
input  CLOCK_TICK, DEV_START, DEV_STOP, SUSPEND,RESUME,
       SET_DEV_PARAM (...),DATA_IN (...),DATA_OUT (...);
output PRESENT_DATA (...),INDICATION (...);

signal DATA AVAILABLE in [
  signal TB_TICK in [
    copymodule BASIC_DEVICE_SYNCHRO
    ||
    copymodule TB_GENERATOR
  ] end    % signal TB_TICK
||
  copymodule END_POINT
] end % signal DATA_AVAILABLE
```

This simple example does not consider error cases. We can note, however, that signals such as INDICATION, SUSPEND, RESUME or SET_DEV_PARAM provide a basis for exception handling, either within the synchronisation manager, or outside of it.

Fig. 2 :

Signals
involved in
device synch.

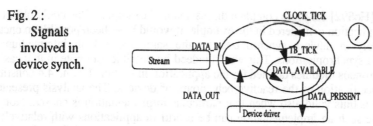

4.6. Intermedia synchronisation

The two devices (video and MIDI) being controlled through a similar datation system, intermedia synchronisation is achieved by combining the two DEVICE modules using the Esterel parallel operator. This combination has the effect of arranging for the TB_GENERATOR modules to be driven by the same physical clock and thus sufficient conditions for synchronisation between the two media are satisfied:

- (i) the two devices are started and stopped together, and

- (ii) when the devices are running, the flow of media proceeds at the correct rate.

```
module VM:
input    START, STOP, SUSPEND, RESUME,
         MIDI_DATA_IN (...), MIDI_DATA_OUT (...),
         VIDEO_DATA_IN (...), VIDEO_DATA_OUT (...),
         CLOCK_TICK;
output   MIDI_PRESENT (...), MIDI_INDICATION (...),
         VIDEO_PRESENT (...), VIDEO_INDICATION (...);

  copymodule VIDEO_DEVICE [signal VIDEO_PRESENT/PRESENT_DATA,
                  START/DEV_START,STOP/DEV_STOP,
                  VIDEO_INDICATION/INDICATION,....]
||
  copymodule MIDI_DEVICE [signal MIDI_PRESENT/PRESENT_DATA,
                  START/DEV_START,STOP/DEV_STOP,
                  MIDI_INDICATION/INDICATION,....
```

5. Conclusion

We have briefly reviewed in this paper two ANSA-based approaches for programming distributed multimdia applications. The CNET proposal is based on a formal hybrid computational model where real-time control is realized by reactive objects (programmed using synchronous languages). The Lancaster platform extends the ANSA engineering structure and provides programmer-oriented features for the organization and configuration of multimedia applications. We have described a mapping between the two sets of abstractions which paves the way for the development of a future integrated platform.

In particular, we have shown how synchronisation management can be realized in the Lancaster platform using synchronous programming. The example presented contained both intramedium and intermedia synchronisation. Intermedia synchronisation in the example, however, was limited to a form of continuous synchronisation. The same principles would of course apply easily well for event-driven synchronisation (see [HoS92] for more complete discussion and example). One could object that, for the application considered in the example, it would have been possible to encapsulate entirely in devices synchronized by the same clock both intramedium and intermedia synchronisations. This would indeed be possible, but note in that case that the synchronous modules defined for the application in section 4.3 and 4.4 constitute a formal specification of the reactive behaviours of devices. The analysis presented in the paper is thus valid also when that particular implementation is chosen. Note also that, while such an implementation can be useful in applications with relatively low level of interactivity and simple multimedia document structure, it is useless if more complex forms of synchronisation are required.

From the analysis presented in the paper, it seems both possible and desirable to construct a distributed multimedia platform which fully integrates the two approaches, and which retains the benefits of both. To be complete, however, the corresponding integrated model should investigate further issues of dynamic configuration in the case of complex multimedia synchronisations, and of QoS support. The development of a complete integrated model is now well underway and will be reported in [CBH92]. We are currently investigating means to formally express QoS requirements, and developing an experimental implementation of our integrated platform using the Chorus micro-kernel [Cho90]. Future work will be targeted particularly at support for performance guarantees and QoS driven resource allocation in the Chorus micro-kernel environment.

BIBLIOGRAPHY

[ATW90]   D.P. Anderson, S.Y. Tzou, R. Wahbe, R. Govindan and M. Andrews, "Support for Continuous Media in the DASH System", Proc. of the 10th International Conference on Distributed Computing Systems, Paris, May 1990.

[ANS89]   ANSA. "ANSA Reference Manual, Release 01.00", APM Cambridge Ltd. Mar. 1989.

[BeG88]   G. Berry, and G. Gonthier. "The ESTEREL Synchronous Programming Language: Design, Semantics, Implementation", INRIA Report No. 842, INRIA, 1988.

[Ber89]   G. Berry, "Realtime programming: special purpose or general purpose languages", Proceedings IFIP Congress 1989, North Holland.

[CBD92]   G. Coulson, G.S. Blair, N. Davies, and N. Williams. "Extensions to ANSA for Multimedia Computing", To appear in Computer Networks and ISDN Systems also available as internal report MPG-90-11, Computing Department, Lancaster University, Bailrigg, Lancaster LA1 4YR, UK, Oct. 1990.

[CBH92]   G. Coulson, G.S. Blair, F. Horn, L. Hazard, and J.B. Stefani. "Supporting the Real-Time requirements of Continuous Media in Open Distributed Processing", in preparation, 1992.

[Cho90]   Chorus systemes. "An Overview of the CHORUS Distributed Operating System", Chorus Systemes CS/TR-90-25, St Quentin en Yvelines, France, 1990.

[DCW91]   N. Davies, G. Coulson, N. Williams, and G.S. Blair. "Experiences of Handling Multimedia in Distributed Open Systems", Presented at SEDMS '92, Newport Beach, CA; also available from Computing Department, Lancaster University, Bailrigg, Lancaster LA1 4YR, UK, Nov. 1991.

[HDJ91]   J.G. Hanko, D. Berry, T. Jacobs, and D. Steinberg, "Integrated multimedia at SUN microsystems" in Proceedings IWNOSSDAV 91, LNCS 614.

[Her90]   R.G. Herrtwitch. "Timed Data Streams in Continuous Media Systems", TR-90-026, International Computer Science Institute, May 1990.

[HHS91]   L. Hazard, F. Horn, and J.B. Stefani. "Notes on Architectural Support for Distributed Multimedia Applications", CNET/RC.W01.LHFH.001, Centre National d'Etudes des Telecommunications, Mar. 1991.

[HHSSS91] D. Hehmann, R.G. Herrtwich, W. Schulz, T. Schutt, and R. Steinmetz. "Implementing HeiTS: architecture and implementation strategy of the Heidelberg High-Speed Transport System" in Proceedings IWNOSSDAV 91, LNCS 614.

[HoS92]   F. Horn, and J.B. Stefani. "On programming and supporting multimedia object synchronization" to appear in special issue of Computer Journal.

[Kop92]   H. Kopetz. "Sparse Time versus Dense Time in Distributed Real-Time Systems", IEEE Proceedings of 12th International Conference on Distributed Computing Systems, Yokohama, Japan, 1992.

[Mei91]   K. Meissner. "Aspects of Multimedia Application support in UNIX", Multimedia in Action, 5th International Protext Conference, Luxembourg, Nov. 1991.

[SHH92]   J.B. Stefani, L. Hazard, and F. Horn. "A computational model for distributed multimedia applications based on a synchronous programming language" in Computer Communications, vol 15, n 2, Mar. 1992.

[Ste90]   R. Steinmetz. "Synchronization Properties in Multimedia Systems", IEEE Journal on Selected Areas in Communications 8, 3, Apr. 1990.

Annex: ESTEREL syntax

This annex presents the basic knowledge of ESTEREL programming necessary to understand the examples presented throughout the paper. Interested readers can find a complete description in [BeG88].

An ESTEREL program can be viewed as a collection of parallel processes which communicate instantly with each other and with the environment, via broadcast valued signals. Broadcast signals are the only communication means provided by ESTEREL: local signals correspond to the internal communication between processes and input/output signals to communication with the environment. There is no absolute time in ESTEREL: it only appears as input signals generated by an external clock.

A signal is broadcast through the instruction **emit** (for example, **emit** SIG (exp) means the broadcast of signal SIG with a value exp. Signals carry 2 types of information: (i) a control information telling whether a signal is present during an execution instant or not present (ii) a value information which is persistent and may be accessed as any other variable or constant. The current value of a signal is accessed through the operator ?SIG.

Reaction to signals relies on a unique basic primitive: **do** *stat* **watching** [exp] SIG, where *stat* is a statement, exp an optional integer expression and SIG a signal. Stat starts immediately; if *stat* finishes before the next (the exp^{th}) occurrence of the signal SIG, so does the "do watching" statement. Otherwise the "do watching" statement is terminated as soon as the next (the exp^{th}) occurrence of SIG occurs. In this case, *stat* is immediately killed (without terminating what has not yet been executed).

The language possesses imperative statements, among which:

-the assignment: X:= *expression*

-the conditional instruction: **if** exp **then** *stat1* **else** *stat2* **end**.

-the infinite loop: **loop** *statement* **end**.

-the halt instruction: **halt**. It is the only instruction that consumes time.

-the sequence operator: *stat1* ; *stat2*. It starts *stat2* as soon as *stat1* is completed.

-the parallel operator: *stat1* || *stat2*. It starts instantly both *stat1* and *stat2*. It terminates, when both *stat1* and *stat2* are completed.

ESTEREL provides a powerful exception handling mechanism: **trap T in stat end** combined with an **exit** T. The trap declares an exception while an exit raises an exception. The body of the **trap** construct is executed normally until an **exit** T is encountered. The body is then immediately terminated.

A statement can be simple or compound: it consists of a block of instructions delimited by the tokens "[" and "]".

Basic statements have been combined to define user friendly compound statements. 3 examples are:

* **do stat1 watching [exp] SIG timeout stat2**, which extends the basic do watching construct: *stat1* started immediately. If *stat1* finishes before the next (the expth) occurrence of the signal SIG, so does the statement. Otherwise *stat1* is immediately killed and *stat2* started. It is equivalent to:
trap T **in** [**do** [*stat1*; **exit** T] **watching** SIG; *stat2*] **end**

* **await** <exp> SIG, which blocks a process until the next (the expth) occurrence of SIG. It is equivalent to:
   do halt watching [exp] SIG

* **every** <exp> SIG **do** *stat* **end**, which is equivalent to:
   loop [**await** <exp> SIG; *stat*] **end**

The standard unit of programming in ESTEREL is the module. ESTEREL possesses a copymodule statement, which provides for in-place expansion, with constant and signal renaming facilities. It has the syntax: **copymodule** module_ident [substitution-list], where a substitution can be either a signal substitution or a constant substitution. The module is then textually copied with the indicated substitutions.

copymodule simple_example [**signal** NEW_SIG1/SIG1, NEW_SIG2/SIG2;
                             constant NEW_CT1/CT1]

This construct promotes the definition of libraries of modules which capture basic and general functionalities e.g. device or medium types.

Scheduling Multimedia Documents Using Temporal Constraints

M. Cecelia Buchanan and Polle T. Zellweger

Information Sciences and Technologies Laboratory
Xerox PARC, 3333 Coyote Hill Road, Palo Alto, CA 94304

Abstract. We describe a scheduling algorithm, based on linear programming, for solving a network of temporal constraints among media in multimedia documents. This algorithm lays out media events and media segments in time so as to conform to a set of constraints specified by the author. This temporal layout is conceptually similar to TEX's spatial layout algorithm, in that it permits time to be stretched or shrunk between events inside media segments to arrive at an "optimal" display for a document. Asynchronous events, such as user interaction, or asynchronous segments, such as programs, can be included.

1 Introduction

Multimedia documents increase the expressive power of documents by allowing the inclusion of dynamic media, such as audio, video, and programs, as well as the traditional static media, such as text and images. The recent proliferation of multimedia hardware and software has made it possible for more people to produce these documents. Unfortunately, the additional power of multimedia documents can be difficult to wield. Two major problems face authors trying to create multimedia documents.

The first problem is *rich media synchronization*. Authors should be able to schedule multimedia events to occur in specific relationships to one another. For example, a bell should ring when its animated clapper touches its side, or a subtitle for a video clip should appear at a certain time.

We believe multimedia documents should be able to include asynchronous information (that is, information whose time of occurrence or duration cannot be determined in advance), such as user interaction and programs. The ability to include programs in documents allows documents to tailor their contents to individual readers or to keep their contents constantly up-to-date. Most existing systems exclude asynchronous information because it makes media synchronization more difficult.

In addition, we believe multimedia documents should support fine-grained synchronization so that authors can synchronize media segments at internal points. Current systems often permit synchronization only at the start or end of a media segment. Finally, authors should be able to specify temporal relationships between asynchronous events and other multimedia events.

The second problem is *authoring ease* during document creation and maintenance. Current systems typically require authors to position their media segments in time manually, and to edit the media segments as needed to satisfy their desired synchronization relationships. Instead, the system should be able to satisfy these relationships automatically by positioning the media, modifying playback rates, discarding data, inserting delays, or performing alternative actions. Several systems provide this capability at runtime to allow media segments to resynchronize following network or operating system delays [2, 15]. In addition, authors should be able to change a multimedia document as easily as possible. In current systems, a small change may require a large effort to re-specify the media synchronization relationships.

To address these problems, we have designed and implemented a multimedia document system called Firefly. Our aim is to enable authors to create richer multimedia documents more easily. Firefly uses an improved multimedia document model that explicitly maintains information about events of interest within media segments and the temporal relationships among those events. Firefly provides a temporal constraint satisfaction algorithm, based on linear programming, that precomputes an "optimal" schedule for controlling the display of a document. During the scheduling process, Firefly can negotiate with the media segments to adjust their durations within an allowable range, if a schedule cannot be computed otherwise.

Firefly addresses the problems of multimedia document specification and scheduling rather than network and operating system level media synchronization and resource allocation. These underlying problems are being investigated by numerous researchers [1, 8, 9, 12, 13]. Firefly's modular architecture will allow it to take advantage of these research results as they become available.

This paper describes Firefly's automatic scheduling algorithm in detail. Section 2 provides an overview of the Firefly multimedia document system. Section 3 describes the scheduling algorithm, and Section 4 presents our initial experiences using it. Section 5 compares Firefly to existing systems. Elsewhere we explain how Firefly eases document creation and maintenance and how Firefly can be used to specify hypermedia documents [3].

2 The Firefly Multimedia Document System

We have designed a system called Firefly for creating, editing, and displaying multimedia documents. A prototype of this system has been implemented in the Cedar programming environment [17] on Sun workstations.

This section presents the Firefly multimedia document system, starting with its document model and continuing with its three main components: authoring tools for creating and maintaining documents, analysis tools for scheduling and debugging documents, and the runtime system for displaying documents.

2.1 The Firefly Document Model

In our model, a document consists of three parts: media items, temporal synchronization constraints, and operation lists.

A *media item* describes the temporal behavior of a piece of information in a document. For example, a media item can describe a video clip, an audio recording, a text file, a program, or a hyper/multimedia document. Media items appear by reference in documents to permit a single media item to be used either in multiple documents or repeatedly in one document. A media item has four parts:

- a *media type*, which indicates the specific kind of medium the media item represents. Firefly uses object-oriented techniques to integrate a variety of media types including text, 2D images, audio, animations, and programs.
- *events*, which represent points at which a media item can be synchronized with other media items. The implementor of each media type determines the granularity at which events can be placed. For example, an event can mark a frame in a video media item or a region in an image media item. Events may be either *synchronous*, meaning their temporal placement is known in advance, or *asynchronous*, meaning their time of occurrence cannot be determined until the media item is displayed.
- *procedures* that operate on a media item and its events. These procedures allow the underlying media items to participate in the Firefly system. They can be classified as: user interface, analysis, control, and runtime. *User interface* procedures support the creation and editing of media items and events. *Analysis* procedures provide information such as the duration between two events or the ordering of events. *Control* procedures affect the display behavior of a media item. All media types must provide control procedures to start, end, pause, and resume the display of a media item. In addition, media types can supply type-specific control procedures, such as increasing the volume of an audio playback or zoom-in on a region of an image. *Runtime* procedures allow media items to communicate with each other based on their current state, and to detect and report the occurrence of asynchronous events. Typically, new procedures must be specified only when new media types are added to the system.
- a *pointer* to the actual data described by the media item. Including the data by reference supports the use of existing media editors and allows the data to be used in other contexts and applications as well.

Temporal synchronization constraints specify the temporal ordering of pairs of events in one or more media items. Each temporal constraint is directed from a *source* event to a *destination* event. Firefly supports two classes of temporal synchronization constraints: *temporal equalities*, such as requiring that two events e_{src} and e_{dest} occur simultaneously or that e_{src} precede e_{dest} by a fixed amount of time, and *temporal inequalities*, such as requiring that e_{src} precede e_{dest} by some (unspecified) duration, that e_{src} precede e_{dest} by at least 15 seconds, or that e_{src} precede e_{dest} by at least 10 and no more than 20 seconds.

Ordered *operation lists* can be associated with any event to control that media item's display behavior, as described above. A media type's control procedures define the allowable operations for all events in media items of that type. Operation lists are kept separately from the media items to allow multiple documents to control shared media items differently.

2.2 Authoring Tools

Firefly provides two direct manipulation tools for multimedia document authoring: a temporal view of a media item and an interactive document editor.

Creating media items. Authors interact with existing media editors to create and edit media items, and to mark points of interest, called *events*, within them. Firefly currently supports four media editors: the Tioga text editor, the Color Trix 2D image editor, the TiogaVoice audio editor [18], and a clock animation editor. The media types supported by these editors are representative of the full range of possible temporal behaviors, allowing us to test the components of our prototype and validate its functionality. New media types, such as video, can be added as they become available.

In response to the author's actions in a media editor, Firefly creates and updates a temporal view of a media item. This temporal view uses a graph notation in which square nodes represent the start and end events, and circular nodes represent internal events. Graph edges represent the temporal adjacency of two events; the length of each edge is proportional to the duration between the events. If a media item contains asynchronous events, those nodes float above its start node because their time of occurrence is unpredictable.

Constructing multimedia documents. Authors use an interactive document editor to specify the temporal behavior of a document. The document editor supports adding media items to a document, placing temporal synchronization constraints between events to specify their temporal ordering, and adding operation lists to events to control the display behavior of a media item.

Figure 1 shows a sample Firefly document, represented using Firefly's graph notation. This document contains an audio recording of a lecture on electricity and magnetism, an animation showing the operation of a battery, a textual technical dictionary defining relevant terminology, and two asynchronous media items: a calendar containing information about scheduled talks and an audio controller supporting audio connections to other parts of the building. The term index has been constrained to appear 10 seconds after the lecture starts, and disappear 30 seconds after the lecture ends. The battery animation has been constrained to display while the lecture describes how batteries operate. The Define Electrolyte asynchronous event occurs when a reader requests the definition of the term "electrolyte". This asynchronous event triggers the display of the "electrolyte" definition. If a related talk starts while the user is viewing the physics document, the calendar pauses this document and opens an audio connection to the appropriate conference room. This connection remains open until the user explicitly asks to leave the lecture. At that point the connection is terminated and the physics lecture is resumed.

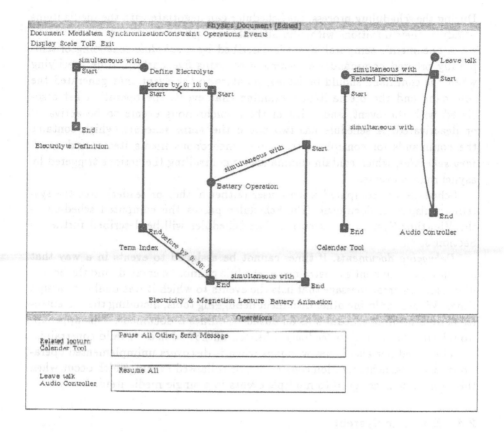

Fig. 1. A sample document represented in Firefly's graph notation.

Certain constraints involving asynchronous events represent impossible temporal behavior and are not allowed. Asynchronous events can only appear as the source event in a synchronization constraint. In addition, constraints involving an asynchronous event cannot specify that the destination precedes the source event. Currently, asynchronous events can start, end, pause, or resume the display of a media item, or send a message to another media item.

2.3 Analysis Tools

Firefly's analysis tools automatically generate schedules for controlling the display of a document and provide a rudimentary document debugger.

Constructing schedules. Firefly includes a Scheduler component that solves temporal constraints to create a temporal document layout, much as a document formatter like TeX solves spatial constraints to create a spatial document layout.

During the scheduling process, the Scheduler can negotiate with the media items to adjust their durations within an allowable range.

A document's temporal layout is described by a *schedule* consisting of temporally ordered commands. A *command* contains five parts: a time specifying when the command should be issued, pointers to the event that generated the command and the media item containing that event, the operation list associated with the event, and a list of the asynchronous events to be activated or deactivated. A schedule has two parts: the *main schedule*, which contains the commands for controlling the main synchronous media items; and *auxiliary schedules*, which contain commands for controlling the actions triggered by asynchronous events.

Schedules are computed when a user (either author or reader) asks the system to display a document. The Scheduler passes the computed schedule to the Viewtime System for execution. The Scheduler will be described further in Section 3.

Debugging documents. If times cannot be assigned to events in a way that satisfies the document's constraints, a schedule cannot be created, and the Scheduler issues an error message that lists the events to which it was unable to assign times. We are exploring other document debugging tools, including the incremental computation of schedules as an author modifies a document. This approach would provide immediate feedback when an author adds an invalid constraint.

The Scheduler also issues warnings when it discovers unimplemented operations, invalid synchronization constraints, or collapsed events, which occur when the same time is assigned to multiple events in a single media item.

2.4   Runtime System

Firefly's runtime architecture consists of three components: the Viewtime System, which manages a document's schedule; the Event Handler, which filters asynchronous events; and the media items, which execute the operation lists and recognize the occurrence of asynchronous events.

Issuing commands. The Viewtime System initiates the display of a document when it receives a schedule from the Scheduler. At that time, the Viewtime System places the commands from the main schedule into the schedule, registers the document's asynchronous events with the Event Handler, and starts the document clock. The document clock is incremented until the display of the document is completed, or until the reader explicitly pauses or terminates the display of the entire document. For paused documents, the document time remains fixed until the reader resumes the display of the document.

The Viewtime System executes commands from the schedule when the document clock reaches the specified time. Commands are executed in two steps. First, the Viewtime System activates or deactivates any specified asynchronous events. Although command data structures permit any event in a media item to activate or deactivate asynchronous events, Firefly's user interface does not yet support this feature. Thus, all asynchronous events associated with a media item are activated at the start of the media item and deactivated at its end. Improved user interface tools are under development. Second, the Viewtime System issues the specified operations by invoking the appropriate control procedures provided by the media item.

Media items can communicate with each other using the SEND MESSAGE operation. To provide the maximum flexibility, media items can dynamically generate messages at runtime. When the Viewtime System encounters a SEND MESSAGE operation, it invokes the sender's GENERATE MESSAGE procedure to determine the appropriate message. Messages may be sent to specific media items or broadcast to all active media items.

Handling asynchronous events. The system cannot know in advance when asynchronous events will occur. Therefore, commands triggered by asynchronous events are maintained in separate auxiliary schedules, which are merged into the schedule whenever the corresponding asynchronous event occurs.

When a media item detects the occurrence of an asynchronous event, it notifies the Event Handler, which maintains a list of active and inactive asynchronous events. If the asynchronous event is inactive, it is ignored. Otherwise, the Event Handler notifies the Viewtime System, which merges the auxiliary schedule for that asynchronous event into the schedule. Commands in auxiliary schedules refer only to *relative times* (i.e., offsets from the asynchronous event occurrence). The Viewtime System converts these to *absolute times* by adding the current document time.

3 Firefly's Automatic Scheduler

The primary component of Firefly's analysis tools is the Scheduler, which automatically produces a schedule for controlling the display of a Firefly document. The scheduling algorithm has four steps: computing media item durations, finding connected components, assigning times to events, and creating commands.

Computing media item durations. The Scheduler begins by querying the media items to obtain the duration between each pair of temporally adjacent events. For each pair of temporally adjacent events, a media item provides the Scheduler with a minimum, optimum, and maximum allowable duration as well as two cost measures: one for stretching the duration beyond the optimum value and another for shrinking the duration. The duration between any two events e_i and e_j must be one of:

known duration. Used by media items with a predetermined temporal behavior, such as audio or video. This duration is specified as:
$$[minDuration_{i,j}, optDuration_{i,j}, maxDuration_{i,j}]$$
where: $0 \leq minDuration_{i,j} \leq optDuration_{i,j} \leq maxDuration_{i,j} \leq \infty$.
If the media type cannot adjust its playback rate, then
$$minDuration_{i,j} = optDuration_{i,j} = maxDuration_{i,j}.$$
variable duration. Used by static media items that do not have a temporal behavior, such as images or text. This duration is specified as:
$$[minDuration_{i,j}, unspecified, maxDuration_{i,j}]$$
where: $0 \leq minDuration_{i,j} \leq maxDuration_{i,j} \leq \infty$.

potentially infinite duration. Used by media items with potentially infinite durations that must be explicitly terminated, such as a radio. This duration is specified as: $[\infty, unspecified, \infty]$.

unknown duration. Used by asynchronous media items with unknown durations, such as programs. This duration is specified as: $[\infty, \infty, \infty]$.

Media items must adjust their durations to reflect the presence of time-altering operations, such as INCREASE PLAYBACK SPEED or PAUSE 10 SECONDS, and to account for media delays, such as the need to spin up video playback heads.

By allowing media items to specify a range of durations rather than a single value, the Scheduler can adjust the durations of media items to generate a schedule when one cannot be created otherwise. This ability to stretch or shrink the duration of a media item to produce an optimal temporal layout is conceptually similar to TEX's ability to stretch or shrink the amount of white space (glue) between items to be typeset (boxes) to produce an optimal spatial layout [10].

The Scheduler is concerned only with the amount by which it can stretch or shrink the duration of a media item. Each media item determines whether its duration can be modified and how that modification is achieved[1]. For example, media such as audio or video can adjust their durations by varying the playback rate. If the playback rate cannot be modified, the duration of a media item could be shrunk by displaying only a portion of the data, or stretched by specifying an alternative action to be performed when there is no more data to display [15].

Finding connected components. The Scheduler uses a Union-Find algorithm to partition a document into *connected components*, consisting of temporally related events. No known temporal relationships exist between two different connected components. Initially, each event is placed in a separate connected component. The Scheduler identifies temporally related events by examining the document's synchronization and duration constraints. In general, two events are placed in the same connected component if they are connected by a temporal synchronization constraint or they are temporally adjacent events in the same media item. However, potentially infinite or unknown durations are ignored because the actual temporal relationship between the two events will not be known until runtime.

For example, the document shown in Figure 1 contains six connected components. The first contains the Lecture, the Animation, and the Term Index (excluding the Define Electrolyte asynchronous event) media items. The second contains the Definition media item and the Define Electrolyte asynchronous event. The Calendar Tool's Start and End events are placed in the third and fourth connected components, respectively. The fifth connected component contains the Related Lecture asynchronous event in Calendar Tool media item and the Start event in the Audio Controller. The sixth connected component contains the Audio Controller's Leave Talk asynchronous and End events.

---

[1] In the future we plan to allow authors to specify the adjustment behavior for any interval.

Assigning times to events. To determine the time at which events within a connected component must occur, the Scheduler combines the duration constraints supplied by the media items with the temporal synchronization constraints specified by the author. The Scheduler solves these's constraints using a linear program whose objective function minimizes the overall cost of stretching and shrinking media item durations.

For each event e_i the Scheduler computes the time of occurrence, $time_i$, and the amount of time this event occurs prior to or later than its optimal time of occurrence, $amtEarly_i$ and $amtLate_i$ respectively. The time of occurrence is relative to the start of the connected component containing that event. For any event e_i, either the $amtEarly_i$ and the $amtLate_i$ are both zero or at most one of these values is positive, non-zero.

The linear program represents the duration between temporally adjacent events e_i and e_{i+1} in a single media item as follows.

Case 1: No optimum duration is specified.

Two constraints are added to ensure that:

1. the duration never falls below the minimum duration
   $$time_{i+1} - time_i \geq minDuration_{i,i+1}.$$
2. the duration never exceeds the maximum duration
   $$time_{i+1} - time_i \leq maxDuration_{i,i+1}.$$

Because no optimal value is specified, there is no cost associated with stretching or shrinking the duration of this interval. Therefore, no new terms are added to the objective function.

Case 2: An optimum duration is specified.

Three constraints are added to specify:

1. an upper bound on amount the duration is shrunk
   $$amtEarly_{i+1} \leq optDuration_{i,i+1} - minDuration_{i,i+1}.$$
2. an upper bound on amount the duration is stretched
   $$amtLate_{i+1} \leq maxDuration_{i,i+1} - optDuration_{i,i+1}.$$
3. the optimal duration constraint
   $$time_{i+1} - time_i + amtEarly_{i+1} - amtLate_{i+1} = optDuration_{i,i+1}.$$

The objective function is updated to include the terms:
$$costShrink_{i,i+1} * amtEarly_{i+1} + costStretch_{i,i+1} * amtLate_{i+1}.$$

The linear program represents a synchronization constraint between two events e_{src} and e_{dest} as follows.

Case 1: simultaneous with: $\quad time_{dest} - time_{src} = 0.0$

Case 2: before by t: $\quad time_{dest} - time_{src} = t$

Case 3: before by at least t: $\quad time_{dest} - time_{src} \geq t$

Case 4: before by no more than s: $\quad time_{dest} - time_{src} \leq s$

Case 5: before by at least t and no more than s: $\quad time_{dest} - time_{src} \geq t$
$$time_{dest} - time_{src} \leq s$$

Creating commands. The Scheduler generates *commands* for each connected component by visiting the events in temporal order, producing at most one command per event. Commands triggered by asynchronous events are placed in that event's auxiliary schedule; all other commands are placed in the main schedule.

The Scheduler must now adjust the media playback rates to obtain the durations computed above. For each event e_i with either $amtEarly_i \neq 0$ or $amtLate_i \neq 0$, the Scheduler adds a CHANGE RATE operation to the operation list for that event. If event e_i is not the end event for this media item, the Scheduler must restore the playback rate to its optimal value. To accomplish this, the Scheduler locates the next event e_k in the same media item for which $amtEarly_k = amtLate_k = 0$ and adds a RESTORE RATE operation to its operation list.

Returning to our sample document, Table 1 contains the commands computed for the six connected components. The Scheduler places the commands for connected components 1 and 3 in the main schedule. The remaining commands are placed in auxiliary schedules.

4 Preliminary Implementation Experience

Our Scheduler has been operational for several months. During that time we have used Firefly to create a variety of multimedia and hypermedia documents.

Algorithm performance. Our informal evaluation of the Scheduler's performance is based on testing and demonstration documents containing approximately 10 constraints and 15 media items each with an average of 5 events. For these documents, the Scheduler computes schedules in well under a minute. A more formal analysis of the algorithm is planned [4].

Representing static media items. We initially represented durations between events in static media items as variable durations. Because no cost is associated with stretching or shrinking variable durations, the Scheduler will attempt to adjust these durations first. Unfortunately, if media items with variable durations do not inherit temporal behavior by being constrained to other events, the Scheduler makes these durations zero and thus the media item effectively disappears. Therefore, we now model static media items as known durations with a minimum duration of zero, an optimum duration that estimates the amount of time an average reader needs to view the material, and a maximum duration of infinity. Very low costs are associated with altering these durations from the optimum so that the Scheduler will attempt to adjust them first.

Specifying documents using temporal constraints. We expect temporal constraint programming to be more tractable for authors than graphical constraint programming because temporal constraints are one-dimensional and are less likely to become over- or under-constrained. Experience with this issue is ongoing.

Table 1. Commands computed by the Scheduler for the sample Firefly document.

Conn. Comp.	Time	Media Item	Event	Async. Events Activate	Deactivate	Operations
1	0 : 0	E&M Lect.	Start	-	-	Start Display
	0 : 10	Term Index	Start	Define Electrolyte	-	Start Display
	20 : 0	E&M Lect.	Battery Oper.	-	-	-
		Battery Animation	Start	-	-	Start Display Change Rate 0.75
	29 : 0	E&M Lect.	End	-	-	End Display
		Battery Animation	End	-	-	End Display
	29 : 30	Term Index	End	-	Define Electrolyte	End Display
2	0 : 0	Term Index	Define Electrolyte	-	-	-
		Electrolyte Definition	Start	-	-	Start Display
	2 : 0	Electrolyte Definition	End	-	-	End Display
3	0 : 0	Cal. Tool	Start	Rel. lecture	-	Start Display
4	0 : 0	Cal. Tool	End	-	Rel. lecture	End Display
5	0 : 0	Cal. Tool	Rel. lecture	-	-	Pause All Other, Send Message
		Aud. Cntl.	Start	Lv. talk	-	Start Display
6	0 : 0	Aud. Cntl.	Lv. talk	-	-	Resume All
		Aud. Cntl.	End	-	Lv. talk	End Display

5 Related Work

In this section we compare Firefly's Scheduler with scheduling mechanisms provided by other systems. Elsewhere we provide an analysis of the strengths and weaknesses of the Firefly temporal model compared to other current models [3]. A more detailed analysis is planned [4].

A variety of mechanisms have been used to manually specify multimedia scheduling including timelines [6, 11], temporal scripting languages [7], and Petri nets [9, 16]. QuickTime provides more power than traditional timeline-based systems [14]. Specifically, media segments can be placed on different timelines, which can be synchronized and scaled in a manner that maintains the desired synchronization relationships.

The CWI Multimedia Interchange Format (CMIF), which addresses the problem of displaying multimedia documents in a heterogeneous hardware environment, uses synchronization constraints that are similar to ours [5]. It appears that this system uses a runtime constraint satisfaction algorithm.

Blakowski et al. also address the problem of displaying multimedia documents in heterogeneous hardware environments [2]. Their Optimizer preprocesses documents to select among alternative presentation forms of the media segments. However, media synchronization is handled by a runtime signalling mechanism. When a media item reaches a synchronization point, it notifies all other media segments that share that synchronization point. Media items that have reached the synchronization point pause their display or perform an alternative "filling" action until they receive signals from all media items participating in that synchronization point. Media items that have not yet reached the synchronization point may perform a skipping action to reach the synchronization point faster. In contrast, Firefly precomputes schedules, which allows it to account for known incompatibilities and delays. As a result, documents presented using Firefly should behave more smoothly than documents presented in systems providing only runtime adjustments.

6 Conclusions and Future Work

We have presented a temporal constraint satisfaction algorithm that automatically computes schedules for controlling the display of multimedia documents. This algorithm can create schedules for documents containing static media, such as text and images, dynamic media, such as audio and video, as well as asynchronous items with unknown durations or times of occurrence, such as programs or user interaction.

Firefly is a work in progress. In the future we plan to add support for hierarchical media items (i.e., documents as media items) and documents generated at viewtime (i.e., documents generated by programs rather than by people). These extensions pose interesting problems for the Scheduler. First, schedules for hierarchical media items cannot currently be computed in a modular manner. We are investigating methods whereby the Scheduler could take advantage of the hierarchy without needing to look at the entire internal structure. Second, to handle documents generated at runtime, schedules must be computed incrementally. This form of incremental scheduling is more complicated than the one envisioned for the debugging tool described in Section 2.3, because it will not be able to reschedule events that have already occurred.

We expect the use of multimedia documents in education, industry, and entertainment to become commonplace in the future. We believe that Firefly represents a significant advance in rich media synchronization and authoring ease.

Acknowledgements

We thank Dennis Arnon, Marshall Bern, Steve Hanks, David Karger, and John Zahorjan for discussing possible approaches to the scheduling problem. We are especially grateful to James Campbell and Dan Greene for their assistance during the design and implementation of this algorithm.

References

1. Anderson, D.: Meta-scheduling for distributed continuous media, UC Berkeley, EECS Dept., Technical Report No. UCB/CSD 90/599, Oct. 1990.
2. Blakowski, G., Hubel, J., Langrehr, U.: Tools for specifying and executing synchronized multimedia presentations, in: R.G. Herrtwich, ed., *Network and Operating System Support for Digital Audio and Video*, Lecture Notes in Computer Science No. 614, Springer-Verlag, 1992.
3. Buchanan, C., Zellweger, P.: Specifying temporal behavior in hypermedia documents, *Proc. European Conference on Hypertext '92*, Milan, Italy, Dec. 1992.
4. Buchanan, C.: Specifying temporal behavior in multimedia documents, PhD dissertation, U. of Washington, 1993 (to appear).
5. Bulterman, D., van Rossum, G., van Liere, R.: A structure for transportable, dynamic multimedia documents, *Proc. 1991 Summer USENIX Conf.*, 137-155.
6. Drapeau, G., Greenfield, H.: MAEstro - A distributed multimedia authoring environment, *Proc. 1991 Summer USENIX Conf.*, 315-328.
7. Gibbs, S.: Composite multimedia and active objects, *Proc. OOPSLA '91*, 97-112.
8. Jeffay, K., Stanat, D., Martel, C.: On non-pre-emptive scheduling of periodic and sporadic tasks, *Proc. 12th IEEE Real-Time Systems Symposium'91*, 129-139.
9. Little, T., Ghafoor, A.: Network considerations for distributed multimedia object composition and communication, *IEEE Network*, 4, 6, Nov. 1990, 32-49.
10. Knuth, D., Plass, M.: Breaking paragraphs into lines, *Software-Practice and Experience*, 11 (1981) 1119-1184.
11. *MacroMind Director: Overview Manual*, MacroMind, Inc., 1989.
12. Nicolaou, C.: An architecture for real-time multimedia communication systems, *IEEE J. Sel. Areas of Comm.*, 8, 3, April 1990, 391-400.
13. Northcutt, D., Kuerner, E.: System support for time-critical applications, in: R.G. Herrtwich, ed., *Network and Operating System Support for Digital Audio and Video*, Lecture Notes in Computer Science No. 614, Springer-Verlag, 1992.
14. Poole, L.: QuickTime in motion, *MACWORLD*, Sept. 1991, 154-159.
15. Steinmetz, R.: Synchronization properties in multimedia systems, *IEEE J. Sel. Areas of Comm.*, 8, 3, April 1990, 401-412.
16. Stotts, D., Furuta, R.: Temporal hyperprogramming, *J. Visual Languages and Computing*, 1, 3, Sept. 1990, 237-253.
17. Swinehart, D., Zellweger, P., Beach, R., Hagmann, R.: A structural view of the Cedar programming environment, *ACM TOPLAS*, 8, 4, Oct. 1986, 419-490.
18. Zellweger, P., Terry, D., Swinehart, D.: An overview of the Etherphone system and its applications, *Proc. 2nd IEEE Comp. Workstations Conf.*, March 1988, 160-168.

The Stratification System
A Design Environment for Random Access Video

Thomas G. Aguierre Smith
SM Media Arts and Sciences

Glorianna Davenport
Assistant Professor of Media Technology
Director of Interactive Cinema Group

The Media Lab
Massachusetts Institute of Technology
20 Ames Street
Cambridge, MA 02139

thomas@media.mit.edu and gid@media.mit.edu

Abstract

Content of a movie is produced in two different types of design environments. The first is the design environment of shooting where a camera is used to capture what is happening at a particular place and time. The second is the design environment of editing where the rushes are interpreted relative to a movie maker's intent. Annotation of the video stream allows the movie maker to make decisions based on specific content of video and in the best case enables a machine to help in that process.

Stratification is a context-based layered annotation method which treats descriptions of video content as objects. Stratification offers an graphical representation of the content of a video stream and enables movie makers to quickly query and view descriptions for any chunk of video. Stratification supports the development of complementary or even contradictory descriptions which result when different researchers access video source material which is made available on a common workstation or over a network.

The Stratification System was implemented on a DECstation 5000 UNIX workstation in Motif. The system was developed under the direction of Glorianna Davenport at the Interactive Cinema Group of the MIT Media Laboratory with partial support from British Telecommunications, Pioneer Corporation, and Asahi Broadcasting.

Introduction

A random access video database system is a challenging information management problem because the way that a particular sequence of images is described effects on how a maker will retrieve it and incorporate it into a movie.

Researchers in Interactive Cinema work with real footage of their own choosing and build systems that demonstrate the relationship between movie making and computational concepts. Stratification focuses on the idea that the movie maker's interpretation of what unfolds in front of a camera is dynamic.

The intention shifts from the shooting environment where the leading question is "What is happening and how can I best represent it with a recorder?" to the editing environment where content is re-evaluated and becomes "How can I arrange moving images with the purpose of communicating something to an outsider?"

Computation becomes a characteristic of the design environment when motion picture editing becomes digital. Digital movie making requires an *environment* that supports how makers use their knowledge of working and interacting with a medium to make content.

Motion picture content is produced in two different types of design environments each having its own set of constraints. Content first emerges in the *Design Environment for Shooting* and then is transformed into the final motion picture in the *Design Environment for Editing*. Between these two processes a log of the content of the video is produced that helps the movie maker remember what was shot and to guide the production of the final movie.

Background

A relational database of keywords (who, what, when, where) was used to describe the content of shots in Glorianna Davenport's *A City in Transition* New Orleans, 1983-86 (1987). A story generation engine relied on complex keyword queries to present the viewer with constrained views of the material that was available on the system. Ricki Goldman Segall in her thesis *Learning Constellations* (1990) edited ethnographic video observations into thematically coherent chunks. In both early experiments, the process of breaking the video into discrete chunks inhibited the dynamic between the maker and the user of the video material.

The development of a computer representation for the content of video called *stratification* represents a first step where a database of descriptions reflects how a camera records with in the context of an environment. Stratification was first developed as theoretical model for how multiple annotations can be applied to a stream of video (Davenport, et al. 1991) and was tested during an ethnographic video production in the state of Chiapas Mexico (Aguierre Smith 1992).

From Discrete to Stream-based Annotations

Most video tape logs consist of content markers which are scribbled on a piece of paper as the movie maker reviews the raw footage. The log depicted in figure 1 represents 26,310 contiguous frames (14 minutes) that were recorded in the Municipio of San Juan Chamula, Chiapas, Mexico while visiting the home of Carmelino Santiz Ruiz. Carmelino shows the video maker a medicinal plant garden and then invites the video maker into the house. Upon entering, the video maker notices five cases of Pepsi bottles stacked in the corner. As the Pepsi cases are video taped, Carmelino begins to pray at the altar in the room. After this, the video maker follows Carmelino into the kitchen where he is asked about the cases of Pepsi bottles.

Figure 1: Video Log of Video Shot at Carmelino's house

```
tape07 | 84793 | and this one also
tape07 | 85163 | for burns ?
tape07 | 85879 | fry it in a comal
tape07 | 86805 | grind it like this
tape07 | 87557 | applied like this
tape07 | 88050 | after four days
tape07 | 88823 | name of burn plant?
tape07 | 89667 | Carmelino's house
tape07 | 90035 | interior w bike
tape07 | 92263 | Pepsi bottles
tape07 | 92957 | lighting candles
tape07 | 93947 | lighting first one
tape07 | 94260 | translation 52.23
tape07 | 94800 | translation 52.41
tape07 | 94847 | beginning the prayer
tape07 | 96720 | squatting and praying
tape07 | 97913 | end zoom INRI cross
tape07 | 99061 | This is the kitchen
tape07 | 99449 | thanks to Dr. Berlin
tape07 | 99819 | hay luz tambien
tape07 | 100615 | grinding corn
tape07 | 103319 | drinking Pepsi
tape07 | 103757 | we only have Pepsi here
tape07 | 104231 | close up of Pepsi cola
tape07 | 107563 | everyone drinks Pepsi
tape07 | 111103 | women walking home
```

Each record in the database consists of the tape name; the frame number; a free text description. The effectiveness of these traditional logs is wholly dependent on the linearity of the medium. For example, when locating "Pepsi bottles" (frame 92263) the video editor uses the shuttle knob on the editing deck to fast forward to that location. In doing so, all the material that has been recorded up to that point appears on the monitor. The editor sees the context that the "Pepsi bottles" -- he sees the shots that come before and after it. He notes that it is taking place in Chamula; in Carmelino's garden; now we are in his house; and in a moment he will begin to pray. When sorted by frame number the free text descriptions provide the context for each content marker. For example, "Pepsi bottles" on the diagram above appears in chronological order between "interior with bike" and "lighting candles."

But when searching the database for the word, "Pepsi", the list of content markers returned cannot provide the context. In Figure 2, a database search for the words "Pepsi" among all the video annotations does not provide the needed context. With a computerized database of content markers one can rapidly find that items that they are interested in. But this is not as useful as expected. The maker needs to see the surrounding annotations.

Figure 2: Database Search for Word "Pepsi"
without Context

tape06	1707	Pepsi or coke
tape07	92263	Pepsi bottles
tape07	103319	drinking Pepsi
tape07	103757	solo hay Pepsi
tape07	104231	close up of Pepsi cola
tape07	107563	everyone drinks Pepsi
tape13	74721	Pepsi bottles
tape14	93083	Pepsi .. Fanta
tape15	28487	Pepsi and orange
tape11	106501	arranging Pepsi
tape23	96843	Pepsi at Antonio's house
tape23	108901	Pepsi delivery

In a random access system the linear integrity of the raw footage is erased. In turn the contextual information that relates to the environment where the video was shot is also destroyed. What is required is a method to record this contextual information so that is can be recovered and re-used at a later time.

Keywords and Context

Keywords provide a more generalized way to create consistent descriptions from one tape to the next. With keywords, one can consistently find related chunks of video among the 27 hours of video tape that were shot. As shown in Figure 3, the keywords of "Chamula" and "Carmelino" remain constant while content markers specifically identify what is happening at a given moment. Keywords provide the context for content markers. Sets of keyword descriptors remain constant while the other descriptions change and evolve.

Figure 3: Content Markers with Keywords Sorted by Frame Number

tape07 I 84793 I and this one also	Carmelino Chamula garden
tape07 I 85163 I for burns ?	Carmelino Chamula garden
tape07 I 85879 I fry it in a comal	Carmelino Chamula garden
tape07 I 86805 I grind it like this	Carmelino Chamula garden
tape07 I 87557 I applied like this	Carmelino Chamula garden
tape07 I 88050 I after four days	Carmelino Chamula garden
tape07 I 88823 I name of burn plant?	Carmelino Chamula garden
tape07 I 89667 I Carmelino's house	Carmelino Chamula house
tape07 I 90035 I interior w bike	Carmelino Chamula house bike
tape07 I 92263 I Pepsi bottles	Carmelino Chamula house Pepsi
tape07 I 92957 I lighting candles	Carmelino Chamula house praying
tape07 I 93947 I lighting first one	Carmelino Chamula house praying
tape07 I 94260 I translation 52.23	<< long text omitted>>
tape07 I 94800 I translation 52.41	<< long text omitted>>
tape07 I 94847 I beginning the prayer	Carmelino Chamula house praying
tape07 I 96720 I squatting and praying	Carmelino Chamula house praying
tape07 I 97913 I end zoom INRI cross	Carmelino Chamula house praying
tape07 I 99061 I This is the kitchen	Carmelino Chamula house
tape07 I 99449 I thanks to Dr. Berlin	Carmelino Chamula house PROCOMITH
tape07 I 99819 I hay luz tambien	Carmelino Chamula house
tape07 I 100615 I grinding corn	Carmelino Chamula house corn
tape07 I 103319 I drinking Pepsi	Carmelino Chamula house Pepsi
tape07 I 103757 I we only have Pepsi	Carmelino Chamula house Pepsi
tape07 I 104231 I close up of Pepsi cola	Carmelino Chamula house Pepsi
tape07 I 107563 I everyone drink Pepsi	Carmelino Chamula house Pepsi
tape07 I 111103 I women walking home	Carmelino Chamula house

Stratification

It becomes evident that descriptions of content have a lot to do with the linearity of a medium. In a random access system we can't rely on the linearity of the medium to provide us with a coherent description. Accordingly we need a new type of descriptive strategy that allows for the annotation of descriptions to have a temporal extent.

When sorted on frame number, the content markers become embedded in patterns of keywords. These patterns illustrate the contextual relationships among contiguously recorded video frames. It also illustrates how context is wedded to the linearity of the medium. We can now trace, in this pattern, what was shot and where. Sorted lists of content markers with keywords produce a layered representation of context. These layers are called *strata* (Figure 4).

Figure 4: Log of Content Markers with Strata.

tape07 I 84793 I and this one also	Carmelino Chamula garden	
tape07 I 85163 I for burns ?	Carmelino Chamula garden	
tape07 I 85879 I fry it in a comal	Carmelino Chamula garden	
tape07 I 86805 I grind it like this	Carmelino Chamula garden	
tape07 I 87557 I applied like this	Carmelino Chamula garden	
tape07 I 88050 I after four days	Carmelino Chamula garden	
tape07 I 88823 I name of burn plant?	Carmelino Chamula garden	
tape07 I 89667 I Carmelino's house	Carmelino Chamula house	
tape07 I 90035 I interior w bike	Carmelino Chamula house bike	
tape07 I 92263 I Pepsi bottles	Carmelino Chamula house Pepsi	
tape07 I 92957 I lighting candles	Carmelino Chamula house praying	
tape07 I 93947 I lighting first one	Carmelino Chamula house praying	
tape07 I 94260 I translation 52.23	<< long text omitted>	
tape07 I 94800 I translation 52.41	<< long text omitted>	
tape07 I 94847 I beginning the prayer	Carmelino Chamula house praying	
tape07 I 96720 I squatting and praying	Carmelino Chamula house praying	
tape07 I 97913 I end zoom INRI cross	Carmelino Chamula house praying	
tape07 I 99061 I This is the kitchen	Carmelino Chamula house	
tape07 I 99449 I thanks to Dr. Berlin	Carmelino Chamula house PROCOMITH	
tape07 I 99819 I hay luz tambien	Carmelino Chamula house	
tape07 I 100615 I grinding corn	Carmelino Chamula house corn	
tape07 I 103319 I drinking Pepsi	Carmelino Chamula house Pepsi	
tape07 I 103757 I we only have Pepsi	Carmelino Chamula house Pepsi	
tape07 I 104231 I close up of Pepsi cola	Carmelino Chamula house Pepsi	
tape07 I 107563 I everyone drink Pepsi	Carmelino Chamula house Pepsi	
tape07 I 111103 I women walking home	Carmelino Chamula house	

Legend for Strata Lines:
Carmelino   Chamula   house   garden   praying   pepsi

This methodological shift is subtle but critical. The linearity of the medium is preserved with stratification because each description has a temporal/linear extent. With other annotation methods, the linearity of the medium is destroyed because we have broken up the footage into *ad hoc* chunks. The keywords form strata that capture changes in descriptive state which the camera recorded.

If you know enough about the environment in which you are shooting you can derive a good description of the images that you have captured using stratification.

Any frame can have a variable number of strata associated with it. The content for *any* set of frames can be *derived* by examining the union of all the contextual descriptions that are associated with it.

Content can now be broken down into distinct descriptive threads or strata. One stratum constitutes a single descriptive attribute which has been derived from the shooting environment. When these descriptive threads are layered one on top of the other they produce descriptive strata from which inferences about the content of each frame can be derived. Stratification is an *elastic* representation of the content of a video stream because descriptions can be derived for chunk of video of any size.

Stratification is a method which produces layers of descriptions that can overlap, be contained in, and even encompass a multitude of other descriptions. Moreover, each additional descriptive layer is automatically situated within the descriptive strata that already exit. Users can create descriptions which are built upon each other rather then worrying about how to uniquely describe each frame independently.

In addition to logging, film makers need tools which will enable them to take segments of raw footage and arrange them to create meaningful sequences. Editing is the process of selecting chunks of footage and sound and rearranging them into a temporal linear sequence. The edited linear sequence may bear no resemblance to the environmental context that was in effect during recording. Conceivably, one can edit the source material in the video database into a documentary movie that will be played back on the computer. Moreover, these "edited-versions" can later be used by someone else to make another movie production.

When editing a sequence, important relationships in the raw footage come to the fore. In an edited sequence causal and temporal relationships in the raw footage are made explicit.

Implementation:

The Stratification system was intended as an experiment to test and work out the details of a stream based annotation system. Users should be allowed to use free text descriptions and more structured keyword type descriptions. A visual mapping of descriptions to a timeline was required to facilitate browsing. In addition the system needed to support multiple users on a network. Researchers needed to apply more than one type of analysis to any shot of video. They might be interested in the transcript of the material in the field, linguistic analysis, narrative style, etc. Good source material should lend itself to be re-employed in different research environments and for different needs. And finally the format for the data files needed to be easily modifiable and accessible over the network. Knowledge about the content of video changes as time progresses. In many cases initial annotations need to be revised as new information becomes available.

Data Representations

The use of keyword classes and a special format for saving descriptions of video called Strata Data Format (SDF) are key features of the Stratification system. The implementation of keyword classes and SDF is designed to complement the file management and text processing utilities currently available in the UNIX operating system.

Each descriptive stratum consists of the source name, begin frame, end frame, free text description field, and keyword classes field. These descriptions are saved in delimited ASCII text files and stored in UNIX directories. SDF files are named in regard to a particular project and owned by an individual or group like any other UNIX file. SDF files can be combined and analyzed for associative browsing of content across projects.

Editing of these files can be accomplished using conventional text editors. In addition, easy to make UNIX shell scripts can be used to parse the files of stratified descriptions (figure 5).

Figure 5: Example of SDF Generated with the Stratification Method. Line 1 - 2 is the header which shows the context of the keyword classes that appear in the file. Line 3 - 11 are content marker annotations. Line 12 - 14 are class keyword strata lines.

```
/mas/ic/src/VIXen/Classes/Maya/places.class|
/mas/ic/src/VIXen/Classes/Maya/people.class
MayaMed|334|334|334|30|corn blowing in the wind
MayaMed|505|505|505|30|path to dominga's house
MayaMed|588|588|588|30|Laguana Pejte'
MayaMed|701|701|701|30|dominga walks down hill
MayaMed|1091|1091|1091|30|chicken's
MayaMed|1267|1267|1267|30|wacking kid
MayaMed|2042|2042|2042|30|pressing boys chest
MayaMed|2264|2264|2264|30|pressing arm
MayaMed|2484|2484|2484|30|throwing plants out
MayaMed|334|2904|1619|30| |places|Chamula
MayaMed|701|1090|895|30| |people|Dominga
MayaMed|1267|2904|2085|30| |people|Dominga
```

The UNIX file system is way to structure and organize annotations and even movie sequences. Ownership can be set for access. The place where a movie is stored can provide important contextual information about the content of a sequence. This of course requires that the user is somewhat rigorous about naming and creating directories. The additional effort pays off when tracing the use of a piece of footage in the system. A consistent format for both raw footage and edited footage enables the researcher to analyze how descriptions of raw and edited footage are built up through use.

Keyword Classes

Key words classes are organized into class hierarchies which are implemented as directory trees in UNIX. Each keyword class is stored as an ASCII text file. If desired, the user can edit a keyword class file with any UNIX text editor. The UNIX file system provides a simple yet useful way to structure different types of knowledge about a video resource. Researchers can have access to each other's keyword class files by setting the permissions on the files accordingly.

Successful perusal of the video database requires knowledge of the descriptive strategies that were used to describe the content. The choice of keywords is related the users intentions; they reflect the purposes and goals of a particular multimedia production. Keyword classes provide a flexible structure that allows for consistency in naming within a particular descriptive strategy. In the end, in the use of keywords consistency will help browsers.

The Stratagraph

The Stratagraph is an interactive graphical display of an annotated video stream. It is a visual representation of the occurrence of annotations of video though time. Keyword classes are displayed as buttons along the y-axis (figure 6).

Each button shows the keywords path name in the UNIX file system. The path name indicates the context for each keyword class. On the y-axis one can inspect where a particular keyword is from and how it is related to other keywords.

These all can be included on the y-axis. For example as shown in figure 6, the first strata displayed belong to the user "morgen"while the others belong to thomas.

Each keyword class has its own color. The keyword classes on the vertical axis are also buttons. When pressed, the graph scrolls to display the first instance of that keyword and the video is cued to the in-point.

The units on the horizontal axis are time code (frame numbers for the laserdisc). Another type of interaction consists of clicking the horizontal axis with the mouse. If the user wants to know about the annotations that are associated with any particular frame. The user can click on a frame number (the horizontal axis) to create a "strata

Figure 6: Key word Classes in Stratagraph

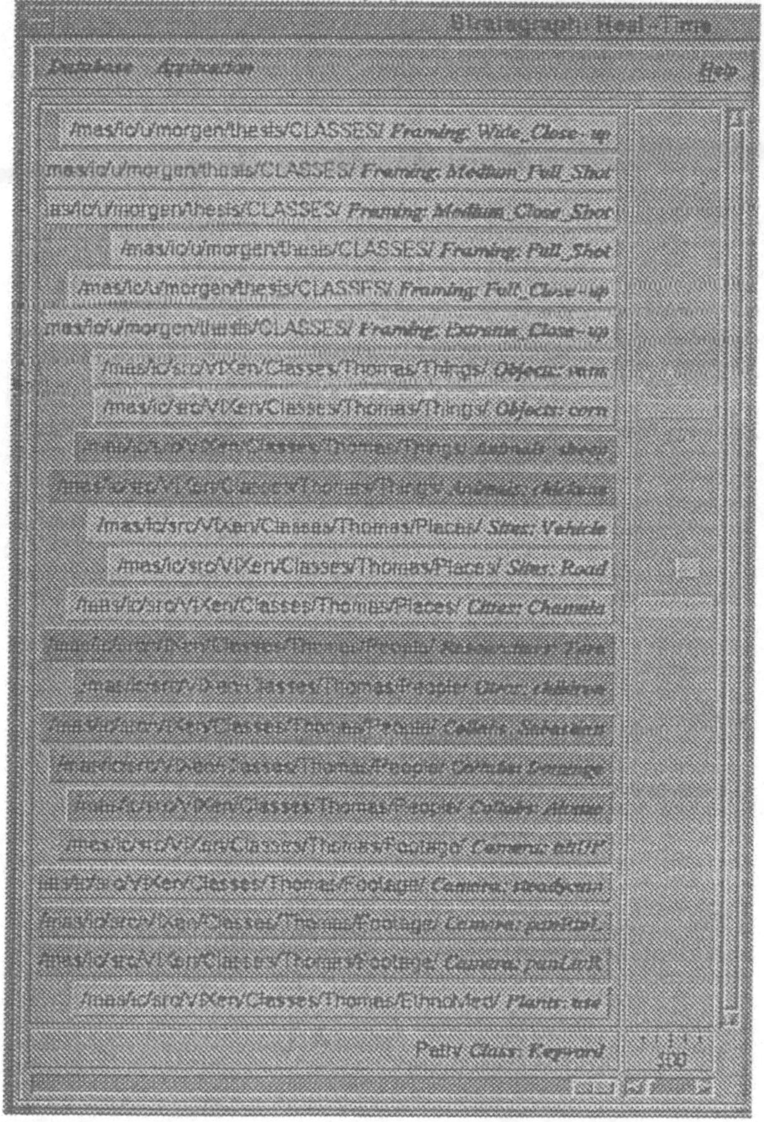

line" that intersects all the strata that are layered on top of that particular frame. The laserdisc is cued to the frame number selected and a report showing all the descriptions that are associated with these strata lines is displayed in the "Strata Content" window.

The strata line can be extended for a chunk of video by clicking the right mouse. The left click and move and right click action is called a "strata rub" (figure 7). This rubbing action displays all descriptions which are in effect for that chunk of video in a "Strata Content" window(figure 8) while the laserdisc plays the shot .

Figure 7: Stratagraph showing strata rubs.

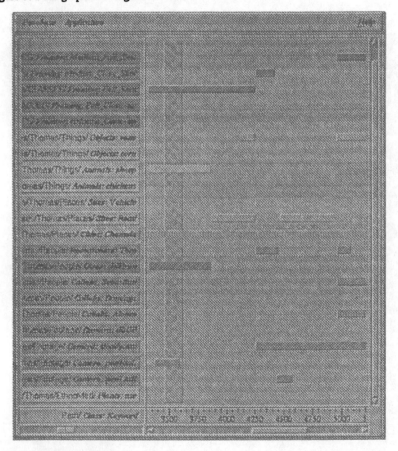

Figure 8: Strata content window.

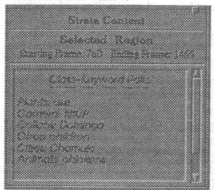

The shot needs to be named and made into an icon so that it can be arranged into sequences. When naming, the maker creates a mnemonic that is used to reference the shot a later time. Although in and out points are important cues for visual continuity, they might not provide an adequate representation of the shot content. A content frame is digitized and made into a Picon (picture icon) its associated frame number is stored as the content marker frame for the shot. A Picon has a title bar which displays the annotation (figure 9).

Arrays of Picons are displayed on the screen and arranged into sequences. When Picons are arranged in a cascade (left to right and up and down) they can be played back in the order that they appear. Ordered groups of PIcons compose a sequence.

Figure 9: Picons represent shots. Sets of Picons represent sequences.

When a satisfactory sequence is arranged, annotations, volume name, in and out frame numbers are saved as an ASCII SDF file with the name of the file as the sequence name (figure 10). Sequences can be saved as play lists and can be re-loaded back into the system at a later time.

Figure 10: SDF file for a sequence "Carmelino.work.movie"

```
Filename: Carmelino.work.movie
MayaMedl13706l13782l13777l30lface
MayaMedl14024l14294l14026l30larranging
MayaMedl14899l15102l15000l30lthe press
MayaMedl15555l16260l16000l30lthe dryer
MayaMedl18012l18507l18400l30lthe garden
MayaMedl18510l19140l19000l30lhouse
MayaMedl19157l19353l19200l30lwith healer
```

A sequence file is a play list that shows how chunks of video are related in a new context. The first generation of annotations reflect the context of where the images were recorded, subsequent annotations reflect shifts in meaning that occur when the chunk appears in a virtual edited sequence. Logging video and assembling sequences collapse into the same process. The source material does not change per se, but the context of the material is what dynamically changes. Since cinematic context is inextricably linked to context, the significance of the source material becomes transformed and refined through use.

Conclusion:

Stratification is a first step which maps and creates a correspondence between the Intentionality which was manifest in shooting to a computational log. This log is designed to enhance the conversational exchange between human and machine via a graphical interface.

Stratification serves as a computer representation of the video stream on random access system computer system. The integrity of the context of where the footage was shot is maintained while additional contexts can be created during the process of sequence assembly.

In a sense, the content of a series of frames is defined during logging. Yet the significance of those frames gets refined and built up through use. This information is valuable for individuals who want to use the same video resources over a network in order to communicate with each other. The stratification method provides a way to represent alternative "readings/significations/edits" of the same video resource to co-exist on the system.

References

Aguierre Smith (1992) If You Could See What I Mean... Descriptions of Video in an Anthropologist's Video Notebook. SM Thesis, Media Arts and Sciences, MIT, September 1992.

Davenport, G. (August 19, 1987). New Orleans in Transition, 1983 -1986: The Interactive Delivery of a Cinematic Case Study. The International Congress for Design Planning and Theory. Boston : Park Plaza Hotel

Davenport, G.,Aguierre Smith, T., & Pincever, N. (1991). Cinematic Primitives for Multimedia: Toward a more profound intersection of cinematic knowledge and computer science representation. IEEE Computer Graphics and Applications(July).

Goldman Segall, R. (1990). Learning Constellations: A Multimedia Ethnographic Research Environment Using Video Technology for Exploring Children's Thinking. Ph.D., Media Arts and Sciences, MIT, August 1990.

Acknowledgments
Special thanks to Erhhung Yuan and Joshua Holden the two undergraduate researchers who have diligently worked on the Stratification modules. We would like to thank Hiroshi Ikeda and Hiroaki Komatsu. This research has been supported by British Telecom, Asahi Broadcasting Corporation and Pioneer.

On the Design of Multimedia Interchange Formats[1]

John F. Koegel
Interactive Media Group
Department of Computer Science
University of Massachusetts--Lowell
Lowell, MA 01854
koegel@cs.ulowell.edu

ABSTRACT

Although it is generally believed that a standard interchange format will play a crucial role in the growth of the multimedia application market, different proposals currently being developed have divergent models and each appears to have primarily evolved using an experimental methodology. In this paper we first provide a survey of these current efforts, discussing goals, architecture, and abstractions of each format. We provide a comparative summary in terms of representational capability and functionality. We classify the composition models as either track oriented or object oriented, and use this distinction to clarify differences in inter-object referencing, compositionality, and access/presentation procedures. We enumerate features that would enhance the ability of a format to support real-time interchange, and conclude with an overview of an approach to rigorously evaluate and compare such format models. This approach is based on a set of benchmark interchange cases and various parameters to be measured in a performance test.

1. Overview

In order for multimedia applications to work together and realize the benefits of distributed computing, a common interchange format for multimedia information is needed. It is not sufficient for the individual media formats to be standardized. The temporal, spatial, structural, and procedural relationships between the media components are an integral part of multimedia information and must also be represented. Today there is a growing realization that lack of a common format is a serious impediment to the development of the market for multimedia applications. Without a representation that is widely adopted and is sufficiently expressive, multimedia content that is created in one application can not be read or reused by another application. Further, in order for multimedia information to be used on several platforms, each application-defined format must have a converter on each platform.

FIGURE 1. Multimedia Interchange Context

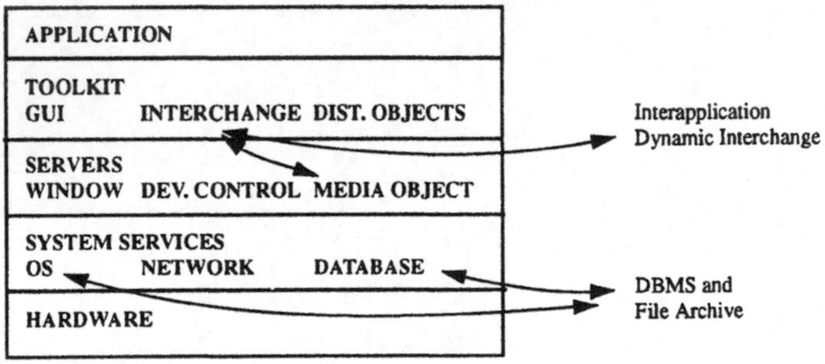

In the architecture of multimedia systems, interchange appears in three different modes (Figure 1):

1. *Interapplication interchange:* two or more applications exchange multimedia information using either an interchange API or a distributed object API.

---

[1] Third International Workshop on Network and Operating System Support for Digital Audio and Video, November 12-13, 1992, La Jolla, California. (c) Copyright 1992 John F. Koegel.

2. *Archive:* an application saves and accesses multimedia information through a file system or DBMS, also using an interchange API. Contemporary multimedia authoring packages follow this case.

3. *Presentation:* A media object server provides media objects to distributed clients in a networked environment. An example application is video-on-demand.

The design of multimedia interchange formats can also be viewed in the context of the interchange format hierarchy (Table 1). In this diagram levels correspond to increasing specificity. At the top level are general container formats. These formats are application independent. At the bottom level are media specific formats. These formats are optimized for a specific media type.

TABLE 1: The Multimedia Interchange Format Spectrum

Category	Examples
General container	GDID, ASN.1, Bento
Metalanguage	HyTime
Multimedia document architecture	HyTime DTD
Special purpose object container	QuickTime Movie File, MHEG, OMFI
Monomedia	MPEG, JPEG Script languages

The major technical issues that must be addressed include:

1. *Multimedia data model:* A data model for structured time-based interactive media (multimedia and hypermedia) including temporal composition, synchronization, multiple media formats, addressing of media objects and composite media objects, hyperlinking, and an input model for interaction.

2. *Scriptware integration:* Many authoring tools integrate multimedia data with specialized procedural scriptware which may be text based or iconic languages. These tools have a tight association of scripts with media objects and media composites, in particular associating input semantics of input objects with script input processing. The interchange formats must retain the associations of the scriptware and the media objects. Further, scripts must be able to reference structured media objects for attribute control, retrieval, and presentation.

3. *Storage efficiency:* An encoding should be efficient for storage, but the container is a small fraction of the information in a typical multimedia presentation.

4. *Access efficiency:* An encoding should be efficient for time-constrained and resource-limited retrieval. Enhanced functions for progressive and multi-resolution delivery, flexible storage organization, media interleaving, index tables, and partial media referencing can support this goal.

5. *Portability:* GUI and platform architecture independence are essential, preferably without penalizing interchange on a single platform. Issues include look and feel independence, input architecture independence, file and object referencing, byte ordering, and data type encoding.

6. *Extensibility:* It should be possible to add new media formats, new media attribute, and other container extensions

In the rest of the paper we first provide a survey of these current efforts, discussing goals, functionality, and abstractions of each format. This review includes composition, time models, hyperlinking and input handling, architecture independence, and extensibility. This part of the presentation leads to an informal comparison to be made in the following section in the form of a function checklist. This is followed by a discussion of the role of interchange formats in supporting realtime interchange. Finally, we propose a benchmark for validating formats which are designed for multimedia interchange.

2. QuickTime Movie File (QMF) Format

QuickTime is a multimedia extension for Apple's System 7 operating system for the Macintosh personal computer. The QuickTime Movie File is a published file format for storing multimedia content for QuickTime presentation. Several QuickTime players are available for other platforms.

QMF uses a track model for organizing the temporally related data of a movie (Figure 2). A movie can contain one or more tracks which can be overlayed. A track is a time ordered sequence of a media type; the media is addressed using an edit list (Figure 3).

FIGURE 2. QuickTime (TM) Abstract Atom Model

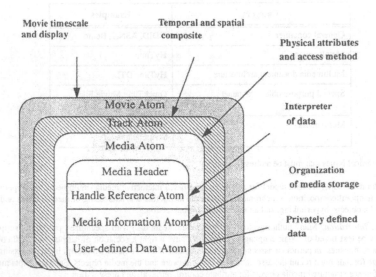

3. MHEG (Multimedia and Hypermedia information encoding Expert Group)

MHEG is an ISO working group that is defining an object-oriented model for multimedia and hypermedia interchange. The interchange model is intended for real-time final-form delivery environments. Here real-time refers to the need for systems to meet time constraints inherent in the media given the resources of the delivery and presentation system. Final-form refers to the content and composition format being oriented towards delivery as opposed to authoring. MHEG uses an object oriented model for composing multimedia (Figure 4).

The composite class is used to associate objects (components) which have temporal or spatial relationships. Two binary temporal relationships are supported: serial and parallel. The link object defines one or more condition-action entries in the context of an interactive composite. This can be used to provide elementary input semantics or implement a hyperlink mechanism. For more elaborate processing, script objects can be associated with specific objects. A GUI-independent input model is associated with the interaction class. Generic input types such as push button and text entry are defined.

FIGURE 3. Components in QMF

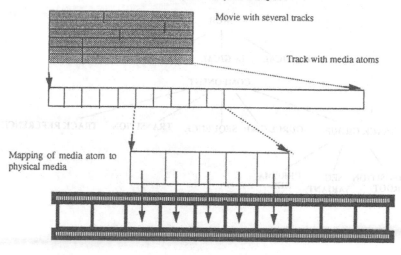

Movie with several tracks

Track with media atoms

Mapping of media atom to physical media

FIGURE 4. MHEG Object Tree

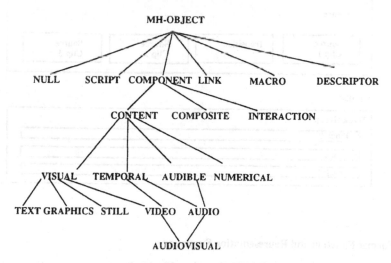

4. OMFI

The Open Media Framework (OMF) is an industry standardization effort being led by Avid Technology to define a common framework and multimedia interchange format (OMFI). The top level composition elements are media objects (MOBS) (Figure 5). The logical MOB is the most important from the standpoint of composition. Media can be organized in parallel time (Track Group) or serial (Sequence). Like QMF, OMFI uses a track model (Figure 6).

FIGURE 5. OMFI media object relationships

MOB
PHYSICAL
LOGICAL
MASTER
COMPONENT
TRACK GROUP
SOURCE CLIP
SEQUENCE
TRANSITION
TRACK REFERENCE
COMPOSITION ROOT
SEG VARIANT
CHROMA KEY

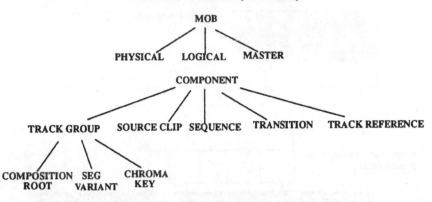

FIGURE 6. OMFI serial and parallel temporal composition examples

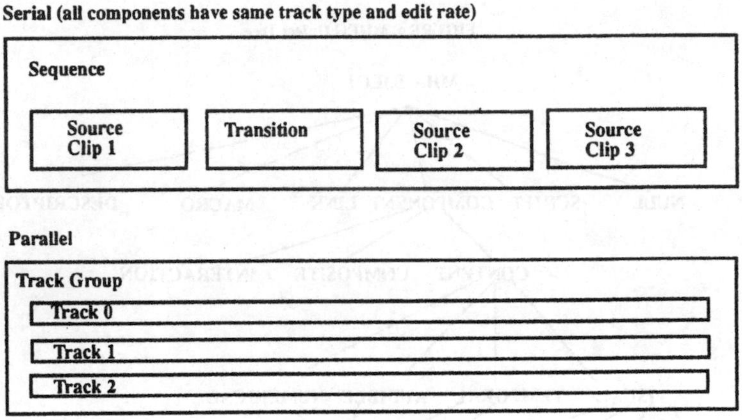

Serial (all components have same track type and edit rate)
Sequence
Source Clip 1
Transition
Source Clip 2
Source Clip 3
Parallel
Track Group
Track 0
Track 1
Track 2

5. Format Function and Representation Summary

TABLE 2: Comparison of the three formats

Feature	QMF	MHEG	OMFI
Model	Track	Object-oriented	Track
Encoding	Unique	ASN.1; others	IFF

TABLE 2: Comparison of the three formats

Feature	QMF	MHEG	OMFI
Media object addressing	File name	Globally unique object id	Globally unique object id
Composition primitives	Movies, tracks, media	Composites, interactors, links	MOBS, tracks
Time composition	Serial and parallel	Both serial and parallel	Sequence for serial, group for parallel
Component reference scope	Local for format units, global for media units	Global: All format units are objects and all have unique id	Nested for format units, global for media units
Media source referencing	No	No	Unique physical MOB
Input model	None	Interactor and link objects	None
Link model	None	Use of link class	None
Scriptware objects	Yes if treated as media with own handler	Yes	Could be treated as media
Architecture independence	No; QuickTime and Mac dependencies	Yes; standard encoding and external referencing	Yes; handles byte ordering, data types, file names
Extensibility	By Apple; user defined atoms available	Yes; private data, private classes, private attributes	By members of OMF
File organization optimization	No specific features	Global object index	No specific features

6. Track Model and Object Model

From the standpoint of media access, the composition model is a presentation list which defines the order in which media are accessed and displayed. In the track model, the primary access sequence is temporal. In an object model, the primary access sequence is through hierarchical descent of the tree; the ordering of the components in a composite would probably be based on display order. In the track model, multimedia presentation is viewed as a sequence of temporally oriented movie segments (Figure 7). In the object model, multimedia presentation is viewed as non-linear hyperlinking between (temporally) composite object trees (Figure 8).

The object model has the feature that all format units (containers, media objects, etc.) are fully visible and addressable units. The track models present provide addressing for media units but not for other container units. The tradeoff here is between the overhead of providing object ids for each format unit versus the possibility of reusing container structure by allowing it to be referenced from several places.

FIGURE 7. Multimedia presentation viewed as a sequence of temporally-oriented movie segments

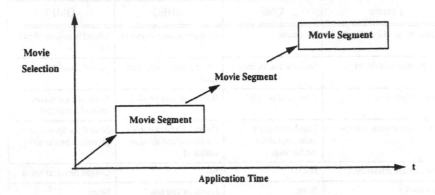

FIGURE 8. Multimedia presentation viewed as non-linear hyperlinking between(temporally) composite object trees

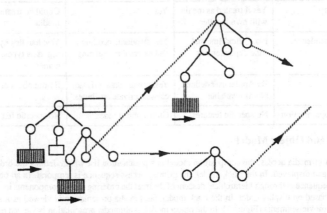

The implied presentation procedure for these two models differs as well (Figures 9 and 10). The two track formats do not model input, leaving interaction support to scriptware or the application. Then the presentation process can be seen as clock based control of a collection of media players (Figure 9). In the object model, the input model allows some predefined input interpretation to be included in each composition, as well as providing links to the application and the script (Figure 10).

FIGURE 9. Abstract presentation procedure for track model

```
// Open
Initialize clock
Initialize movie, track and media indices
Initialize handlers
// Players
Repeat
    Get next segment of media call handlers
    Advance clock
Until input condition or end of movie
```

FIGURE 10. Abstract presentation procedure for object model

```
// Open
Get table of contents, object tables
Initialize presentation processes
//Present
Repeat
    Repeat
        Get set of objects to be presented
        Invoke players for each object
    Until media event or input condition
    Pass input condition to application or script or interpret
    Identify next set of objects to present
Until exit condition
```

7. Realtime Interchange

7.1 Definitions

For interchange during presentation to satisfy realtime, the temporal nature of the systems, objects, etc. are delivered or executed in a way that follows the original time-base as specified by the designer of the system. This implies that object retrieval, presentation, and response to user input must meet certain deadlines to be considered "realtime". However, minimal configurations are allowed to present subsampled time-based data (e.g., reduced rate video) and still satisfy this "realtime" criteria.

The interchange format can provide support for realtime by providing file and data organization techniques which enable a delivery engine to have faster access to objects. Some selected techniques are listed in the next section. Some of these techniques are independent of the format; all could potentially be available to the delivery system by preprocessing.

7.2 File Format Techniques for Supporting Realtime Interchange

1. *Object Placement Optimization:* Objects are stored so that objects which are likely to be accessed simultaneously are adjacent from the standpoint of the access mechanism.

2. *Partial Object Retrieval:* Large objects can be retrieved in sections since in many cases the entire content of such an object will not be presentable at one time.

3. *Object Sequencing:* The order in which objects are expected to be presented is maintained for use by the access mechanism.

4. *Global Object Index:* A table of all object ids and their position in the object set is provided to support fast lookup of objects.

5. *Object Interleaving:* Objects which are to be retrieved simultaneously are interleaved so that large objects don't cause delays for other objects.

6. *Separate Retrieval of Object Description and Object Content:* The object description can be retrieved without necessarily retrieving the content so that the system can use information about a set of objects to optimize the access for this set and so that resources needed for the access can be prepared.

7. *Progressive access of objects:* images can be retrieved and presented in increasing resolution for systems in which presentation delay is significant. Scalable versions of objects can be represented and retrieved for systems with insufficient resources for full fidelity presentation.

8. *Resource recommendations:* The resource requirements for retrieval and presentation by the target system are available by lookup rather than by derivation.

8. Evaluation Criteria

Although each of the interchange formats has similar features, the benefits of a rigorous comparison would be to enhance future versions of the formats and assist developers in using the formats efficiently. There are three ways in which these models can be evaluated and compared:

1. Representational power: define the composition, synchronization, media, interaction, and other primitives that are intrinsic to the domain and verify that each format can express these

2. Functional checklist: define a functional requirements list and identify which requirements are met by each format

3. Benchmarks for representative operations and application scenarios

8.1 Example Test Cases

Test cases are needed in order to obtain quantitative results from a benchmark of the different formats. Test cases should naturally cover a broad range of the domain and should emphasize functionality that is most important. We list several test cases as examples, and the comparisons are brief. Each of the following cases represents a distinct category from the perspective of the authoring environment and the composition characteristics. Each test case should be used in both a small and a large example.

1. Edited video: a sequence of video segments (analog and video) interspersed with various transitions and incorporating subtitles

2. Hypertext: a hypertext document with a large number of links; multiple links to and from specific objects.

3. Slide-show style presentation: simple branching structure with small number of media objects per screen

4. Script-based interactive multimedia presentation: complex presentation, as in a model-based interactive lesson, controlled by a script

8.2 Benchmarks

The value of benchmarks rests in selecting performance measures that are relevant to the intended uses. The following list of measurements would most likely be performed across several platforms. For each test case, the following measurements can be made:

1. *Storage cost:* this measures the encoding efficiency

2. *Retrieval cost:* for a presentation delivery sequence of accesses, what is the retrieval cost for accessing the necessary objects

3. *Random access cost:* for a random sequence of object accesses, what is the cost to retrieve these objects

4. *Update cost:* what is the cost of changing/adding/deleting an object or composition during an edited operation in terms of file reorganization

5. *Conversion cost:* what is the cost of converting from/to one or more proprietary formats to/from the common format

9. Summary

Interchange is important issue that will effect market for multimedia products. Important technical issues relate to the representational power of the data model. There is no systematic design evaluation process.

In this paper we have given a comprehensive overview of the significant features of a number of current formats under development. Several of these formats are under design, and the description given here could change. We have summarized and compared the major contributions of each approach in a function checklist. We have given an overview of various approaches for evaluating the design of such formats which covers both representational power and performance issues.

When these formats are finalized, we anticipate opportunity for further work in performing such an evaluation.

10. Acknowledgements

The following individuals provided information about formats referenced in this paper: Dave Grandin at Avid, Bob Gordon of IMA, and Rita Brennan of Apple. The author appreciates the collaboration of colleagues on the ISO MHEG committee.

QuickTime is a trademark of Apple Computer Inc. Open Media Framework is a trademark of Avid Technology.

11. References

1. Apple Computer. Proposal to Standardize a Temporal Media Movie Format. Aug, 1991.

2. Avid Technology, Open Media Framework (OMF) Concepts and Overview, June 1992.

3. Avid Technology, Open Media Framework (OMF) An Introduction to OMF Interchange (TM), June 1992.

4. Avid Technology, Open Media Framework (OMF) Interchange Format, Sept. 1992.

5. Interactive Media Association. IMA Architectural Framework (Draft). Oct. 1992.

6. Interactive Media Association. IMA Compatibility Project Proceedings. Sept. 1991.

7. ISO/IEC JTC1/SC18/WG8 IS 10744 Hypermedia/Time-base Structuring Language (HyTime). Aug. 1992.

8. ISO/IEC JTC 1/SC 29, "Coded Representation of Multimedia and Hypermedia Information Objects, version S.7, October 1992.

9. Koegel, J., MHEG and Realtime Interchange. ISO/IEC JTC 1/SC 29/WG 12 N 92/388, July 1992.

10. Ortiz, G. "QuickTime 1.0". Dialog, Summer 1991.

Session 7: Multimedia Workstations and Platforms

Chair: Daniel C. Swinehart, Xerox Palo Alto Research Center

The theme of this session was that hardware and system software partnerships can be created to produce an environment better suited for high-performance continuous media than conventional workstations with only user level software support. Each of the three papers described methods for managing limited system resources involved in transport and rendering. The first focused on I/O bus bandwidth management, the second on system and user memory management, and the third on CPU and network management. All were concerned with marshaling the available resources in order to provide an acceptable level of service for continuous media.

Wall et al addressed the real-time requirements of video workstations. The experimental environment was the High-Resolution video workstation developed jointly by Sun and Sarnoff Laboratories. Several general and special purpose processors, large memories, network, file, and video hardware interfaces are interconnected by a high-speed (320 Mbyte/sec) bus, a common arrangement for emerging high-performance systems.

The problem is to manage the bus in such a way that the real-time demands of all the running applications can be met, and to favor the higher-priority applications when the constraints cannot be met. The bus hardware can implement a scheduling mechanism, in the form of an event list of transfers, but cannot make the scheduling policy. This role is served by software in the modified OS kernel.

Several increasingly sophisticated protocols are discussed for managing the demands of such applications as HDTV rendering, NTSC film loops, and remote medium-bandwidth conference-quality video, giving priority to the more demanding applications while retaining the flexibility of a software implementation. The Time-Division Resource Management protocol, essentially a deadline-scheduled approach which honors priorities when resources are tight, outperforms purely priority-scheduled and purely deadline-scheduled approaches.

Druschel et al address a different part of the performance problem. They point out that since the speed of buses and memories are not keeping up with speedups in CPUs and communication networks, memory operations in continuous media applications need to be managed very carefully. Conventional approaches to memory performance, such as data caching, have limited value since there is little locality of reference.

The situation gets worse when the need to communicate across domain boundaries (I/O board to kernel, kernel to user) is considered. The straightforward approach involves copying the data at each such transition. Copy-avoidance schemes, using memory mapping approaches, have drawbacks, especially if they are presented directly to users.

The authors propose a buffer abstraction, presented as an applications program interface (API) to provide a framework for using a number of copy-avoidance and cache-optimization techniques to improve performance. Data as it arrives or is created is put into newly-allocated buffers, which can be shared across multiple domains. The operations which actually reference or modify the buffer information are heavily constrained; other operations use a "buffer-editing" abstraction to combine, split, and modify buffers without actually referring to the contents. Through these mechanisms, high performance can be guaranteed without special effort by the applications programmers.

Dannenberg et al. concern themselves with the software capabilities to manage scheduling of the computing resources of workstations and network continuous media transmission, via a general time-driven time scheduler, well integrated with an extended X service (including video and audio and MIDI). The mechanism is a client toolkit and a temporal events server (Tactus server). The server, at the presentation site, is default timing manager for services that have no notion of real time, such as X. The server can synchronize information streams arriving from where in the network. Client programs use the toolkit to provide information in a timely fashion, sufficiently in advance of the time when it will be rendered at the server. Tactus includes a number of novel approaches to media synchronization and management, including a "cut" capability assuring smooth transitions from one sequence to another. The prototype implementation runs on top of a Mach microkernel, achieving kernel-level performance.

Bus Bandwidth Management in a
High Resolution Video Workstation

Gerard A. Wall, James G. Hanko, and J. Duane Northcutt
Sun Microsystems Laboratories, Inc.
2550 Garcia Avenue, MTV29-110
Mountain View, CA 94043

Time-Critical Computing encompasses that set of applications to which correctness of operation is a function of time. Applications that manipulate digital audio and video frequently have timeliness requirements associated with their correct function. These applications must perform operations such as the acquisition, processing, and delivery of audio or video data within a specified set of time constraints (as defined by the user, the application, or the media itself) or the value of the computation is significantly diminished. Furthermore, a system's effectiveness may be reduced if it permits computations to consume resources when they can no longer provide value to the user (e.g., after their deadlines are missed). A technique known as Time-Driven Resource Management (TDRM) has been developed to manage system resources, and this paper describes the application of this technique to the management of i/o bus bandwidth within an experimental integrated video workstation.

Introduction

Considerable attention has recently been focused on the problem of integrating digital audio and video into the workstation environment. Digital audio and video are instances of *time-critical media* — i.e., data streams that are meant to be consumed by the human sensory system, and whose correctness depends as much on timing relationships in the presentation as on the accuracy of the bit stream [Northcutt 90b]. If time-critical media are not presented with the proper timing, the data can lose most, or all, of its informational content for the human consumer. For example, significant variations in the rate of presentation of digital audio can make speech unintelligible, or variations in the delivery of frames of video from a scientific visualization application can distort the results of the experiment. Furthermore, the policies for dealing with errors in presentation can be both media- and application-specific. For example, in some applications the outright loss of a frame of video might be better than displaying it late, and causing subsequent frames to be late. Finally, digital audio and video applications typically present continuous demands for large amounts of processing and data transfer resources, and this makes the task of supporting them even more challenging.

Because of these requirements, careful resource management is essential for providing effective system-level support for many types of time-critical media applications. At Sun Microsystems Laboratories, the Time-Critical Computing project is employing a technique known as Time-Driven Resource

Management (TDRM) to support time-critical applications (including time-critical media applications). TDRM attempts to manage all system resources in such a fashion as to meet as many as possible of a system's application-defined time constraints. While much attention has been given to the problems associated with managing processor cycles (e.g., [Locke 86], [Northcutt 87], and [Clark 90]), relatively less effort has been put into the management of the other system resources. In many time-critical systems, the importance of effective management of shared i/o buses is rivaled only by processor cycles in terms of its impact on the value delivered to users. In this paper, the effectiveness of applying the principles of TDRM to the management of i/o bus bandwidth is explored in context of an actual system that supports the manipulation of a considerable volume of time-critical media.

This paper begins with a statement of the problem of supporting time-critical media applications, and focuses on the issue of managing i/o bus bandwidth in particular. Then, a broad overview of the technical approach being proposed to address these problems is given. This is followed by a description of the experimental context within which this approach was investigated — including a description of the hardware and software testbed, as well as the test programs, used to generate a realistic applications environment for the experiments. The results of the experimentation are then presented, and the effectiveness of TDRM is evaluated with respect to other common resource management techniques.

Problem Statement

In order to effectively support time-critical media applications, a system must work to meet the timeliness requirements imposed on it by its current application mix. With time-critical media, the individual data items that are manipulated by the applications may have end-to-end time constraints associated with their use. In such a case, the failure to meet time constraints may be manifested in incorrect behavior of the applications. For example, a simple NTSC video player program must digitize, buffer, process, transfer, and display video frames 30 times per second to avoid lost frames, noticeable inter-frame jitter, "tearing" of frames, or other display inconsistencies. Note that while some forms of time-critical media (such as audio or video) have inherent time constraints, applications can impose additional time constraints on the media (e.g., for purposes of interactive control and synchronization [Hanko 92]).

With this class of applications, many different types of system resources are needed to perform the desired operations. For time-critical media applications, the system must ensure that all of the resources needed to complete a computation's manipulation of time-critical media are made available to the proper portions of the computation, at the proper time. This calls for the careful (run-time) management of all system resources, as a function of application-defined time constraint inputs.

In a video-capable workstation, i/o resources are of considerable importance, since much of a time-critical media application's problem involves the transfer of large volumes of continuous information within the system. Careful management of i/o resources is desirable because of their shared nature, and the fact that they are frequently not specifically designed to handle the needs of time-critical media. Most workstation i/o subsystems are designed to support bursty, low data rate devices (e.g., 1-2MB/s from disks or networks), and a single stream of raw video (e.g., 30MB/s of continuous data) can easily overload most i/o buses. The major issue being explored in this paper is whether it is worthwhile to consider the time constraints associated with individual i/o operations when servicing requests for i/o resources.

Technical Approach

The primary function of an operating system is resource management, and the degree to which a given system supports the needs of time-critical applications depends on the policies it applies in the management of its resources. To meet the needs of time-critical computing, a new form of system resource management has been developed known as Time-Driven Resource Management (TDRM) [Northcutt 88]. The fundamental concept of TDRM is that the system attempts to maximize the user-defined value by managing all system-level resources (or as many as practical) in a manner that permits all of the time constraints of the current applications mix to be met, when sufficient resources exist to do so. When all of the given time constraints cannot be met, the TDRM approach dictates that resource requests are granted in a manner that maximizes the delivered value to users by deferring servicing the requests of computations that are of lesser importance, in favor of those with greater user-defined importance. The key concepts behind TDRM are described below.

Providing Time Constraint Information

The existence of time constraints on computations is the basis of the TDRM approach's effective management of resources on behalf of time-critical applications. First, the system uses application-provided time constraints associated with the outstanding requests for a given resource to develop a request servicing order that meets all of the deadlines. When a feasible ordering cannot be found, the system uses application-defined (relative) importance levels to choose which requests to defer in order to maximize the system's delivered value. Therefore, applications must provide both the time constraint and importance information to the system for each time-critical portion of their execution stream. For time-critical media, the time constraints are generally derived by the application (or support libraries) from the intrinsic characteristics of the media itself (e.g. "natural" video frame rate).

Predicting Overload Conditions

In order to properly predict overload conditions, a system based on TDRM needs both an indication of the time constraints of all active computations (provided by the applications), and an estimate of the resources required by these computations to meet their time constraints. While some resource requirements are obvious to applications, other requirements are fairly opaque and may vary from implementation to implementation of a given architecture. In the same sense that the working-set model of virtual memory management attempts to dynamically construct a memory usage model for each virtual address space, the system must attempt to model the expected resource demands for each of its time-critical activities. These system-maintained resource requirement estimates are used to determine when the system is in overload.

Deferring the Servicing of Requests for Resources

When overload is detected, the user-defined importance of each request is considered and requests are deferred (i.e., moved to a later part of the order), starting with those of the least importance and moving upward in importance until the maximum possible number of deadlines can be met. Note that deferring a time-critical computation until later in a schedule (or service order) does not rule out that possibility that the deferred computation's time constraints can be met. It is possible for a deferred computation to complete on time due to an overestimation of the resources required by other (higher importance) computations, or due to an underestimation of the available resources.

Handling Overload Indications

Whenever time constraints are placed on computations, the system may not have sufficient resources to meet them. Therefore, TDRM also involves the notification of applications when time constraints cannot be met, so that appropriate actions can be taken. The system should permit applications to specify the earliest possible point at which the system should initiate the notification process when it predicts or detects overload — e.g., immediately upon issuing the time constraint (so that an application can renegotiate requests that have a low probability of being met), prior to the deadline (so that an application has sufficient time to assume a fall-back position and still take meaningful action before the deadline expires), at the deadline (so that an application can stop consuming resources that will not contribute value to the user), or some period of time following the deadline (so that an application can continue to execute even though the ideal time to have completed has passed — i.e., a "soft" deadline). Applications that make use of time-critical information must be written to expect and rationally deal with overload exceptions.

The TDRM approach differs from fixed policies such as static priority or resource reservation schemes that enforce a First Come, First Served (FCFS) policy on the servicing of resource requests. Static priority and FCFS are inappropriate when different computations may have varying time constraints and different intrinsic values (as is often the case in real applications that deal with time-critical information). TDRM is more flexible in that it takes into account (at run-time) the user-defined, (dynamic) time constraint and value of each computation — not simply the arrival time of their requests for resources.

Research Testbed

The testbed upon which this experimentation was carried out is the HRV (High Resolution Video) Workstation, which is the product of a DARPA-supported research project, in which Sun Microsystems Laboratories (in conjunction with David Sarnoff Research Center and Texas Instruments) designed and constructed a workstation prototype which integrates high resolution digital video as a first class data type within the system [Northcutt 92]. That is, it is possible for programs to manipulate video data as though it were any other form of data in the system (this precluded the addition of video to the workstation through some dedicated path).

The HRV Workstation was designed to support a substantial load of multimedia applications, representative of the kind of application mix that may appear on a future class of workstations — i.e., workstations that are primarily communications engines, as opposed to compute engines. In this scenario, the workstation must support as many simultaneously active windows of video as it does text or graphics windows today. Furthermore, while a stream of today's (i.e., NTSC-quality) video represents 30 MB/sec of bandwidth, a single stream of future (i.e., HDTV-quality) video represents a continuous load of 240 MB/sec.

The HRV workstation consists of a set of boards which support the acquisition, storage, processing, transportation, and display of raw (i.e., uncompressed) high resolution (i.e., HDTV-quality) video. The HRV workstation also has a well integrated operating system, window system, and video support software that give the user and programmer the full environment of a workstation (as opposed to that of an attached image/video processor).

HRV Workstation Hardware

The HRV workstation contains a number of general-purpose RISC processors and several video-related i/o devices, all of which are interconnected by a high performance bus. This primary interconnection mechanism is known as the High Speed Data Bus (HSDB) and provides a 320 MB/sec (peak rate) transport path for data (including digitized video) among the boards in the HRV system. The HSDB performs variable-size block transfers with a software-managed (centralized) controller initiating transfers, selecting the participating boards, clocking the data across the bus, and terminating the transfer operation.

Video acquisition in the HRV system is performed by the Video Input Processor (VIP), which is capable of digitizing, and delivering over the HSDB, digital video at full frame rates (up to SMPTE240M/HDTV rates). The Video Output Processor (VOP) contains double buffered sets of 8-bit and 24-bit frame buffers, each of which is two megapixels in size. The VOP also contains a processing element, and all the control logic for the HSDB. The system also contains a Bulk Memory (BM) board with 256 MB of RAM. The final board, the Algorithm Accelerator (AX), contains four processing elements and performs most of the significant video data computations in the system.

HRV Workstation System Software

The HRV workstation system software is derived from standard Sun Microsystems software components. The RISC processors on the AX boards run a Solaris 2.0 mini-kernel, modified to include a lightweight interprocess communications (IPC) facility (to complement the kernel's lightweight processing structures), a new process scheduling framework (to provide TDRM-based processor management), and new i/o handling facilities (to support the needs of the new devices added to the system). In addition, the HRV workstation runs the Sun OpenWindows Version 3 window server, with extensions to provide programming interfaces for video in windows.

Using the Solaris loadable module structure, a number of additional system service functions come standard with the basic HRV workstation. Included among the system service modules is a High Speed Data Bus transfer agent that manages requests for access to the HSDB.

Experimental Context

It was realized early in the design of the HRV Workstation that the HSDB would be a limited, shared resource, and that the effective management of it would be essential. To provide the flexibility to explore different approaches to achieving this goal, the HSDB was designed according to the principles of policy/mechanism separation — i.e., only the basic mechanisms of data transport were placed in the hardware, and all of the resource management policy was left up to the operating system.

Managing Bandwidth on the HSDB

The HSDB has a peak data rate of 320 MB/sec, but the actual rate achievable in a given instance may vary considerably. This is because, in addition to the fixed amount of work that the global HSDB hardware must perform at the start and end of each transfer, there is a per-transfer overhead which varies from board to board in the system and is dependant on the addresses selected and the state of the boards involved. For example, the VOP's frame buffer is implemented with triple-ported VRAM, and its secondary shift registers must be set up at the start of a transfer and then stored at the end of the transfer. If another operation (e.g., a screen refresh) is being performed via the VRAM's random port, the VOP will stall the HSDB's operation for several (HSDB) clock periods. As a result, the amount of overhead per transfer to the VOP varies from 250 nanoseconds to 1.5 microseconds. Furthermore, since each horizontal scan line segment in a window requires a separate HSDB transfer, the amount of bandwidth available for actual transfers depends on the size and shape of the target windows. Although the maximum bandwidth available for 1920 pixel-length (i.e., full HDTV screen width) transfers is 298 MB/sec, the achievable bandwidth for shorter transfers can be substantially lower (e.g., 240 MB/sec for 256 pixel-length line segments). Finally, software must keep the transfer command queues loaded for full utilization of the HSDB, otherwise bus bandwidth is wasted while new transfer commands are being issued.

Video Application Programs

In the experiments described here, a suite of different video application programs were used to provide the type of loading on the HSDB that can be expected from actual, realistic programs. The traffic generated from these programs was used in experiments that measured the resulting bus utilization and effective frame delivery rates, under a wide range of different loading conditions (ranging from moderately loaded, to heavily overloaded). The application suite consists of the following applications, each of which uses the HSDB to transfer streams of video information:

- The *HDTV Capture* application moves fields of digitized video from the VIP to a window on the VOP. The goal of this application is to move 30 frames of 1920 by 1000 HDTV video to the display each second.

- The *HDTV Magnifier* application transfers a 128 by 128 section of the digitized HDTV video stream from the VIP (at a point selected by the user), scales it up by a factor of four, and then moves the resulting 512 by 512 image to another window on the VOP.

- The *Network Video* application program receives up to 30 frames/sec of compressed video from an ATM network, decompresses the video, scales the video frames up by a factor of four, and then moves the resulting 512 by 480 image into a window. Multiple copies of the Network Video application can run simultaneously on the HRV system, and can be used to provide a variable degree of loading.

- The *Cine Loop* program stores sequences of video frames in the bulk memory unit, and displays the sequence, at up to 60 frames/sec.

With these sample application programs, it is possible to investigate a wide range of different loading conditions, each with a variety of different time constraints and user-defined importance levels. It should be noted that each of these applications is asynchronous with respect to the 72 Hz HRV Workstation display (the HDTV Capture and HDTV Magnifier are bound to the 60 Hz HDTV source, the Cine Loop runs off the Sun 4/670 system clock and the Network Video is completely uncoupled). The applications were also constructed with a high-level, end-to-end flow control mechanism which monitors the delivery time of transfer requests, falls back in the event of massive overload, and then attempts to slowly return to the target frame rate. In addition, the processing load was manually distributed over a set of processors to ensure that each task had sufficient processing power to complete on time. Therefore, the management of the shared i/o bus became the deciding factor in the success or failure of each application to present good quality video.

For the purpose of these experiments, the HDTV Capture program's requests were given the highest importance, the HDTV Magnifier application was given the second highest importance, the Network Video applications all shared a lower importance, and the Cine Loop application was given the lowest importance. All of the transfer requests generated were assigned time constraints based on the target frame rate of the application that issued the request (e.g. 33 milliseconds per cycle for the HDTV Capture).

Finally, a flexible framework was developed within the HRV operating system kernel that allowed the simple substitution of a wide variety of different bus scheduling policies for the HSDB. Currently, four different scheduling policies are available in the HSDB transfer management facility. A *First Come, First Served* policy was implemented to provide a baseline measure, by illustrating the behavior of typical i/o buses. A *Static Priority* policy was implemented to show what happens in more sophisticated systems, where a limited number of arbitration priorities are available on the i/o bus. A *Dynamic Deadline* policy was also provided, which allows applications to express directly their time constraints and illustrates the behavior of a more complex bus management facility (without overload detection and handling). Finally, a TDRM-based policy was implemented, which attempts to service requests by their deadline, predict overload situations, and defer requests of lower importance when necessary.

Results of the Experiments

Using the previously defined programs and bus scheduling policies, a number of different test runs were made and the results examined. Measurements were obtained by keeping running totals in the HSDB transfer agent of key variables, including: requested data rate, achieved data rate, number of transfers performed on the bus, number of missed deadlines, and the amount of bus bandwidth lost through transfers that missed their deadlines. These measurements (given in the Appendix) are used to determine the effectiveness of the individual bus scheduling policies, and a summary of the behavior of each of the tested policies in underload and overload is shown in Figure 1 and Figure 2, respectively. These figures graphically illustrate the major discontinuity that exists over the continuum from light underload to heavy overload, which is most apparent in the performance of the deadline scheduling policy.

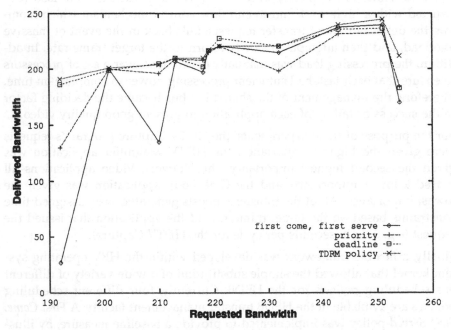

Figure 1 Bus Bandwidth Utilization in Underload

The following subsections summarize the results of these experiments for each of the bus scheduling policies explored, and associate the observed effects with the general characteristics of each policy.

First Come, First Serve (FCFS)

Across the board, the FCFS policy produces obviously unacceptable results, and the effect of missing delivery deadlines was clearly visible on the workstation's display. Even in underload, the Network Video application suffered from a high degree of jitter, and some amount of jitter was also noticed in the other applications (particularly when two consecutive HDTV capture requests appeared in the FCFS queue, due to the relative phasing of request arrival at the transfer agent). In overload, there was considerable jitter in the displays of all applications and there was frequent tearing visible in the HDTV capture window when its transfer took longer than the time available to move the data before the VIP would overwrite it (i.e., 50 msec — 16.67 msec per field, times the three other field buffers on the VIP).

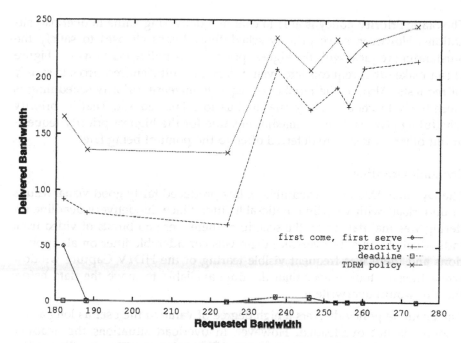

Figure 2 Bus Bandwidth Utilization in Overload

The first limitation that must be placed on the use of a FCFS policy is that resource being managed must not be overloaded; as seen in Figure 2, the performance of a FCFS policy approaches zero in overload situations. This is due to FCFS simply serving the oldest request first, which builds an increasing queue of newer requests. This policy ensures that, as requests miss their deadlines, subsequent requests within the queue will be later still. Second, even in underload situations (as shown in Figure 1), FCFS can fail to deliver a high degree of effective resource utilization whenever there is contention. In general, the greater the competition for a resource, the more likely it is that FCFS will waste resources by allocating them to requests which are (or will be) late and which, therefore, cannot deliver value to the user.

Static Priority

The Static Priority scheduling policy is obviously better than FCFS in all situations. However, even in underload, priority scheduling resulted in some degree of observable jitter and burstiness. These effects were due to this scheduling policy selecting the highest priority request, regardless of the display time constraints of other requests. This often defers a request one or more display times beyond what would be optimal for the video stream. In overload, more of the bursting and jitter effects were noticeable, since lower priority requests would frequently be deferred and then several of them would be satisfied at once.

The Static Priority policy is able to continue delivering value in overload situations. However, since priority scheduling always chooses to satisfy the outstanding request with the highest priority, this policy (as shown in Figure 1) can make other requests late even if there are sufficient resources to satisfy all requests. Also, as the number of requests increase, priority scheduling is more likely to choose to allocate the bus to a late request. That is, priority scheduling always delivers maximum value for the highest priority requestor but other requests are deferred often to the point of being late.

Dynamic Deadline

The Dynamic Deadline scheduling policy produced fairly good visual results in underload with very little noticeable jitter. Since the delivery deadline ordering was maintained by the scheduler there were no bursts of video from individual streams. In overload, there was considerable jitter on all applications and there was frequent visible tearing of the HDTV Capture window when transfer took longer than the time available to move the data before the VIP would overwrite it.

The deadline policy delivers a high degree of value to the user as long as the resource is not overloaded. However, in overload situations the deadline scheduling policy performs very poorly. This is because the deadline policy always grants the request with the earliest deadline, regardless of load. In overload situations, this means that deadline scheduling tends to service the requests that are most likely to be late. As a result, this policy can waste bandwidth and cause all requests to be late until the overload is removed. Note that even in situations where the long term average load can be satisfied with available resources, deadline is suspectable to instantaneous overload conditions which cause deadline scheduling to waste bandwidth.

TDRM-Based Scheduling

The TDRM scheduling policy had good visual results; in underload there was no jitter and in overload jitter was limited. There was no tearing of the HDTV Capture application's display window. These results are consistent with our expectations, due to the policy's continual attempts to meet all feasible deadlines.

The TDRM policy wasted very little bandwidth and achieved high utilization of the bus. The TDRM policy showed highly effective allocation of the bus with an increasing load and an increasing number of applications making requests. Occasionally, this policy can defer a request during underload due to a falsely predicted overload, which can result in slightly lower performance than deadline scheduling (as shown in Figure 1). In overload, TDRM outperforms static priority scheduling (as seen in Figure 2) since it will select a lower importance request if both the lower and higher importance requests can be satisfied.

Analysis of the Results

With only a limited number of applications making video transfer requests, there was no opportunity for the law of large numbers to smooth out anomalies. When a policy made an incorrect resource management decision, the general trend was to continue to make incorrect decisions because previous errors reduced the usable bandwidth. This is most obvious with the priority scheduling policy where after satisfying the highest priority request, the scheduler often only had bad choices left.

FCFS and dynamic deadline policies had difficulty dealing with an increasing number of requestors and increasing load. Both policies have a greater likelihood of making the wrong resource allocation decision as the load on the HSDB (and number of applications) grows.

The TDRM policy had some difficulty detecting (or predicting) overload due to an incomplete analytical model for computing the cost of a transfer request. Keeping statistics on the results of previous transfers to the same window probably would have filled in the gap between the analytical model and reality.

Note the applications' high level end-to-end flow control did not help as much as it could, since it was based on long-term rate calculations instead of direct feedback from the dispatcher. More graceful overload handling requires additional information from the resource manager than a simple indication of success or failure. The flow control modules must know why their constraint was missed in order to make more intelligent choices as to which type of corrective actions should be taken.

One major issue with i/o buses is the level of granularity of transfers. Preemption of the active transfer on the HSDB has a limited effect in these application scenarios due to length of time preemption takes (1 millisecond) and the relatively small size of most of the transfers (2 to 4 milliseconds).

Conclusions

The results of these experiments have been consistent with our expectations based on our previous experiments with processor scheduling [Northcutt 90a]. These results suggest that in low utilization situations (i.e., less than 50% loading) all four policies produced roughly equivalent results. As load increased, FCFS delivered consistently poorer results than the other three policies. Finally, as overload occurred Dynamic Deadline scheduling began to perform worse than FCFS. In overload situations, the TDRM policy outperformed Static Priority scheduling in terms of total effective bus utilization. In addition, the TDRM policy produced qualitatively superior visual results across the entire range of scenarios.

References

[Clark 90]　　　Clark, R. K.
Scheduling Dependent Real-Time Activities.
Ph.D. Thesis, Technical Report CMU-CS-90-155, School of Computer
Science, Carnegie Mellon University, August 1990.

[Hanko 92]　　　Hanko, J. G., Kuerner, E. M., Neih, J., Northcutt, J. D., and Wall, G. A.
"System Support for Media Synchronization."
Computer Networks and ISDN Systems, To Appear, Winter 1992.

[Locke 86]　　　Locke, D. C.
Best-Effort Decision Making for Real-Time Scheduling.
Ph.D. Thesis, Department of Computer Science, Carnegie-Mellon
University, May 1986.

[Northcutt 87]　Northcutt, J. D.
*Mechanisms for Reliable Distributed Real-Time Operating Systems:
The Alpha Kernel.*
Perspectives in Computing Series, Volume 16. Academic Press, 1987.

[Northcutt 88]　Northcutt, J. D.
The Alpha Operating System: Requirements and Rationale.
Archons Project Technical Report #88011, Department of Computer
Science, Carnegie-Mellon University, January 1988.

[Northcutt 90a]　Northcutt, J. D., Clark, R. K., Maynard, D. P., and Trull, J. E.
Decentralized Real-Time Scheduling.
Final Technical Report to RADC, RADC-TR-90-182, School of
Computer Science, Carnegie-Mellon University, August 1990.

[Northcutt 90b]　Northcutt, J. D. and Kuerner, E. M.
*System Support for Time-Critical Media Applications: Functional
Requirements.*
HRV Project Technical Report SMLI 92-0077, November 1990.

[Northcutt 92]　Northcutt, J. D., Wall, G. A., Hanko, J. G., and Kuerner, E. M.
"A High Resolution Video Workstation."
Signal Processing: Image Communication 4(4-5), Elsevier, August 1992.

Appendix

Experiment Number	1	2	3	4	5	6	7	8	9	10
HDTV Capture	30	30	30	30	30	30	30	30	30	30
HDTV Magnify	—	—	15	—	—	15	15	15	15	15
Cine Loop	30 H	40 H	—	—	60 H	30 H	60 H	30 H	60 H	30 N
Network Video 1	—	—	—	15	—	—	—	—	—	—
Network Video 2	—	—	—	—	—	—	—	—	—	—
Network Video 3	—	—	—	—	—	—	—	—	—	—
Requested Bandwidth (MB/sec)	243	251	221	213	274	237	253	232	256	218

Experiment Number	11	12	13	14	15	16	17	18	19	20
HDTV Capture	30	30	30	30	30	30	30	30	30	30
HDTV Magnify	15	—	—	—	—	—	—	15	15	15
Cine Loop	60 N	—	30 LB	60 H	30 N	60 N	—	—	30 N	60 N
Network Video 1	—	15	15	15	15	15	15	15	15	15
Network Video 2	—	15	—	—	15	15	15	15	15	15
Network Video 3	—	—	—	—	—	—	15	15	15	15
Requested Bandwidth (MB/sec)	254	201	219	260	210	246	192	182	188	224

Table 1 Individual Experiment Configurations

Note that experiments 6, 7, 8, and 9 differ in window stacking order and therefore represent different data points. Also, the Cine Loop was run with an HDTV format window (H), a letter-boxed NTSC format (i.e., 640 by 210) window (LB), as well as a full NTSC format window (N).

The table below shows achieved data rate in units of MB/sec (A), bandwidth lost due to late transfers in units of MB/sec (L), transfers dispatched per second (D), and the number of transfers missing their deadline per second (M).

Expr. #	FCFS				Static Priority				Dynamic Deadline				TDRM			
	A	L	D	M	A	L	D	M	A	L	D	M	A	L	D	M
1	238	0	55	0	232	5	55	5	236	0	55	0	240	0	60	0
2	240	11	64	10	233	14	64	12	237	7	63	6	245	2.4	58	2
3	221	0	60	0	221	0	66	0	228	0	64	0	221	0	60	0
4	207	0	39	0	209	0	41	0	210	0	42	0	210	0	42	0
5	0	229	76	76	215	34	63	26	0	236	76	76	245	0	51	0
6	4.4	233	81	73	208	28	81	24	5	232	72	81	235	0	83	0
7	0	218	84	84	192	44	82	33	0	219	85	85	234	0	80	0
8	215	6	75	8	198	23	75	20	221	0	75	0	221	0	75	0
9	0	207	86	86	175	47	76	42	0	208	87	87	216	1.2	70	1
10	197	6	75	9	208	5	77	4	202	5.5	76	4	210	0	75	0
11	170	68	100	50	184	53	100	40	183	54	100	44	212	1.3	80	1
12	200	0	59	0	200	0	59	0	199	0	58	0	201	0	60	0
13	217	0	68	0	215	1.6	68	2	217	0	68	0	217	0	68	0
14	0	230	89	89	209	30	77	29	0	233	86	86	230	0	65	0
15	135	72	90	37	196	4	82	3	205	0	84	0	204	0	85	0
16	4.4	218	101	100	172	45	97	34	4.5	223	102	101	207	0	85	0
17	27	157	71	48	130	63	71	21	186	0	69	0	191	0	71	0
18	50	114	77	40	92	74	78	39	0	158	74	74	165	0	81	0
19	0.8	159	97	96	80	85	99	53	0	161	99	99	136	0.2	92	1
20	0	169	108	108	70	97	102	67	0	177	109	109	134	0	89	0

Table 2   Results of Experiments — Raw Data

Analysis of I/O Subsystem Design for Multimedia Workstations

Peter Druschel   Mark B. Abbott   Michael Pagels   Larry L. Peterson *

Department of Computer Science
The University of Arizona
Tucson, AZ 85721

Abstract. This paper argues that the CPU/memory data path is a potential throughput bottleneck in multimedia workstations, and analyzes the implications for the design of the I/O data path. It identifies known techniques for optimizing the data path, and highlights subtle interactions that occur when these techniques are used in combination. The paper concludes by proposing an abstract type for data buffers, called *IOData*, and argues that this abstraction provides a flexible basis for the end-to-end optimization of the I/O data path.

1   Introduction

With the emergence of fiber optics technology, local area networks have recently seen an order-of-magnitude increase in bandwidth. Another order-of-magnitude leap—to 1Gbps—is likely to occur in the near future. Suitable network interfaces can deliver this increased bandwidth to desktop workstations. At the same time, dramatic increases in microprocessor performance continue. These technological advances make it feasible to support powerful multimedia applications, including continuous media (CM) such as audio and video.

The substantial increases in processor performance and network bandwidth contrast with only moderate improvements in main memory and bus performance of workstations. We expect this trend to continue because cost considerations will prevent the use of dramatically faster main memory and interconnect technology in this class of machines. In particular, this paper argues that the CPU/memory bandwidth of workstations will remain within the same order of magnitude as their network bandwidth, and as a consequence, care must be taken to minimize the number of trips that CM data takes across the CPU/memory data path.

To achieve a well-balanced system that minimizes bus traffic, an end-to-end design that integrates the entire data path—from source device, through the operating system and one or more application programs, to the sink device—is required. Existing I/O systems typically fail to do this. For example, software

* This work supported in part by National Science Foundation Grants CCR-9102040, NCR-9204390, and DARPA Contract DABT63-91-C-0030

data manipulation steps are performed serially, resulting in poor locality of reference and thus increased memory traffic; data is often copied from one main memory location to another when crossing the user/kernel protection boundary; and additional copies may occur when the data passes through a user-level server, such as a window manager. We show that the CPU/memory bandwidth will remain within an order of magnitude of network bandwidth. Therefore, the CPU/memory data path will form a bottleneck in this type of system.

This paper analyzes the I/O data path from an end-to-end perspective, focusing on data dependent costs. (To support CM applications, appropriate scheduling and IPC mechanism such as the ones described in [8] are required in addition to the techniques presented in this paper.) Unlike previous work on this topic, which concentrates on alternative network interfaces and OS buffering techniques, we consider the entire data path, highlighting the subtle interactions that occur when various techniques and mechanisms are used in combination. Moreover, our analysis considers the trend towards kernelized operating systems and third-party user-level servers. We take the position that, in order to enable software innovations, applications should have access to CM data, which precludes the use of low-level streaming. The paper concludes with the introduction of a new abstract type for data buffers, called *IOData*, that serves as a flexible basis for an integrated I/O data path.

2 Performance of Workstation Hardware

This section analyzes several performance characteristics of desktop workstation hardware that affect CM data throughput. It includes both measurements of commercially available workstations, and projections of the relevant characteristics of next-generation workstations.

Table 1 shows, for each of four commercial workstations, the SPEC performance rating (integer and floating point), the peak memory bandwidth of the memory subsystem (taken from the hardware specification), and the results of several CPU/memory bandwidth measurements described below. The ratio of measured bandwidth to the peak memory bandwidth is shown in parentheses.

	SPECInt89	SPECfp89	Memory (Mbps)	CPU/Memory (Mbps)		
				Copy	Read	Write
IBM RS/6000 340	28.8	88.8	2133	405 (.19)	650 (.30)	590 (.28)
HP 9000/720	39.5	78.5	1600	160 (.10)	450 (.28)	315 (.20)
Dec 5000/200	19.5	26.7	800	100 (.13)	300 (.38)	570 (.71)
Sun SPARC 2	21.7	27.4	640	90 (.14)	180 (.28)	140 (.22)

Table 1. Performance of some current Workstations

The CPU/Memory bandwidth numbers reported in Table 1 were obtained using a simple benchmark that measures the sustained bandwidth during a series of read, write, and copy operations. The benchmark was coded in C and compiled with each workstation vendor's native C compiler at the highest level of optimization. No attempt was made to tune the C source or the generated machine code for a particular machine. The read and write columns were measured by reading (writing) each element of an array. The copy column was measured using two methods: Element-wise assignment of two arrays, and invoking the bcopy() library function. The benchmark used arrays of type int (32bits) and double (64bits). The numbers reported are the best that could be achieved in each case; i.e., int versus double for the read and write columns, and int versus double versus bcopy() for the copy column. Finally, the size and alignment of the arrays was chosen to eliminate effects of the data cache.

The main result is that the measured bandwidths are only a fraction of the peak memory bandwidth. In particular, the read bandwidth ranges from 28 to 38%, and the copy bandwidth from 10 to 19% of the peak memory bandwidth. Note that the maximal possible copy bandwidth is one half the peak memory bandwidth. This is because copy involves moving the data across the CPU/memory bus twice.

This bandwidth degradation is the combined effect of two limitations. First, the compiler-generated machine code, and the vendor-provided implementation of the bcopy() function may be less than optimal for the intended task of the benchmark program. We have deliberately not tried to eliminate this limitation, since it equally affects real programs. Second, there is a hardware imposed limit to the bandwidth that the CPU can sustain. All memory subsystems use some form of pipelining (interleaving/page mode) to achieve high bandwidth despite the relatively high access latency of dynamic RAM. The average bandwidth drops off for small transfer sizes, where the startup latency contributes significantly to the transfer time. Data is transferred between cache and main memory at the size of a cache line, which is too small to achieve a large fraction of the peak memory bandwidth. We conclude that the CPU/memory bandwidth of current workstations is within the order of magnitude of their network bandwidth—hundreds of Mbps.

Next generation workstations will support Gbps network adapters that can transfer data into main memory at network speeds. The continuing increases in CPU speed will soon permit software data manipulations at this speed. For example, the first implementation of DEC's Alpha processor [5] permits the execution of 12 to 24 CPU instructions for every machine word in a 1Gbps data stream. However, dramatic increases in CPU/memory bandwidth cannot be expected. Consider what designers of the next generation of workstations might do to improve CPU/memory bandwidth. Peak memory bandwidth can be improved by increasing the memory width, but cache line size must be increased proportionally to achieve a substantial increase in CPU bandwidth. Cache line sizes that result in an optimal hit rate, however, are typically too small to achieve a substantial fraction of the peak memory bandwidth. Another approach is to reduce transfer latencies, but DRAM access times are considered to be near their technological limit. Several recently announced components integrate some form of a cache with a dynamic RAM to reduce the average access latency [4]. These integrated second level caches use large cache lines and are connected to the DRAM using wide data paths. As for any cache, the hit rate of these components depends on locality of reference. As we shall discuss in Section 3, accesses to CM data exhibit poor locality. Consequently, we expect CPU/memory bandwidth to be of the same order of magnitude as network bandwidth for at least the next generation of desktop workstations.

3   Effects of Data Caches

Workstations employ caches to bridge the gap between CPU and main memory speeds. The idea is to keep a portion of the main memory's data in a high speed memory close to the CPU. A cache increases system performance by reducing the average access latency for data and instructions. It also reduces contention for main memory access in shared memory multiprocessors. However, the effectiveness of caches is influenced by a number of factors, such as size and organization of the cache, locality of data access, and processor scheduling.

Assuming an integrated system that supports full application-level programmability of CM data—i.e., ignoring workstations with a built-in TV set—processing CM data requires the CPU to inspect and possibly modify every word of a CM data unit, potentially multiple times. (The next section identifies several reasons why CM data might be transferred across the CPU/memory data path.) This section argues that data caches are not effective in eliminating these CPU/memory transfers. Consider the following factors, ordered according to importance.

Processor Scheduling: CPU scheduling may cause the execution of other programs to be interleaved with the processing of a data unit. By the time processing resumes, cached portions of the data unit have most likely been replaced. On a multiprocessor, the processing may even be resumed by a different CPU with its own data cache. There are a number of situations where scheduling occurs during the processing of a data unit. When the data unit is passed on to another thread (i.e., queued), a processor must be scheduled to run that thread. Queuing typically occurs at the user/kernel boundary, in certain protocols, and between the device driver's interrupt handler and the top half of the driver. In the worst case, additional queuing may occur between protocol layers. Moreover, hardware interrupts, and the events they indicate, trigger processor rescheduling.

Cache Write Policy: Uniprocessor data caches often employ a write-through policy, meaning that every store operation requires a write to main memory. Write buffers are typically used with write-through caches to decouple the CPU from the timing of memory writes. However, many consecutive writes—such as would occur when reading and writing every word of a data unit—could still cause the CPU to stall on store instructions.

Cache Lookup Strategy: Many operating systems employ virtual memory techniques—e.g., virtual copying—that cause physical page frames to be mapped into different virtual address spaces. With a physically addressed cache, cached portions of such pages remain cached even after a processor reallocation to another address space. However, some machines use virtually indexed caches tagged with an address space identifier. With these caches, accesses to cached data from shared pages cause cache misses after a processor reallocation.

Cache Size: Data caches, particularly fast on-chip caches, are limited in size. For data to remain cached during a data inspection/modify step that involves loading and storing every word, the cache must be at least twice the size of the data unit. In practice, cache size requirements are further increased by cache line collisions due to limited associativity of the cache, and by accesses to program variables during and between data manipulations.

4   Avoiding Data Transfers

The previous two sections document our argument that the CPU/memory bandwidth of workstations is within an order of magnitude of the network bandwidth, and that data caches have a minimal effect in reducing CPU/memory traffic resulting from CM data access. It follows, therefore, that in order to preserve the bandwidth on the data path from the source (e.g., network device), through the OS and application, to the sink (e.g., a display device), multiple transfers of the data between the CPU and memory must be avoided.

Each of the following subsections identifies a potential data transfer along this path, briefly describes one or more technique for efficiently handling or avoiding it, and discusses the assumptions and limitations of these techniques.

4.1   Device/Memory Transfers

Data must be moved between main memory and network/device adapters. The techniques most commonly used are *Direct Memory Access* (DMA) and *Programmed Input/Output* (PIO). DMA implies that the I/O adapter transfers data directly from/to main memory, without involving the CPU. PIO requires the processor to transfer every word between main memory and I/O adapter in a programmed loop.

With DMA, the data transfer can proceed concurrently with other activity by the processor(s), allowing a degree of parallelism. Moreover, it is generally possible to transfer large blocks of data in a single bus transaction, thereby achieving transfer rates close to the limits of main memory and I/O bus. On the downside, DMA requires some complexity in the device adapters. Also, data caches are often not coherent with respect to DMA transfers, that is, after a DMA transfer from an I/O adapter to main memory, the data cache may be stale. Consequently, there are costs due to the necessary flushing of the data cache, and subsequent cache misses when the data is accessed by the CPU.

PIO maintains data cache coherency. On the downside, CPU cycles are consumed, and a transfer between main memory and an I/O adapter requires two bus trips, compared to one for DMA. In general, data in memory mapped I/O locations cannot be cached, meaning that each load/store instruction causes an

I/O bus transaction. Consequently, only a small fraction of the peak I/O bandwidth is typically achieved using PIO.

Network adapters may support some packet demultiplexing prior to data transfer to main memory, allowing filtering and selective placement of data units in memory. In the simplest case, the adapter allows the host to peek at a network packet's header. The host makes the demultiplexing decision and initiates the data transfer to the appropriate location in main memory, using DMA or PIO. More elaborate adapters can be programmed by the host CPU to automatically recognize network packets by matching their headers, and place them into appropriate memory locations using DMA [9].

4.2 Cross-Domain Transfers

Protection necessitates the transfer of data between protection domains. In the simplest case, CM data is handled by a single application process running on top of a conventional monolithic kernel. In this case, each data unit must cross the user/kernel boundary. In general, additional user processes, such as window managers and multimedia servers, and the trend in operating system design towards a kernelized system structure introduces additional domain boundaries into the I/O data path.

Software data copying as a means of transferring data across domain boundaries exacerbates the memory bottleneck problem. A number of techniques exist that rely on the virtual memory system to provide copy-free cross-domain data transfer. *Virtual page remapping* [2, 11] unmaps the pages containing data units from the sending domain and maps them into the receiving domain. *Virtual copying* [7] shares the transferred pages among the sending and receiving domain, and delays copying until one of the sharing domains attempts to write the shared data unit, in the hope that the data won't be written at all. *Shared virtual memory* employs buffers that are shared among two or more domains to avoid data transfers.

Virtual page remapping has move rather than copy semantics, which limits its utility to situations where the sender needs no further access to the transferred data. Virtual copying has copy semantics, but it can only avoid copying when the data is not written by either the sender or the receiver after the transfer. Both techniques require careful implementation to achieve low latency. The time it takes to switch to supervisor mode, acquire necessary locks to VM data structures, change VM mappings, perform TLB/cache consistency actions, and return to user mode poses a limit to the achievable performance of both techniques.

Shared virtual memory avoids data transfer and its associated costs altogether. However, its use may compromise protection and security between the sharing protection domains. Since sharing is static—a particular page is always accessible in the same set of domains—a priori knowledge of all the recipients of a data unit is required. *Dynamic shared memory* [6] is a variation of this technique that attempts to overcome these shortcomings.

4.3 Data Manipulations

Data manipulations are computations that inspect and possibly modify every word in a data unit. Examples include encryption, compression, error detection/correction, and presentation conversion. Data manipulations can be performed in hardware or software. Hardware support for data manipulations can reduce CPU load, and when properly integrated, reduce memory traffic. For certain manipulations like video (de)compression, hardware support may be necessary in the short term, due to the computational complexity of the task. To rely on hardware for all data manipulations, however, seems too constraining for multimedia applications.

Since software data manipulations are often associated with distinct program modules, they are typically implemented in series, resulting in poor locality of data access. This can lead to serious bandwidth degradation due to repeated transfers across the CPU/memory data path. Techniques for avoiding this bottleneck in the implementation of network protocol suites are known as *Integrated Layer Processing* (ILP) [3, 1]. ILP combines several serial data manipulation steps, so that they can be performed while a particular data item is in registers or cache. The basic approach is to restructure the software so that all data manipulations are performed during one traversal of a data unit. This results in a much improved locality of data access, so that each data item can remain in the data cache or in registers until all data manipulations have been performed on that item. A major limitation of ILP is that data manipulations performed by program modules in different protection domains cannot be effectively integrated.

4.4 Buffer Management

Buffer editing—which we distinguish from data manipulations that require the inspection and/or modification of each word of the data—can be expressed as a combination of operations to *create, share, clip, split, concatenate,* and *destroy* buffers.

A buffer manager that employs lazy evaluation of buffers to implement the aforementioned primitives can facilitate copy-free buffer editing. The manager provides an abstract data type that represents the abstraction of a single, contiguous buffer. An instance of this abstract buffer type might be implemented as a sequence of not necessarily contiguous fragments.

Since buffer editing occurs frequently in network protocol implementations, buffer managers are used in the network subsystem of many operating systems [10]. However, the scope of these managers is restricted to a single protection domain, typically the kernel. Consequently, an evaluation of buffers—that is, a software copy into a contiguous buffer—is necessary when a data unit crosses a protection domain boundary.

4.5 Interface Design

The *Application Programming Interface* (API) defines the semantics for data transfer to and from the application program. The semantics defined by this interface can be a significant factor in the efficiency of the implementation. Consider, for example, the UNIX read() and write() system calls. Applications programs may choose a contiguous data buffer in their address space with arbitrary address, size, alignment, and have unconstrained access to that buffer. This low-level representation of data buffers makes it difficult for the system to avoid copying of data.

An API that lends itself to efficient data transfer methods should use an abstract data type to represent data buffers. Application programs access the buffer type indirectly through its operations. They ask the system to create an instance when they need a buffer, pass/receive instances as arguments/results in system calls, and ask the system to deallocate the instance when the associated buffer is no longer needed. The system then has complete control over buffer management, including address, alignment, method of transfer, and so on. A proposal for an API of this kind is presented in Section 6.

5 End-to-End Design

This section analyzes the design space for an I/O subsystem that optimizes data-dependent costs along the entire data path from source to sink device. The goal is to minimize the number of CPU/memory transfers. We analyze several representative sample points in the design space. For each of sample point, we discuss tradeoffs, determine the optimal data path, and select appropriate implementation techniques to achieve the optimal data path.

5.1 Hardware Streaming

One approach to avoiding the CPU/memory bottleneck is to remove both the CPU and main memory from the data path. The data path may be set up and controlled by software, but the data itself is transferred from source to sink device without CPU involvement.

In the simplest case, both the source and sink device are integrated on the same I/O board. Data never reaches the I/O bus, memory, or CPU. Data can only be exchanged among the devices intergrated on the same board, and functionality is restricted to that provided by the board. The second approach uses peer-to-peer I/O bus transactions to transfer data directly from source to sink device, bypassing both main memory and the CPU. This method offers some limited hardware configurability. Device adapters must support peer to peer bus transfers and have the capability to perform demultiplexing, and any necessary data format conversions.

Both methods can be characterized by a lack of integration with the workstation computer system. The result is a tradeoff between complexity in the multimedia hardware on one hand, and flexibility and programmability on the other. In either case, software programs have no access to the data, and are constrained by the functionality provided in hardware. Thus, hardware streaming offers little room for innovation in multimedia applications. No specific techniques are required for data transfer in the I/O subsystem; however, many of the issues discussed in this paper may apply to the *internal* design of the multimedia adapter boards.

5.2 DMA-DMA Streaming

A second approach is for data to be transferred using DMA from a source device to buffers in main memory, where some buffer editing may occur under control of the operating system, and then, the data is transferred to a sink device using DMA. The CPU controls data transfers, may change the size of the data units and control information (headers), but remains removed from the data path. Unlike the previous approach, DMA-DMA streaming requires no special hardware other than DMA support on both the source and sink device adapters. Consequently, generic devices (e.g., disks) can be used to support multimedia.

Two DMA I/O bus trips are required by this approach. It follows that the throughput is bounded by one half of the I/O bus bandwidth. In practice, the sustainable throughput is lower, since main memory accesses caused by concurrent CPU activity compete with the DMA transfers for main memory access, even if the I/O bus and memory bus are separated.

Two key techniques are required to keep the CPU removed from the data path in this approach: scatter-gather DMA support on the device adapters, and a buffer manager that supports copy-free buffer editing. Support for gather DMA in the sink device adapter is critical. Recall that a lazily evaluating buffer manager may cause a buffer to consist of multiple discontiguous fragments. In the absence of a gather capability, it is necessary to copy data units into contiguous space prior to a transfer to the sink device. A DMA scatter ability of the source device is not strictly necessary, but it allows the use of a fixed buffer size, which may lead to a more efficient implementation of the buffer manager.

5.3 OS Kernel Streaming

Now consider systems where software data manipulations are performed, but all the manipulations are executed within the privileged kernel protection domain. In other words, the data is not accessed in any user-level protection domain. Clearly, the data must pass through the CPU/data cache. The goal is to keep the resulting CPU/memory data traffic to the minimum, namely two transfers. The solution is to integrate data manipulations—if there is more than one—using ILP. Note that data manipulations may include data movement from and to devices via PIO.

If a device supports both DMA and PIO, it may be beneficial to use PIO in this type of system, since programmed data movement can be integrated with other data manipulations. That is, instead of first using DMA and then loading/storing data to/from main memory, the CPU could directly load data from the source I/O adapter, and store data to the sink I/O adapter, bypassing main memory. This approach saves two DMA bus trips, which would otherwise compete with the CPU for memory access. However, it also trades memory accesses for potentially much slower load and store operations across the I/O bus. Which approach results in a more efficient data path depends on the relative performance of memory accesses and DMA transfers, versus PIO on the target hardware.

Unlike the previous methods, OS kernel streaming offers full programmability of the data path. However, all data manipulations must be performed in the kernel of the operating system. Applications and user-level multimedia servers are restricted to the set of data manipulations implemented by the kernel.

5.4 User-Level Streaming

Next consider systems where data passes through the kernel, plus one or more user-level protection domains. These user-level domains could be part of a kernelized operating system, implement third-party servers for multimedia support or windowing, or be part of a multimedia application. In addition to the issues discussed in the previous subsections, the designer is faced with protection boundaries. Protection domain boundaries are an obstacle to the integration of data manipulations. We conclude, therefore, that designers should make every effort to locate all data manipulation functions in the same domain. Protection also requires an efficient method for transferring data between domains.

Section 4.2 discusses several VM techniques for cross-domain data transfer. The use of such a facility is critical for data transfer between two user-level protection domains. Otherwise, two software copies are required for user-to-user transfers on most systems—from the source user domain to a kernel buffer and from the kernel buffer to the target user domain.

As a special case, data transfer between a user domain and the kernel domain can be implemented without extra cost, if the transfer is combined with the data movement between I/O adapter and main memory. That is, data is transferred directly between the I/O adapter and the user buffer. However, there are several problems with this approach when applied to network I/O. On the receiving end, the data transfer must be deferred until the appropriate user buffer can be determined, which may depend on the behavior of the user program. This can lead to contention for buffers in the network adapter. Support for packet demultiplexing is required in the network adapter, since the receiver must be determined, prior to the data transfer to a user buffer. On the transmitting side, network protocols may need to retain a copy of the data for retransmission, but the only copy of the data is in a user buffer. Finally, a system that relies on this approach as the sole means of cross-domain data transfer is constrained to a monolithic OS structure, and requires that data passes only through a single user domain.

5.5 Need for Integration

API, cross-domain data transfer facility, and buffer manager must be integrated in a manner that takes into account their subtle interactions. Consider the effects of naively applying efficient cross-domain data transfer and lazily evaluated buffer management to a system with a UNIX-style API. The buffer manager is restricted to the kernel domain. A virtual copy facility is used for cross-domain data transfer; due to per-operation latencies, it works most efficiently for large, contiguous buffers. The system API specifies a pointer to a contiguous buffer in the user domain's address space.

In this case, data units from the source device are placed in main memory buffers, and some buffer editing occurs as part of the in-kernel I/O processing; e.g., reassembly of network packets. When a data unit represented by a lazily evaluated buffer reaches the user/kernel boundary, it must be evaluated (copied into contiguous storage), despite the use of a virtual copy facility. The reason is that the interface defines data buffers to be contiguous. Since the API allows applications to specify an arbitrarily aligned buffer address and length, the buffer's first and last address may not be aligned with page boundaries. Consequently, the data transfer facility may be forced to copy the portion of the first and last page that is overlapped by the buffer.

Once in the user domain, more buffer editing may need to be performed. Since the buffer management tool is not available at user-level, the application must either perform data copying, or employ its own lazily evaluating buffer management. In the latter case, another copy is required when data crosses the next domain boundary. After data is transferred back to the kernel, the semantics of the API allow the user program to reuse the data buffer instantly, which is likely to force the virtual copy facility to copy parts of the buffer. We conclude that, despite the use of copy-avoiding techniques, multiple copies occur along the data path, and these copies are an artifact of the poor integration of these techniques.

One problem is that the implementation of the buffer manager is local to the kernel domain; a global implementation is necessary to maintain the lazily evaluated representation of data buffers along the entire data path. A global abstract buffer type has the additional benefit that all domains (including applications) can perform copy-free buffer editing. A second problem is the API, which commonly does not permit a non-contiguous representation of buffers, and as a consequence, stands in the way of efficient data transfer. A potential third problem is the cross-domain data transfer facility's inability to efficiently support the transfer of non-contiguous buffer fragments.

In conclusion, it is necessary to maintain a lazily evaluated representation of data buffers along the entire data path. This implies that all programs must deal with this representation of buffers. Consequently, a global buffer manager is needed that is integrated with the API and cross-domain transfer facility. The choice of a cross-domain data transfer method may further influence the design of a network adapter. For example, virtual shared memory requires demultiplexing prior to the data transfer from adapter to main memory.

6 An Integrated I/O Data Path

This section proposes an API based on an abstract buffer type that permits the integration of the API, buffer manager, and cross-domain data transfer facility to achieve an efficient end-to-end data path. The approach hinges on an abstract data type called *IOData*. This section gives an overview of the IOData type design, discusses its use, and briefly sketches how it can be implemented using different buffer management schemes and cross-domain data transfer facilities.

The use of the IOData type for data buffers has important advantages. First, it ensures that a single buffer representation can be maintained along the entire data path, permitting lazy buffer evaluation. Second, it isolates applications, user-level servers, and large parts of the kernel from details of buffer management and cross-domain data transfer. This increases portability of both applications and operating system, and permits the use of the most efficient buffer management and data transfer on different platforms. Third, the IOData type gives applications access to efficient buffer manipulation operations, and eliminates the need for separate application-level buffer management.

An instance of the IOData type represents an abstract data buffer of arbitrary length. It encapsulates one or more physical buffers that contain the data. At any given time, the physical buffers may not be contiguous, mapped in the current domain, or even in main memory. The IOData type is *immutable*, i.e., once an instance is created with an initial data content, the content cannot be subsequently changed. IOData instances can be manipulated using a well-defined set of operations. An implementation of this abstract type—i.e., code that implements its operations—is included as part of a library in each protection domain. The exact form and syntax of IOData's operations depends on the programming language used, which may vary from domain to domain.

The IOData type supports the following operations. An *allocate* operation creates a new instance of the requested size and allocates an appropriate number of physical buffers. During an initialization phase, the client is provided with a list of pointers to these physical buffers, for the purpose of initialization. A *share* operation creates a logical copy of an IOData instance; it does not actually copy the physical buffers. *Clip*, *split*, and *concatenate* operations implement the necessary buffer editing operations. A *retrieve* operation generates a list of pointers to the physical data buffer fragments, thereby allowing the client to access the data. A *mutate* operation is a combination of retrieve and allocate. It allows a client to read the data from an IOData instance, and store the (perhaps modified) data into a new IOData instance. The operation generates a list of pointer pairs, one referring to a fragment of the source, the other pointing to a physical buffer of the target. Finally, a *deallocate* operation destroys an IOData instance, and deallocates the physical buffers if no logical copies of the data remain.

Consider an implementation of the IOData type. One key feature is that the implementation has complete control over the size, location, and alignment of physical buffers. Consequently, a variety of buffer management schemes are feasible. All buffers may be part of a system-wide pool, allocated autonomously by each domain, located in a shared VM region, or they may reside outside of main memory on an I/O adapter. Physical buffers can be of a fixed size to simplify and speed allocation. The other key feature of the IOData type is its immutability. It allows the transparent use of page remapping, shared virtual memory, and other VM techniques for the cross-domain transfer of IOData instances. Virtual copying can be used with increased efficiency since physical buffers are guaranteed not to be written after a transfer.

It is possible to extend an existing API (such as that of UNIX) to include input/output operations based on the IOData type. New applications that depend on high bandwidth (such as multimedia) can use the new interface. The conventional interface can be maintained for backward compatibility.

Acknowledgement

We would like to thank Steve Scott for providing technical information on IBM RS/6000 workstations and for running our benchmark on these machines.

References

1. M. B. Abbott and L. L. Peterson. Automated integration of communication protocol layers. Technical Report TR 92-25, Department of Computer Science, University of Arizona, Tucson, Ariz., Dec. 1992.
2. D. R. Cheriton. The V distributed system. *Commun. ACM*, 31(3):314–333, Mar. 1988.
3. D. D. Clark and D. L. Tennenhouse. Architectural Considerations for a New Generation of Protocols. In *SIGCOMM Symposium on Communications Architectures and Protocols*, pages 200–208, Philadelphia, PA, Sept. 1990. ACM.
4. R. Comerford and G. F. Watson. Memory catches up. *IEEE Spectrum*, 29(10):34–57, Oct. 1992.
5. Digital Equipment Corporation, Palo Alto, California. *Alpha Architecture Technical Summary*, 1992.
6. P. Druschel and L. L. Peterson. High-performance cross-domain data transfer. Technical Report TR 92-11, Dept. of Comp. of Sc., U. of Arizona, Tucson, AZ (USA), Mar. 1992.
7. R. Fitzgerald and R. F. Rashid. The integration of virtual memory management and interprocess communication in Accent. *ACM Transactions on Computer Systems*, 4(2):147–177, May 1986.
8. R. Govindan and D. P. Anderson. Scheduling and IPC mechanisms for continuous media. In *Proceedings of 13th ACM Symposium on Operating Systems Principles*, pages 68–80. Association for Computing Machinery SIGOPS, October 1991.
9. H. Kanakia and D. R. Cheriton. The VMP Network Adapter Board (NAB): High-Performance Network Communication for Multiprocessors. In *SIGCOMM Symposium on Communications Architectures and Protocols*, pages 175–187, Stanford, CA, Aug. 1988. ACM.
10. S. J. Leffler, M. McKusick, M. Karels, and J. Quarterman. *The Design and Implementation of the 4.3BSD UNIX Operating System*. Addison-Wesley, 1989.
11. S.-Y. Tzou and D. P. Anderson. The performance of message-passing using restricted virtual memory remapping. *Software—Practice and Experience*, 21:251–267, Mar. 1991.

Tactus: Toolkit-Level Support for Synchronized Interactive Multimedia

Roger B. Dannenberg, Tom Neuendorffer, Joseph M. Newcomer, Dean Rubine

Information Technology Center, School of Computer Science, Carnegie Mellon
University, Pittsburgh, PA 15213 USA

Abstract. Tactus addresses problems of synchronizing and controlling
various interactive continuous-time media. The Tactus system consists of
two main parts. The first is a server that synchronizes the presentation of
multiple media, including audio, video, graphics, and MIDI, at a workstation. The second is a set of extensions to a graphical user interface
toolkit to help compute and/or control temporal streams of information
and deliver them to the Tactus Server. Temporal toolkit objects schedule
computation events that generate media. Computation is scheduled in
advance of real time to overcome system latency, and timestamps are
used to allow accurate synchronization by the server in spite of computation and transmission delays. Tactus supports precomputing branches
of media streams to minimize latency in interactive applications.

1 Introduction

Recently, many proposals have emerged for extending graphics systems to support multimedia applications with sound, animation, and video [17, 18, 16, 7].
Other research has been directed toward real-time transmission of multimedia
data over networks [13, 2] and standards for the representation and exchange
of multimedia data [14]. New capabilities for real-time interactive multimedia
interfaces [5] create new demands upon application programmers. In particular,
programmers must manage concurrent processes that output continuous media.
Timing, synchronization, and concurrency are among the new implementation
problems.

Traditionally, object-oriented graphical interface toolkits have presented a
high-level programming interface to the application programmer, hiding many
details of underlying graphics systems such as X or Display Postscript. However,
timing is usually overlooked in these systems. Programmers usually add animation effects by ad-hoc extensions, and synchronization at the level of milliseconds
needed for lip-sync, smooth animation, and sound effects is not generally possible.

We have extended an existing toolkit with new objects, abstractions, and
programming techniques for interactive multimedia. We also implemented a synchronization server that supports our toolkit extensions. Intuitively, our synchronization server does for time what a graphics server does for (image) space. In
our terminology, the application program is the *client*, which calls upon the

server to synchronize and present data. We call the combined toolkit and server the Tactus System.

The Tactus System has a number of novel and interesting features. It works over networks with unpredictable latency, and it can maintain synchronization even when data underflows occur. The techniques are largely toolkit-independent, and the Tactus Server is entirely toolkit independent. Data is computed ahead of real time to overcome latency problems, but the initial latency of a presentation is due only to computation and bandwidth limitations. Tactus is organized so that pre-existing graphical objects acquire real-time synchronizing behavior without changes to the existing code. The Tactus System also offers a new mechanism called *cuts*, whereby precomputed media can be selected with very low latency.

1.1 Assumptions

Before describing Tactus, we will present some assumptions and ideas upon which it is based. First, we are interested in distributed systems, and thus we assume that there will be significant transmission delays between servers and clients. Second, we assume that multimedia output will require the merging of multiple data streams; we want more than just "canned" video in a window. By *data stream*, we mean any set of timed updates to an output device. Data streams include video, audio, animation, text, images, and MIDI.

The assumption that delays will be present imposes limitations on the level of interaction we can expect. Network media servers may take seconds to begin presenting video even though the presentation, once started, is continuous. We intend to support applications where media start-up delays and latency due to computation of 10 to 1000 ms are tolerable. This includes such things as multimedia documents, presentations, video mail, and visualizations. It also includes more interactive systems such as hypermedia, browsers, and instructional systems where user actions determine what to view next. Although we rule out continuous feedback systems such as video games, teleconferencing, and artificial reality, we want to support rapidly altering the presentation at discrete choice points.

1.2 Principles

To deal with transmission and computation delays, it is necessary to start sending a data stream before it is required at the presentation site. Because of variance in computation, access, and transmission delays, it is also necessary to have a certain amount of buffering in the Tactus Server at the presentation site. When multiple streams are buffered, it is necessary to synchronize their output. With Tactus (see Figure 1), all data streams are timestamped, either explicitly or implicitly, so that Tactus can determine when each component of a stream should be forwarded to a device for presentation. We assume a distributed time service that can provide client software with an accurate absolute time, with very little skew between machines [12], although this assumption is not critical for most applications.

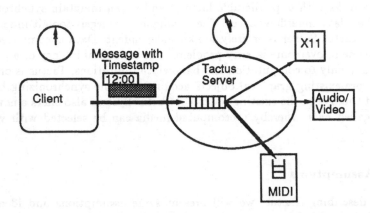

Fig. 1. The Tactus System. Clients send timestamped data (heavy lines) to the server ahead of real time. Data is buffered and then delivered to various presentations devices. Some presentation devices (e.g. MIDI as shown here) may accept data early and provide further buffering and more accurate timing than can be provided by the Tactus Server. The clock on the left shows logical time as seen by the client, while the clock on the right shows real time as seen by the Tactus Server.

Multiple presentations may be buffered at the Tactus Server. At any time, one is being presented while the others are potential responses to user choices. This avoids the latency of transmitting a presentation over the network after the user makes a choice. Transition points are marked so that smooth cuts are possible (see Section 4).

Input is handled in mirror image to output. There is latency between real input at the device level and the arrival of the input data at the application process, so all input must be timestamped. Input events can then be related back to the output that was taking place at the time of the input. It is up to the application to deal with the delay between input and output, for example, by "rewinding" to an indicated stop point or reflecting the input in future output.

The task of synchronizing output in a distributed environment is simplified by pre-computing or pre-transmitting data streams and timestamping them. Without additional support, however, this would complicate the work of the application, which then must compute data streams in advance of real time. One way to reduce this problem is to schedule application activity by a clock that is ahead of real time. A good analogy is that if you set your watch ahead by 5 minutes, you are more likely to show up on time for meetings.

In summary, the three most important principles of Tactus are (1) compute data streams in advance of real (presentation) time, (2) use a server at the presentation site to buffer and synchronize data streams, and (3) buffer responses to user choices to minimize response times. Buffering data at the presentation site can greatly increase the timing accuracy with which data is presented.

1.3 Previous Work

Few of these *principles* are original, but their integration and application are new. Tactus was inspired by David Anderson and Ron Kuivila's work on event buffering for computer music systems [3, 4]. This work is in turn related to discrete-event simulation. Later, Anderson applied these ideas to distributed multimedia [1], but not to interface toolkits. Active objects have long been used for animation [11] and music [6] systems, but have only recently gained attention in multimedia circles [10]. To our knowledge, we are the first to extend an object-oriented application toolkit with support for managing latency through precomputation and event buffering. CD-ROM based video systems have used buffering of images at choice points to allow for seek time. Our work focuses more on the implications of all these techniques for application toolkits.

Recently, many commercial multimedia systems have been introduced, including Apple's Quicktime [17], Microsoft's MPC [18], and Dec's XMedia [7]. These systems emphasize storage, playback, and scalability. HyTime [14] provides a standard representation for hypermedia but no implementation is specified. These systems could benefit from the synchronization and latency management techniques we propose, and our work suggests how a graphical interface toolkit might be extended to take advantage of commercial multimedia software.

2 The Tactus Toolkit Extensions

The Tactus Server could be used without the Tactus Toolkit, but this would require the user to compute data in advance of real time, implement various protocols (described in Section 3), and interleave computation for various streams. The toolkit simplifies these programming tasks. Our toolkit extensions include clock objects for scheduling and dispatching messages, active objects that receive wake-up messages and compute media, and stream objects that manage Tactus Server connections and timestamping (see Figure 2).

2.1 Active Objects

Active objects form the base class for all objects that handle real-time events and manage continuous time media in the Tactus extensions to ATK [15]. Each active object uses a clock object (set via the UseClock method) to tell time and to request wake-up calls. The RequestKick method schedules the active object to be awakened at some future time (according to its clock), and the Kick method is called by the system at the requested time.

Active objects are intended to take the place of light-weight processes, and often perform tasks over extended periods of time. This is accomplished by having each execution of the Kick method request a future Kick.

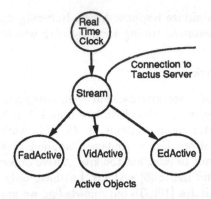

Active Objects

Fig. 2. A Clock Tree. Objects, including clocks, request a wake-up message from their parent in the clock tree. RealTime is at the root of the tree and interfaces with the operating system timing facilities. Stream is a subclass of Clock and manages connections to the Tactus Server. The leaves of the tree are subclasses of Active which produce and control multimedia data. Kick messages flow in the direction of the arrows, while RequestKick messages are sent in the opposite direction.

2.2 Clock Objects

Clocks are a subclass of Active. Each clock object keeps track of all the active objects (users) that have attached themselves via the UseClock method. Since clocks are active objects, they too can be attached to other clocks. Clocks are useful not only for their wake-up service but also because they manage mappings from one time system to another. Mappings are linear transformations, meaning that a clock can shift and stretch time as seen by its users. When a change in the mapping of time occurs, users of the clock are notified (whether or not they are waiting for a Kick). We call the time seen by users of Clocks *logical time*, as opposed to the *real time*. Logical time allows active objects to compute in "natural" time coordinates. Meanwhile, clocks can be adjusted to achieve "fast forward", "rewind", "pause", and "continue" effects.

2.3 The RealTime Object

Clocks form a "clock tree" whose leaves are active objects, whose internal nodes are clocks, and whose root is a special subclass of clock called RealTime. A RealTime object serves as the true source of time for the entire clock tree. It should be noted that the clock tree is entirely independent of the graphical view tree typically found in graphical user interfaces [8].

2.4 Stream Objects

Stream objects are a subclass of Clock. In addition to scheduling and kicking users, stream objects communicate with the Tactus Server and establish timestamps for Tactus messages. Stream objects also schedule their children ahead of real time by the worst-case system delay called *Latency*, a number which is presently determined empirically.

Recall that the Tactus Server expects all messages to have timestamps which serve as the basis for synchronization of multiple media. It might seem logical to use the kick times of active objects, but because kick times are the composition of perhaps several mappings at different levels for the clock tree, the active object kick time may have no simple relationship to real time or to the kick times of other active objects.

Rather than use active object kick times, timestamps are based on the idealized real time of the kick, that is, the requested kick time mapped to real time. Before a kick, the clock tree is inactive. When the kick time arrives, a Kick message is propagated from the RealTime object through the tree to an active object at a leaf of the tree. On the path from root to leaf, a stream object is kicked. The stream sets a globally accessible timestamp and stream identifier before propagating the Kick message. If the active object performs an output action, the output function called accesses the timestamp and stream identifier in order to compose a message for the Tactus Server. In this way, timestamps are implicitly added to client output.

Together, these classes and their specializations serve to insulate the application programmer from the detailed protocols necessary to send streams of data to the Tactus Server. The extensions do such a fine job of hiding details that existing programs can use Tactus without modification. (Tactus libraries are linked dynamically.) Although this provides no benefits to existing applications, it means that *existing application components can be given real-time synchronization capabilities*. For example, objects that formerly displayed text or images can now be called upon to deliver output synchronously with other media.

3 The Tactus System

As described in the introduction, the Tactus System consists of a Tactus Server as well as a set of extensions to an object oriented toolkit. In this section, we will describe how the two work together.

3.1 Steady State Media Delivery

Steady-state on the client side consists of active objects waiting for wake-up messages. At each wake-up, an object computes data such as a packet of audio or a frame of animation and sends it to the Tactus Server. The wake-up message is scheduled by a stream object that forces the computation to happen ahead of real-time. When no more computation is pending, the stream object computes when the next wake-up will occur and sends a null message to the Tactus Server with that timestamp. This tells the Server not to expect more messages until that time.

In a slight variation of the above, Tactus may choose to deliver the data to the presentation device slightly ahead of time, relying on the hardware or device driver to delay the presentation until a given timestamp. For example, our MIDI driver maintains buffers of timestamped packets of MIDI data and outputs data at the designated time. In contrast, X11 (our "graphics device driver") has no buffering or timestamping capability yet, so Tactus provides all timing control for graphics. These differences are invisible to clients.

Time is used to regulate the flow of data from clients to Tactus, thus alleviating the need for explicit flow-control messages. The client simply produces "one second of data per second" and sends it to Tactus. When the client is behind, it computes as fast as possible in order to catch up. If the client falls too far behind, the Tactus Server buffers will underflow and a recovery mechanism must be invoked. (See below.)

3.2 Stream Start-Up

We anticipate that the worst-case delay from client to device will be quite large (perhaps seconds). This is too large to be acceptable for the normal stream start-up time. Tactus clients typically will start streams with the goal of delivering media to the user as soon as possible. Therefore, the stream object advances logical time, causing the client to run compute-bound until it catches up. To further facilitate rapid start-up, each device has a *minimum* amount of buffering (measured in seconds) required before it can start, and the Tactus Server rather than the client determines when to start a presentation.

3.3 Underflow

An underflow is caused by the stream buffer running out of data. More precisely, underflow occurs when it is time to dispatch a data packet at time T, but there is no packet containing data at a time greater than T. Since data arrives in time order, a timestamp greater than T is desired because it indicates that all data for time T has arrived. (It is the current policy of Tactus to halt all media presentation at time T until all media for time T can be updated, but we believe other policies should be supported as well [1]).

No immediate feedback to the client is necessary upon underflow (presumably, the client is already compute-bound trying to catch up). When Tactus resumes data output, it sends a message to the client indicating the amount by which the presentation was delayed. This information can be used to control the total delay between computation and presentation. The default behavior is for the client to keep a constant presentation latency; if the Tactus Server stops the presentation for 2 seconds, then the client holds off on computation for 2 seconds as well.

Generally, this protocol takes place only at the stream level, and the clocks and active objects beneath the stream remain oblivious to the time shifts. On the other hand, active objects can attempt to avoid underflow by noticing or predicting when computation falls too far behind real time. For example, our animation object drops frames, maintaining a constant number of frames per second, when the logical time rate (playback speed) is increased.

4 Cuts

Because of various latencies, multimedia systems are often unable to respond to input without obvious "glitches" where, for example, the video image is lost, digital audio pops, and graphics are partially redrawn. This usually happens because there is a time delay between taking down one stream and starting up another. These annoying artifacts could be hidden if the new stream could be started before the the old one is stopped. Tactus supports this model, and a switch from one stream to another is called a *cut*.

In Tactus terminology, a *cut* is made from a primary stream to a secondary stream. To minimize latency, cuts are performed by the Tactus Server on behalf of its client. The client requests a cut, but the request may or may not be honored, depending upon whether the secondary stream is ready to run.

There are two attributes that describe a cut (see Figure 3). The first determines whether a cut may be taken at any point in time or only at certain time points, and the second describes whether the cut is made to the beginning of a secondary stream or to the current time.

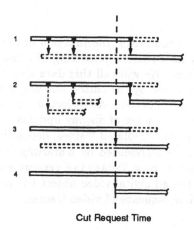

Cut Request Time

Fig. 3. There are four (4) types of cuts. The top two cuts are restricted to discrete time points, whereas the lower two can take place at any time. The first and third cuts here cut to a stream already in progress while the second and fourth types cut to the beginning of a stream.

Cuts must be anticipated by the application. Since the application runs ahead of the real presentation time, it will naturally come to choice points before the user has a chance to make a choice. For example, the application will generate graphics or video for an intersection before knowing whether the user will say "turn left" or not. At this point, the application will create a cut object to arbitrate between the current (primary) stream and a new "turn left" (secondary) stream.

Within the Tactus Server, the secondary stream will perform a normal stream start-up. If the user requests "turn left", the application[1] sends a cut message with a timestamp to the Server. If the message arrives before the time indicated by its timestamp, and if the secondary stream is ready to run, then Tactus switches to the secondary stream at the designated time. A message is returned to the application indicating success or failure. If the cut was a success, then the objects generating the no-longer useful primary stream will be freed. Clients can also specify an initial set of (initialization) commands to be issued when a cut takes place.

5   An Example Application

It is now time to see how clocks, streams, active objects, and the Tactus server work together to produce a synchronized multimedia presentation. We will describe an application we have actually built: a time-line editor for sequencing video and animation.

5.1   The Editor

As far as this discussion is concerned, the function of the editor (see Figure 4) is merely to produce a data structure consisting of a list of animations to run and video segments to show. We will call this data structure the *cue sheet*. An animation sequence represented by the editor consists of a file name, a starting frame, an ending frame, and a duration. An object of class FadActive takes these parameters and generates a sequence of display updates showing the sequence of frames and some number of interpolated frames, depending upon the duration. Similarly, a video segment is represented by a starting frame, an ending frame, and a duration. An object of class VidActive generates control commands for a laser videodisc player (an all digital video object has also been implemented) to generate the appropriate sequence of video frames.

5.2   Active Objects and the Clock Tree

The structure of the application was shown in Figure 2. An editor creates three active objects: FadActive for graphical animation control, VidActive for video

---

[1] User cut requests are processed by the application, not by the Tactus Server; this requires a round-trip message to the application, but keeps the input-processing model uniform.

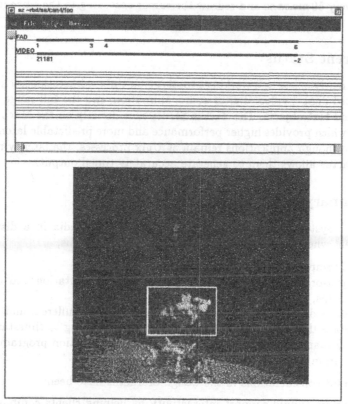

Fig. 4. The Editor Application. The top window is a time-line for animation, video, and music. The presentation below shows a video image and an animated graphical overlay, in this case highlighting the Lunar Excursion Module as it lifts off from the moon.

control, and **EdActive** for sequencing through the editor data structure. All three of these use a stream object as their clock, and the stream object uses **RealTime** for its clock.

5.3 Adding an Animated Cursor

Suppose we want to modify the timeline editor to display a vertical cursor that travels across the timeline during playback. The change is trivial since during playback, any graphics commands are automatically synchronized with other media, and the active object architecture supports interleaving the cursor computation with other output. The changes required are: Subclass **EdActive** to create a **CursorActive** object, and add it to the clock tree as a child of the stream. The **CursorActive** object will automatically be notified when the stream starts and stops. When the stream starts, use the **RequestKick** and **Kick** messages to wake up every 50 ms or so and redraw the cursor according to the current logical time.

6 Current Status

Tactus currently exists as a working prototype on Unix workstations; it supports synchronized digital video, digital audio, animation, and the full range of ATK graphics and text objects. The Tactus Server runs directly on the Mach 3.0 Microkernel, which provides higher performance and more predictable latency than Unix. Even though applications remain as Unix processes, the latency management of Tactus allows them to generate accurately timed output.

7 Summary and Conclusions

Tactus is a system for synchronizing multiple digital media in a distributed environment where latency is an important factor. In Tactus,

 - data is computed ahead of real time,
 - an object-oriented system is used to schedule computation and compute timestamps,
 - the data is delivered to a synchronization server and buffered, and
 - the data is dispatched to presentation devices according to timestamps.
 - Latency management is carefully hidden from application programs, which benefit from a simple-to-use interface.

Important considerations in the design of Tactus have been:

 - The system should support interactivity by helping clients compute multimedia information, The server should handle synchronization at the point of presentation,
 - Both discrete events and continuous media should be supported, and
 - Existing graphics support libraries should be usable with a minimal amount of change.

Consequently, a large effort has gone into designing the client side of Tactus as well as the server. This has paid off by simplifying the task of writing multimedia applications.

Tactus has important implications for the design of multimedia systems. Few device drivers or devices support timestamped data, yet this seems to be a useful technique for solving current synchronization problems. Designers of network time protocols should keep in mind that abruptly setting clocks ahead or behind can cause problems for multimedia and other real-time systems. Media delivery subsystems should be designed with facilities for external control and synchronisation, whereas most current designs are closed and self-regulating. Finally, Tactus raises some interesting and unaddressed scheduling problems [9] not solved by traditional periodic scheduling models. Tactus events are aperiodic, but the precise time and content of a Tactus event dispatch is known significantly in advance.

8 Acknowledgments

The work was entirely sponsored by the IBM Corporation. We would also like to thank Jim Zelenka, and Kevin Goldsmith for implementation assistance, and Carol Krowitz and Joy Banks for work on this manuscript.

References

1. D. P. Anderson, R. Govindan, and G. Homsy. Abstractions for continuous media in a network window system. Technical Report UCB/CSD 90/596, Computer Science Division (EECS), U.C. at Berkeley, 1990.

2. D. P. Anderson and G. Homsy. A continuous media i/o server and its synchronization mechanism. *Computer*, pages 51–57, October 1991.

3. D. P. Anderson and R. Kuivila. Accurately timed generation of discrete musical events. *Computer Music Journal*, 10(3):48–56, Fall 1986.

4. D. P. Anderson and R. Kuivila. A system for computer music performance. *ACM Transactions on Computer Systems*, 8(1):56–82, February 1990.

5. M. M. Blattner and R. B. Dannenberg, editors. *Multimedia Interface Design*. ACM Press, 1992.

6. P. Cointe and X. Rodet. Formes: an object and time oriented system for music composition and synthesis. In *1984 ACM Symposium on LISP and Functional Programming*, pages 85–95, New York, 1984. ACM.

7. Digital Equipment Corporation. Xmedia tools, version 1.1a. Software Product Description SPD 36.55.02, 1992.

8. B. J. Cox. *Object-Oriented Programming: an evolutionary approach*. Addison-Wesley, Reading, Mass., 1987.

9. R. B. Dannenberg. *Real-Time Scheduling and Computer Accompaniment*, pages 225–262. System Development Foundation Benchmark Series. MIT Press, 1989.

10. S. Gibbs. Composite multimedia and active objects. In A. Paepcke, editor, *OOPSLA '91 Conference Proceedings*, pages 97–112, New York, 1991. ACM/SIGPLAN, ACM Press.

11. K. Kahn. Director guide. Technical Report MIT AI Laboratory Memo 482B, MIT, December 1979.

12. R. Kolstad. The network time protocol. *UNIX Review*, 8:58–61, December 1990.

13. T. D. C. Little and A. Ghafoor. Spatio-temporal composition of distributed multimedia objects for value-added networks. *Computer*, pages 42–50, October 1991.

14. S. R. Newcomb, N. A. Kipp, and V. T. Newcomb. The hytime multimedia/time-based document structuring language. *Communications of the ACM*, 34:67–83, November 1991.

15. A. J. F. Palay, M. Hansen, M. Kazar, M. Sherman, M. Wadlow, T. Neuendorffer, Z. Stern, M. Bader, and T. Peters. The andrew toolkit - an overview. In *Proceedings USENIX Technical Conference*, pages 9–21. USENIX, Winter 1988.

16. G. D. Ripley. Dvi - a digital multimedia technology. *CACM*, 32(7):811–822, July 1989.

17. P. Wayner. Inside quicktime. *Byte*, 16, December 1991.

18. T. Yager. The multimedia pc: High-powered sight and sound on your desk. *Byte*, 17, February 1992.

Short Paper Session I: Scheduling and Synchronization

Chair: Jim Kurose, University of Massachusetts at Amherst

First paper: **An Analytical Model for Real-Time Multimedia Disk Scheduling** by *James Yee and Pravin Varaiya*

Second paper: **Real-Time Scheduling Support in Ultrix-4.2 for Multimedia Communication** by *Tom Fisher*

Third paper: **Continuous Media Synchronization in Distributed Multimedia Systems** by *Srinivas Ramanathan and P. Venkat Rangan*

An Analytical Model for Real-Time Multimedia Disk Scheduling

James Yee, Pravin Varaiya

EECS Dept., University of California at Berkeley,
Berkeley, CA 94720, USA

Abstract. Traditionally, storage systems have been designed to provide good average performance without requiring explicit knowledge of the application requirements. However, with real-time multimedia applications, the knowledge of the requirements can be used by a storage system to schedule the real-time access of multiple streams. In this paper, we present an analytical model for a simple class of multimedia system, and examine classes of deterministic open-loop and closed-loop disk scheduling policies suggested by existing results from manufacturing system scheduling problems. The buffer requirements, system stability and convergence dynamics of these disk scheduling policies are presented, accompanied by numerical and simulation results.

1 Motivation

Sustained high bandwidth real-time access of various media types is required in many real-time multimedia (MM) applications (e.g., *video teleconferencing*, *collaborative work* applications, and *authoring* applications). To support such access, the characteristics of the network, storage system, and presentation requirements of the MM application need to be considered. In this paper we focus on the class of real-time MM systems with which the media types are characterized by constant rate requirements, and the storage device behaves deterministically (e.g., single platter disk with contiguous file allocation and tightly bounded seek time). Furthermore, we assume each file is infinite in length, and the storage device has a fixed nonzero seek time and contiguously allocated data can be read out at a constant rate.

2 Problem Description

Consider the task of scheduling a disk server to read N files stored contiguously on a disk requested by N real-time MM applications, where each application requests a file to be read at the rate μ_i. At any t, one file, say file i, may be read from the disk at a rate of R into a buffer b_i located on the disk server. Each b_i, in turn, is read out at a constant rate of μ_i to meet the media types' demand. The disk server may switch from reading one file to another, but a fixed [1] switching

---

[1] We can equally well handle switching delays which vary with time and with the pair of buffers between which switching occurs, provided the switching delays are bounded.

delay (seek time) of $\delta \geq 0$ is incurred each time it switches, during which no data is read into the buffers.

Such a system can be modeled using a deterministic fluid model as in figure 1a, where an input server writes into the buffers b_i one at a time at rate R,

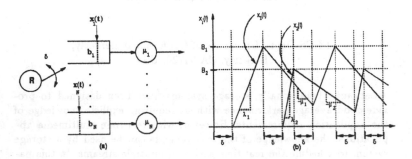

Fig. 1. a) Switched Arrival System (SAS), b) a typical trajectory for $N = 2$

and each b_i is emptied by a constant rate output server μ_i. $x_i(t)$ is the buffer length of each b_i at time t, and the total buffer length of the system is denote by $x(t) = \sum_{i=1}^{N} x_i(t)$. This model, referred to as a switched arrival system (SAS), is described by the set of parameters (N, R, μ_i, δ). We also define a *scheduling policy* for a SAS by the following three sequences of parameters:

- *order* $\{\pi_i\}$ - π_k is the index of the k_{th} buffer served.
- *switching delay* $\{\delta_i\}$ - $\delta_k \geq \delta$ is the switching delay prior to serving the k_{th} buffer.
- *service duration* $\{l_i\}$ - l_k is the amount of time for which the k_{th} buffer is served.

Starting at $i = 1$ and $t = 0$, a policy is executed by repeating the process of seeking for duration δ_i then writing into b_{π_i} for duration l_i, and then incrementing the value of i.

The buffer length of each b_i at time t is denoted by $x_i(t)$. We assume that the buffer size B_i is at least as large as $\sup_t x_i(t)$ so that $0 \leq x_i(t) \leq B_i$. Each $x_i(\cdot)$ is a piecewise linear function with slope $\lambda_i := R - \mu_i$ if b_i is being written into and $-\mu_i$ if it is not. The vector $\bar{x}(t) = [x_1(t), \ldots, x_N(t)]'$ is referred to as the *trajectory* of a SAS policy. See figure 1b. A policy is said to be *stable* if $\exists M < \infty$ s.t. $\sup_t x(t) \leq M$. A non-decreasing *starvation function* for each b_i is define by a $s_i(t) = \int_{S_i(t)} \mu_i d\tau$, where $S_i(t) = \{\tau | x_i(\tau) = 0, \forall 0 \leq \tau \leq t\}$. We also define $s(t) = \sum_{i=1}^{N} s_i(t)$ as the starvation function of the system. At time t, we say b_i is starving if $s_i(t)$ is increasing, and the system is starving if $s(t)$ is increasing. An obvious necessary condition for bounded $s(t)$ is:

$$\rho := \frac{\sum_{i=1}^{N} \mu_i}{R} < 1 \tag{1}$$

Unless otherwise specified, we assume (1) for all systems.

In addition, a policy is called *open loop* if both π_i and l_i are predetermined, and *closed loop* otherwise. With closed loop policies, the buffer state information $\bar{x}$ is used to determine the policy dynamically. We also distinguish between *non-idling* (if $\delta_i = \delta, \forall i$) and *idling* (otherwise) policies. An idling policy permits the server to serve other jobs, possibly non-realtime, during the idle intervals.

2.1 Previous Work

For certain open-loop and closed-loop SAS policies, analytical and simulation results concerning buffer and initial delay bounds can be found in [GC92] and [AOG91]. However, these work provide no analysis of the buffer utilization dynamics or system stability.

A model similar to the SAS that has been studied extensively is the switched server system (SSS). A SSS is just the SAS with all its flows reversed, and insightful analysis of SSS policies can be found in [PK89, SCG91].

3 Periodic Policies

We define a *periodic* policy as a policy with $\pi_i = \pi_{i+N}, \forall i$, and define *cycle time* $= \sum_{i=1}^{N} l_i$. In the following, we examine a class of *open loop* policies where l_i has period N, and a class of *closed loop* fill policies. W.l.o.g., we assume $\pi_1 = 1$ unless otherwise specified.

3.1 Open Loop Periodic Policy

If an open loop periodic policy has a periodic $\bar{x}$, then for every b_i, we require

$$\lambda_i l_i = \mu_i (N\delta + \sum_{j \neq i, j \in \{1, \dots, N\}} l_j) \tag{2}$$

A periodic trajectory for the case of $N = 2$ is shown in figure 2a. From this observation we obtain the following theorem:

Theorem 1. *Given a periodic open loop SAS policy, a periodic trajectory exists iff $\rho < 1$.*

If we let $B_i = l_i \lambda_i$, then for any $\bar{x}(0)$, it can be shown that $x_i(t_0 + \delta + l_i) = B_i, \forall i$, therefore we have the theorem:

Proposition 2. *Given a periodic open loop SAS policy satisfying (2), $B_i = \lambda_i l_i$, and any initial position $\bar{x}(0)$ with $0 \leq x_i(0) \leq B_i$, the trajectory $\bar{x}$ will converge to the periodic trajectory within 1 cycle time.*

The main advantage of such a policy is the quick convergence, but it comes at a cost of requiring exact knowledge of the μ_i. Also, the recomputation of $\bar{l}$ is required whenever a file is added or removed, and in order to keep track of the l_i dynamically, precise timers are required in implementation. Finally, this policy is non-idling.

3.2 A Closed Loop Policy

We define a *fill* policy as a closed loop policy that serves each buffer b_i until $x_i(t) = B_i$, for some fixed positive B_i, and a *periodic fill* policy is defined as a fill policy with $\pi_i = \pi_{i+N}$. We show that B_i can be chosen such that a fill policy will have bounded starvation, and the trajectory of such a policy converges to a unique limit cycle in finite t.

By assuming starvation is bounded, we first derive expressions for the system trajectory when there is no starvation, and show the convergence of the trajectory to a unique limit point. We then show the conditions on B_i for ensuring bounded starvation. By construction, we can show:

Theorem 3. *Given a periodic fill SAS policy, if $\exists t_0 \geq 0$ s.t. $(s(t) - s(t_0)) = 0, \forall t \geq t_0$ (i.e., no starvation after t_0), then the trajectory $\bar{x}$ can be expressed by:*

$$\bar{x}(t_{k+1}) = A\bar{x}(t_k) + B\bar{b} - \delta\bar{c} \tag{3}$$

where t_k is the beginning of the k_{th} cycle after t_0, A, B are $N \times N$ matrices, $\bar{c}$ is a vector of dimension N, and $\bar{b} = [B_1, \ldots, B_N]'$.

Following from the fact that $B, \bar{c}, \bar{b}$ are constant, we have:

Corollary 4. *If $\exists t_0 \geq 0$ s.t. $(s(t) - s(t_0)) = 0, \forall t \geq t_0$, the SAS policy with trajectory described by (3) converges to an unique limit cycle $\Leftrightarrow$ the homogeneous system:*

$$\bar{x}(t_{k+1}) = A\bar{x}(t_k) \tag{4}$$

converges to a unique limit point $\Leftrightarrow$ the same SAS policy with $\delta = 0$ converges to a unique limit point.

From Corollary 4, we know that in order to show (3) converges, we only have to show that the homogeneous system (4) converges to a limit point. Taking advantage of this equivalence, we can show:

Theorem 5. *Given a periodic fill SAS policy and t_0 as defined above, a sufficient condition for $\bar{x}$ to converge to a unique limit cycle with $t \geq t_0$ is: $\rho < 1$.*

Furthermore, it can be shown that this unique limit cycle is simply a translation of the unique limit cycle obtained using the periodic open loop policy. That is:

Corollary 6. *Let $\bar{x}'$ be the unique non-starving, non-overflowing periodic trajectory for the open loop periodic policy with buffer size $\bar{b}'$. Let $\bar{x}$ be the unique non-starving limit cycle to which a periodic fill policy converges for the same SAS with buffer size $\bar{b}$. Then there exists a τ s.t.:*

$$x(t) = x'(t - \tau) + (\bar{b} - \bar{b}')$$

i.e., $\bar{x}$ is just $\bar{x}'$ translated, as shown in figure 2.

We have so far assumed the existence of a t_0 beyond which no starvation occurs, next we present the conditions for guaranteeing such a t_0 and for zero starvation.

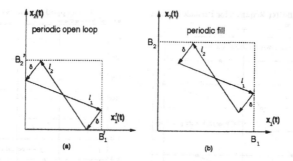

Fig. 2. Trajectories for Periodic Open Loop and Fill Policies

Theorem 7. *Given a periodic fill SAS policy and any initial position $\bar{x}(0)$, $\exists t_0 < \infty$ s.t. $(s(t) - s(t_0)) = 0, \forall t \geq t_0$ (bounded starvation) if and only if:*

$$\lambda_i l_i \leq B_i < \infty \tag{5}$$

where the l_i satisfy (2). Given (5), if $x_i(0) \geq B_i - l_i\lambda_i + n_i\delta\mu_i$, where $n_i = min_j\{j|\pi_j = i\}$, then $s(t) = 0, \forall t \geq 0$.

With this policy, a unique limit cycle can be achieved without explicit and precise knowledge of the file parameters.

4 Numerical and Simulation Results

For accessing multiple files stored on different types of disks using a periodic policy, the minimum total buffer required as calculated by (2) are shown in figure 3a. These values may be compared to those obtained by simulation in [AOG91].

To validate the analytical model in the presence of disk access block size, disk sector/track size, and to compare against existing heuristic scheduling algorithms, simple simulation (based on the work in [AOG91] and [Nic92]) has been developed to examine the performance of different scheduling policies in a client-server video application. In figure 3b, we notice the robustness of the periodic fill policy results in lower frame loss than the periodic open-loop policy, and how favorably they performed compared to a fcfs disk access policy.

5 Conclusions and Future Work

In this paper we showed the application of some control theory results to scheduling problems in MM systems. Specifically, we proposed a SAS model for MM systems, derived the buffer requirements for both periodic open-loop and fill policies, and examined their trajectory convergence dynamics. We emphasize the fact that simplifying assumptions have been made in characterizing the storage system, communication system, and presentation requirements. To meet a

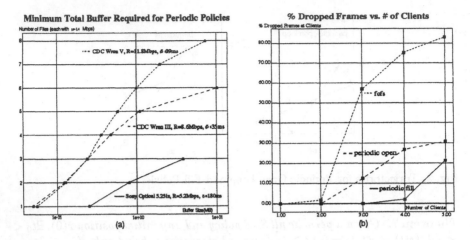

Fig. 3. Numerical and Simulation Results for Periodic Policies

wider range of MM system requirements, further characterizations of real-time performance requirements (e.g., synchronization mechanisms), buffer and bandwidth requirement of media types, and real-time communication protocols, are needed.

Motivated by [PK89], we are also examining a class of aperiodic closed loop Fill-A-Fraction (FAF) policies that selects and fills a buffer k if $x_k(t) \leq \epsilon x(t)$, for some $0 < \epsilon < 1$. Extending the SSS results from [PK89], some preliminary results on the existence of lower bounds on $\liminf x(t)$ and upper bounds on the the time average of $x(t)$ have been derived [Yee93].

References

[AOG91] David P. Anderson, Yoshimoto Osawa, and Ramesh Govindan. Real-Time Disk Storage and Retrieval of Digital Audio/Video Data. Technical report, U.C. Berkeley, EECS Dept., Report No. UCB/CSD 91/646, August, 1991.

[GC92] Jim Gemmell and Stavros Christodoulakis. Principles of Delay-Sensitive Multimedia Data Storage and Retrieval. *ACM Transactions on Information Systems*, January 1992.

[Nic92] K. Nichols. Performance Studies of Digital Video in a Client/Server Environment. In this proceeding, November, 1992.

[PK89] James R. Perkins and P.R. Kumar. Stable, Distributed, Real-Time Scheduling of Flexible Manufacturing/Assembly/Disassembly Systems. *IEEE Transactions on Automatic Control*, February 1989.

[SCG91] Ali Sharifnia, Michael Caramanis, and Stanley B. Gershwin. Dynamic Setup Scheduling and flow control in Manufacturing Systems. *Journal of Discrete Event Dynamic Systems, pg. 149-175*, September 1991.

[Yee93] James C. Yee. Real-Time Scheduling Policies for Multimedia Systems, Ph.D. Thesis. Technical report, UC Berkeley, in preparation, 1993.

Real-Time Scheduling Support in Ultrix-4.2 for Multimedia Communication*

Tom Fisher
fisher@tenet.Berkeley.EDU

The Tenet Group
Computer Science Division
Department of Electrical Engineering and Computer Sciences
University of California
Berkeley, California 94720
and
International Computer Science Institute
1947 Center Street
Berkeley, California 94720

Abstract. Advanced workstations are now being equipped to handle various types of multimedia applications such as audio and video. The quality of these images and sounds depends highly on the timely management of multimedia data. In order to provide such management, support for real-time process scheduling must exist. In this paper we describe the fundamental mechanisms implemented in the Ultrix4.2 kernel that provide the real-time scheduling support needed for multimedia applications. Our primary goal is to reduce and bound the overall delay in the response times to both internal as well as external events. To achieve this goal we have modified the Ultrix4.2 kernel to be preemptible at highly specific locations and have added a small subset of data structure locks to maintain data consistency. Work has also been done towards reducing and bounding the time spent resolving lock conflicts. The end result is that we have a kernel capable of the real-time processing necessary for multimedia applications. Finally, we demonstrate and validate through experimentation that the above claim is true.

1 Introduction

When we talk about real-time in the Tenet Group in a broad sense we are referring to guaranteed performance, however in this paper, real-time as in real-time scheduling, refers to scheduling processes under bounded response times. In

* This research was supported by the National Science Foundation and the Defense Advanced Research Projects Agency (DARPA) under Cooperative Agreement NCR-8919038 with the Corporation for National Research Initiatives, by AT&T Bell Laboratories, Hitachi, Ltd., Hitachi America, Ltd., Pacific Bell, the University of California under a MICRO grant, and the International Computer Science Institute. The views and conclusions contained in this document are those of the authors, and should not be interpreted as representing official policies, either expressed or implied, of the U.S. Government or any of the sponsoring organizations.

order to achieve such bounded response times there needs to exist greater control over scheduling behavior. To achieve this greater control several areas of real-time functionality have been added to the Ultrix4.2 kernel. First, we discuss in detail the implementations in each of these areas. Then we proceed to validate the additional functionality through measurements. We note that in order to achieve valid and accurate measurements, all areas of the entire system must be vigorously and continuously exercised. It is only through this type of exhaustive measurement process do we establish guarantees about the work.

In addition, it must be understood that we cannot totally solve this problem of bounding the existing delays by software alone. We do not attempt to reduce nor bound the delays caused by interrupt processing and/or DMA processing, even though both types of high priority activities in the lower half of the kernel may produce very noticeable delays. To adequately bound such delays would require hardware support.

The rest of this paper is organized as follows: Section-2 states our objective. Section-3 describes and discusses our implementation. Section-4 describes our performance measurements. And, section-5 gives our conclusions.

2 Objective

Scheduling latency is defined in the context of this paper to be the delay between the time when an event (or request) takes place and the time when execution to handle (or respond) to that event actually occurs. The primary goal of our work is to reduce and bound the duration of this delay. In particular, the target is to reduce this delay to below *3ms* for close to %100 of the time while the machine is operating under loaded conditions.

3 The Implementation

The fundamental mechanisms that make up the real-time scheduling features added to the Ultrix4.2 kernel are included in 5 areas: real-time process establishment, preemption points, data structure locks, an implemented scheme for resolving lock conflicts, and a separate real-time callout mechanism. This combined functionality provides for the scheduling control needed to reach the above stated objective.

3.1 Real-Time Process Establishment

The notion of a real-time user process has been introduced into the Ultrix kernel. In this context, a real-time process is simply a normal user process that has explicitly been accelerated to real-time status. It is this real-time status, which any user-process in the system is equally able to gain, that we would like to describe. In order to obtain this real-time status several things must occur:

Admission control. First of all, the kernel uses a simple admission control scheme to accept or reject a process's request to obtain real-time status. The scheme, at this point, is straight forward and simply allows only 4 (user-requested) real-time processes to exist at any one time, and rejects all others. It is possible, however, for the system to accelerate specific processes, and exceed this limit. This is done to reduce and limit the duration of time spent handling lock conflicts and will be described later (see section 3.4). In the future, we expect this admission control scheme to take into account multiple user-defined real-time priorities as well as user-defined resource requirements (or requests). The user process, under this new scheme, would then negotiate and haggle with the kernel until an agreement, if any, is made about the nature of the real-time status to be allocated.

Initialization. After an approval (or agreement) has been made, the process's pages are locked in memory, an entry in the system wide real-time process table is allocated and initialized, and the user/kernel process priorities are changed to a static real-time UNIX priority (see next section). And finally, a global flag is set signaling the kernel that execution of a real-time process is now pending and that preemptions are now enabled. Once a preemption has occurred and the real-time execution is under way, the global flag is turned off, preemptions are then disabled and the kernel proceeds to operate as normal.

Assignment of Priorities. A real-time process has associated with it several priorities. These are a static UNIX priority and 3 kinds of real-time priorities.

When a process is first accelerated to real-time status it is assigned a static UNIX priority. This priority, referred to as the process's real-time UNIX priority, defines the overall position of the priority of real-time processing in relation to all other high priority activities in the system.

In order to distinguish between real-time processes of various urgency we have adopted the notion of global, inherited and effective priorities from SunOS5.0 [Khanna92]. The use of these priorities is further described in section 3.4.

3.2 Preemption Points

Preemption points are highly specific locations in the kernel where a process may choose to give up its execution of the CPU in favor of a higher priority process. This approach is much simpler for us than putting locks on all data structures and making the entire kernel fully preemptible. We do not say that this is the correct approach, but only that it's the simplest way for us to achieve our needs which are to satisfy the requirements of multimedia applications. Other work has successfully been done to make the kernel fully preemptible, for example in OSF-1.0, SunOS-5.0, and HPUX.

Initially our goal was to see how far we could get (how many effective points we could add and far we could reduce and bound the scheduling delay) without having to implement locks over any data structures. As a result, the first generation of preemption points we added totaled 11 and our measurements reflected a major improvement in the overall average and maximum scheduling response times (see fig. 1).

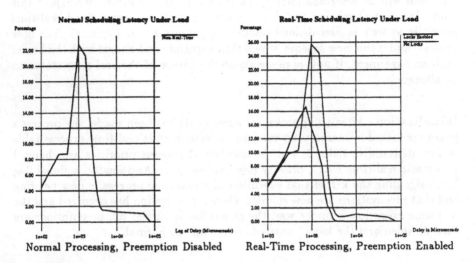

Normal Processing, Preemption Disabled   Real-Time Processing, Preemption Enabled

Fig.1. Average Delay Times Between Execution Request and Execution.

After having implemented locks over a small set of data structures (see next section) we were able to, one by one, add 31 more preemption points. It was at this time where we made our final measurements (described in section 4.2).

3.3 Data Structure Locks

As stated above, our goal in the beginning was to avoid locking any data structures and to reach the highest level of kernel preemptibility possible without using these them. We soon realized though that this goal was unrealistic (see fig. 1) and proceeded further with the new goal of implementing locks over as few a number of data structures as possible while still achieving our above stated target.

Our next move was to implement a mechanism in the kernel to determine exactly which data structures needed locking for any given point within the kernel. As a result, we were able to come up with a small subset of data structures, totally 27, that would need protection when all the preemption points were in place and operational. This small subset of total kernel data structures then became the set of structures to which we applied our version of real-time locks to.

The real-time locks we've implemented are, at this point, only simple mutual exclusion locks (mutexes). In the near future we intend to implement multiple readers, single writers (readers/writer) locks as well. These mutexes are primarily in place to prevent races on access to each of the shared data structures in the small subset mentioned above.

An important point is that under normal sleep situations, locks are never held. But during real-time processing when a lock conflict is discovered, the process can go to sleep, blocked waiting for the lock while holding onto other locks. This is ok in this situation. Work has been done to ensure that locks are never acquired out of order.

3.4 Priority Inheritance

To limit the duration of time spent handling lock conflicts (situations where lower priority processes hold locks needed by higher priority ones) a strategy referred to as the *basic priority inheritance protocol* was adopted from SunOS-5.0 [Khanna92]. A more elaborate description may be found in [Sha90]. In short, the priority inheritance protocol comes into play only when higher priority processes block while waiting to obtain locks. It attempts to limit the duration of this time spent blocking by having the high priority process propagate (will) its priority to all lower priority processes that currently block it. In our particular case, the lower priority process is ramped up to real-time status and the priority that it inherits is the real-time process priority of the process(es) blocked waiting for it. When lower-priority processes cease to block a high priority process, the lower priority ones revert back to their original priority, which in our particular case may mean to giving up the newly acquired real-time status. The strategy was modified and developed for use with processes instead of threads and to operate in the context of a kernel only preemptible through specific preemption points.

Under normal operations the amount of time spent blocking is potentially very long—in fact, it is unbounded since the amount of time a high priority process must wait for a lock to be freed depends not only on the duration of some critical sections, but possibly on the duration of the complete execution of some processes. However, with the use of priority inheritance this time is bounded. It is described in [Rajkumar88] how, if we have m lower priority processes holding onto k distinct locks, the maximum number of critical sections that a real-time process will have to wait is at most $MIN(m, k)$.

3.5 Internal Event Processing

Callout processing is the traditional UNIX way of handling internal timeout events. In a standard UNIX kernel like Ultrix4.2 this processing is all done at software interrupt level. This processing, however, can produce unexpected, variable delays even while the a real-time process is in the middle of execution. In fact, in SunOS 4.0 and 4.1 this delay has been measured to be up to $5ms$. This is unacceptable for real-time processing.

Currently, what we've done is implement a separate real-time callout queue for processing especially time-critical internal events. And, we've taken the approach of blocking out all other normal callout processing during the execution of real-time processes.

4 Measurements

As we have noted in the introduction, we need to vigorously and continuously exercise all pathways throughout the entire kernel. Otherwise, we may miss the affects of exercising certain pathways and later suffer either previously undetected race conditions and/or unexpected delays. Therefore, we have taken care to develop scripts that spawn processes of all types to load our machines in the best way we can.

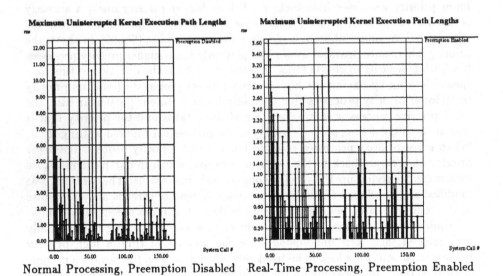

Normal Processing, Preemption Disabled   Real-Time Processing, Preemption Enabled

Fig.2. System Call Uninterrupted Kernel Execution Path Lengths.

Uninterrupted kernel execution times. This experiment was designed to observe where in the kernel we would experience the greatest uninterrupted execution path lengths. The initial idea was to determine where preemption points needed to be placed and to gain a better understanding of where the greatest scheduling delays would arise from. As a result, we can see the affects the preemption points have on reducing the length of uninterrupted execution path lengths (fig 2). Note, that in the graph with preemption enabled, we can see that all system calls now have uninterrupted execution path lengths almost entirely under *3ms*.

Delay between wakeup and actual execution. This experiment was conducted to determine the exact scheduling delays that take place under actual execution between when a process wakes up and when it actually runs. Measurements were taken with a normal process and real-time processes before and after locks and many preemption points were added (fig 1). Note, that in order to have reasonably legible graphs here, we've displayed the log of the X values in both graphs. ***Also note, that after our final generation of preemption points were added (those added after locks were enabled) that our original objective of bounding this delay to under *3ms* has been achieved***.

5 Conclusions

Work has been done in the Ultrix Kernel to help reduce and bound scheduling latencies so as to better support the real-time processing requirements of multimedia applications. In particular, work has been done to add preemption points and data structure locks and a scheme has been implemented to help resolve lock conflicts in a straight forward way. In particular, we have been able to bound the scheduling latency to be almost entirely under *3ms*, which was our original target. We hope to continue this work in the area of supporting multiple, simultaneous, co-existing real-time processes.

References

[Eykholt92] . R. Eykholt, S. R. Kleiman, S. Barton, R. Faulkner, A. Shivalingiah, M. Smith, D. Stein, J. Voll, M. Weeks, and D. Williams, "Beyond Multiprocessing: Multithreading the SunOS Kernel", *Summer 1992 USENIX*, June 8-12, 1992, San Antonio, Texas.

[Herrtwich90] . G. Herrtwich, "An Introduction to Real-Time Scheduling", *ICSI Technical Report, TR-90-035*, Jul. 1990.

[Khanna92] . Khanna, M. Sebree, and J. Zolnowsky, "Realtime Scheduling in SunOS 5.0", *Winter 1992 USENIX*, San Francisco, California.

[Leffler89] . J. Leffler, M. K. McKusick, M. J. Karels, and J. S. Quarterman, *The Design and Implementation of the 4.3 BSD UNIX Operating System*, Addison-Wesley, 1989

[Nakajima91] . Nakajima, M. Yazaki and H. Matsumoto, "Multimedia/Realtime Extensions for the Mach Operating System", *USENIX Proceedings Summer '91 - Nashville*, Jun. 1991, p183-198.

[Rajkumar88] . Rajkumar, L. Sha, and J. P. Lehoczky, "Real-Time Synchronization Protocols for Multiprocessors", *Proceedings of the Real-Time Systems Symposium*, Dec. 6-8, 1988, Huntsville, Alabama.

[Sha90] . Sha, R. Rajkumar, and J. P. Lehoczky, "Priority Inheritance Protocols: An Approach to Real-Time Synchronization", *IEEE Transactions on Computers*, Vol. 39, No. 9, September 1990.

[Tokuda87] . Tokuda, J. W. Wendorf and H. Y. Wang, "Implementation of a Time-Driven Scheduler for Real-Time Operating Systems", *Proceedings - IEEE Real-Time Systems Symposium*, Dec. 1987, p271-280.

Continuous Media Synchronization in Distributed Multimedia Systems

Srinivas Ramanathan and P. Venkat Rangan

Multimedia Laboratory
Department of Computer Science and Engineering
University of California at San Diego
La Jolla, CA 92093-0114
E-mail: {sramanat, venkat}@cs.ucsd.edu

Abstract. Future integrated networks are expected to offer a variety of multimedia services, some of which may involve recording and playback of multiple media strands such as video and audio. Media synchronization, which refers to the temporal coordination of the playback of multiple media strands, is the subject matter of this paper. We compare the requirements for media synchronization against those of clock synchronization and argue that clock synchronization may not be necessary, sufficient or desirable for ensuring media synchronization of pre-recorded media strands. We present a synchronization technique targeted for environments in which non-deterministic variations exist in network delays and in the rates of recording and playback. In this technique, at the time of recording, temporal relationships among media strands are recorded in the form of relative time stamps. In order to facilitate synchronization, at the time of playback, the display sites transmit lightweight *feedback units* back to synchronizer nodes, concurrently with playback of media units. Using these feedback units, the synchronizers estimate playback times of media units of different media strands and readjust playback so as to enforce synchronization.

1 Introduction

The advent of high-speed networks is stimulating the development of a wide variety of multimedia services, such as multimedia entertainment, mail, news distribution, tele-conferencing etc. All such applications usually involve simultaneous playback of multiple media, such as video and audio. Media synchronization, which refers to the temporal coordination of the playback of multiple media strands (a strand is a sequence of media units, such as audio samples and video frames), is the subject matter of this paper.

2 Problem of Media Synchronization

In a distributed multimedia system, media strands may be recorded at different media capture sites (such as telephones, cameras etc.), transmitted over integrated networks, possibly stored at information servers, and subsequently played back at different display sites (such as audio speakers, videophones etc.). During simultaneous playback of multiple media strands, it may be necessary that playback of these strands must remain mutually synchronized throughout [7]. Since temporal relationships between media strands are continuous and on-going, media synchronization in this case, is termed as *continuous synchronization* (this is in contrast to synchronization at discrete points, referred to as *discrete event synchronization*). In the rest of this paper, we use the term "synchronization" to refer to "continuous media synchronization".

Media synchronization is automatic if the media strands are transmitted as analog signals over synchronous channels, as is the case in cable TV networks. However, when media strands are transmitted digitally over an integrated network, synchronization is no longer automatic because the network itself may introduce non-deterministic delays during multimedia transportation, owing to varying queueing delays at intermediate nodes. Furthermore, absence of synchronization between clocks of the media capture and display sites, may also lead to loss of synchronization.

Depending on whether or not media strands are recorded prior to the commencement of their playback, multimedia applications can be classified into: *stored media* applications (e.g. multimedia on-demand retrieval services), and *live media* applications (e.g. tele-conferencing). In Section 3 we propose feedback techniques for synchronization in stored media applications. In Section 4, we contrast the synchronization needs of live media applications with those of stored media applications.

3 Synchronization in Multimedia Storage and Retrieval

3.1 Problem Formulation

In order to formulate the problem of synchronization during retrieval of stored media strands, consider the simultaneous recording of a pair of media strands S_1 and S_2. Let θ_1^r and θ_2^r represent the periods of recording of media units of the media strands S_1 and S_2, respectively (therefore, the recording rates of the media strands are $\frac{1}{\theta_1^r}$ and $\frac{1}{\theta_2^r}$, respectively). At the time of retrieval, the playback rates of the media strands may differ significantly from the recording rates, as is the case during fast-forward or slow-motion playback. Inter-media synchronization is ensured if playback of the individual media strands commence simultaneously and remain mutually coordinated throughout the duration of their retrieval. Usually, loss of synchronization is perceived by humans only when positional relationships existing between media strands at the time of recording are not maintained at the time of playback. Based on this observation, *playback can be said to be synchronous if and only if media units that were recorded simultaneously are played back simultaneously*[1]. Specifically, if media units μ_1 and μ_2 of media strands S_1 and S_2 are recorded simultaneously, then

$$\mu_1 * \theta_1^r = \mu_2 * \theta_2^r \tag{1}$$

If μ_1 and μ_2 are also played back simultaneously, then

$$\mu_1 * \theta_1^p = \mu_2 * \theta_2^p \tag{2}$$

---

[1] Throughout this paper, we assume that additional mechanisms, such as those proposed in [4] are used to ensure continuity of playback of the individual media strands.

where θ_1^p and θ_2^p are the playback periods of media units of media strands S_1 and S_2, respectively. From Equations (1) and (2), we derive the condition for synchronization to be:

$$\frac{\theta_1^p}{\theta_2^p} = \frac{\theta_1^r}{\theta_2^r} \tag{3}$$

Equation (3) requires that the relative periods of playback of the media strands match the relative periods of recording of these media strands. Hence, media synchronization of this type is termed *Relative Synchronization*.

A special case of relative synchronization occurs when $\theta_1^r = \theta_1^p$. In this case, playback will be synchronous only if $\theta_2^r = \theta_2^p$. Since playback of media strands will exactly mimic their recording, synchronization of this type is termed *Absolute Synchronization*. Unless explicitly mentioned, our efforts in the remainder of this paper will focus on the problem of ensuring relative media synchronization between media strands.

3.2 Techniques for Media Synchronization

In this section, we present a feedback technique for ensuring synchronization during retrieval of pre-recorded media strands from a multimedia server on to display sites on an integrated network. If the clock rates at the recording sites are deterministic and are known prior to the commencement of playback, it is simple to enforce relative synchronization; all that is required is to adjust the playback periods such that relative periods of playback match the relative periods of recording. However, in actual practice, both the periods of recording and the periods of playback may vary from time to time due to absence of synchronized clocks. Usually, it is possible to provide bounds on these variations. Let ρ represent the bound on the fractional drift in periods of recording and playback. The actual period of recording of a media strand S_i with nominal period θ_i^r, may vary between $[\theta_i^r * (1 - \rho), \theta_i^r * (1 + \rho)]$; similar variations may also exist in the periods of playback.

Synchronization can be enforced simply by bounding the buffering at the destinations and transmitting corresponding media units to all the destinations simultaneously. Since the maximum extent by which one display site may lead another is at most equal to the buffering capacity of a lagging display site, all that is necessary to ensure that the asynchrony between display sites never exceeds its maximum tolerable value is to appropriately size the buffering at the display sites. The dependence of maximum asynchrony on the buffering capacity at a display site makes this scheme inflexible. Furthermore, once the asynchrony reaches a maximum value, it may remain close to the maximum value, resulting in greater average asynchrony. We now propose a synchronization technique that overcomes these drawbacks.

In order to enforce relative synchronization between media strands, we propose a feedback-based synchronization technique that forces a match between the fluctuations in the relative playback periods of the media strands and the relative recording periods of those strands. This technique assumes that the networks used for media communication are asynchronous; although the network delays may vary non-deterministically, their maximum variations are bounded by means of admission control and intelligent packet scheduling at the network[8]. Let Δ_{min} and Δ_{max} represent the bounds on delays offered by the network[2]. At the time of recording, media units recorded by the media sites are transmitted to the multimedia server for storage. Given the bounds on network delays, the multimedia server, when it receives a media unit, μ at time $\tau(\mu)$, can estimate the earliest and latest instants at which this media unit, μ could have been recorded to be:

$$r^e(\mu) = \tau(\mu) - \Delta_{max} \tag{4}$$

$$r^l(\mu) = \tau(\mu) - \Delta_{min} \tag{5}$$

In order to indicate the positions of media units of a media strand relative to the those of other media strands, the multimedia server assigns relative time stamps (RTSs) to all media units of the media strands being recorded[5]. In the absence of a global clock, RTS assignment is based on a master clock, which could be the clock of one of the media sites or even that of the multimedia server itself. Since exact recording times of media units are unknown to the multimedia server (owing to network jitter and uncertainties in recording periods of media units), media units whose recording times lie within a time window W are deemed to be recorded simultaneously. All media units (of all the media strands) that are recorded simultaneously are assigned the same RTS. The precision in RTS assignment is determined by the choice of the window W; smaller the value of W, greater the precision. The choice of W is, however, limited by the minimum uncertainity, $\Delta_{max} - \Delta_{min}$ in determining recording times. Additionally, W may also be restricted by the need to assign RTSs in terms of media units; W must be less than half the period of a media unit to guarantee unique RTS assignment and W must be at least equal to half the period of a media unit in order to guarantee that integral RTSs can be assigned to each media unit.

Playback is synchronous only if media units assigned the same RTS are played back simultaneously. Since RTSs are maintained at the multimedia server, it is best suited for functioning as the synchronizer. In the synchronization technique we propose, at the time of playback initiation of media units, the display sites transmit *feedback units* back to the multimedia server. Feedback units contain only RTSs of the media units whose playback was initiated concurrent with their transmission, and hence, feedback transmission is lightweight. Since feedback transmission is concurrent with playback initiation of media units and since network delays are bounded, it is possible to use the arrival time of feedback units to estimate the playback times of the corresponding media units in a manner similar to that in Equations 4 and 5. The synchronization scheme that we propose employs a master-slave approach: one of the media strands is chosen as the *master* strand, all other strands that are to be synchronized with it being *slave* strands. The master strand is played back continuously at the master display site, whereas playback of the slave strands are periodically speeded up or slowed down to synchronize with the master. Using estimates of playback times of media units at master and slave display sites, the multimedia server can determine media units that are being played back simultaneously. Differences in RTS values of such media units are symptomatic of asynchrony between the master and slave. If the RTS of a master media unit exceeds the RTS of a slave media unit that is being played at the same time, the multimedia server speeds up the slave, by skipping an appropriate number of media

---

[2] All times are measured on the clock of the multimedia server.

units from the slave media strand. If the RTS of a master unit is less than that of the corresponding slave unit, the multimedia server slows down the slave, by duplicating media units in the slave media strand. Even after resynchronization, the sites may once again drift apart and therefore, periodic resynchronizations may be necessary.

A number of extensions of this basic scheme are possible:

- *Predictive Resynchronization*: Since the fractional drift of media sites is bounded, it is possible to use an estimate of playback time of a media unit to predict the playback times of future media units. For example, using the estimates, $p^e(\mu)$ and $p^l(\mu)$, of playback time of media unit μ of media strand S_1 with nominal playback period θ_1^p, the playback time of $\mu + \nu$ can be predicted to be:

$$p^e(\mu + \nu) = p^e(\mu) + \nu * \theta_1^p * (1 - \rho) \qquad (6)$$

$$p^l(\mu + \nu) = p^l(\mu) + \nu * \theta_1^p * (1 + \rho) \qquad (7)$$

Given the maximum tolerable bound on asynchrony, the multimedia server can actually predict the time in future when asynchrony could exceed the tolerable bound, and can initiate remedial action *in advance* to synchronize their playback.

- *Selective Feedback Transmission*: Feedback transmission for every media unit played back, although it enables good estimation of playback times of media units, imposes additional overheads on the network, on the display sites and on the multimedia server. Since the drift between any pair of display sites is bounded, feedback transmission can be selective; the minimum frequency at which feedback units should be transmitted, so as to still restrict the maximum asynchrony to within tolerable limits can be either statically pre-computed prior to the commencement of playback, or *adaptively* controlled during playback, based on the maximum tolerable asynchrony. Implementation of adaptive feedback control is made possible by using a binary flag in headers of media units to distinguish media units corresponding to which feedbacks must be transmitted by the display sites. In [6], we compute the minimum feedback frequency, which is defined as the ratio of feedback units transmitted to media units played back, for the case when θ represents the nominal periods of recording and playback of all the media strands to be

$$\mathcal{F} = \frac{1}{\left\lfloor \frac{(A_{max} * \theta * (1 - \rho) - W - 2 * \theta * \rho) * \frac{1 - \rho}{2 * \rho} - 3 * \Delta_{max} + \Delta_{min}}{2 * \theta * (1 + \rho)} \right\rfloor}$$

where $\mathcal{A}_{max}$ is the maximum tolerable asynchrony between media strands.

The advantages of the feedback technique are manifold. Synchronization is ensured even in the absence of globally synchronized clocks. Initial performance simulations indicate that the overheads entailed are very small (for an environment in which $\Delta_{max} - \Delta_{min} = 30ms$, $\theta = 33ms$, $\rho = 0.001$, and maximum tolerable asynchrony $\mathcal{A}_{max} = 5$ media units, the minimum feedback frequency was about 1 feedback unit for every 1000 media units played back). The feedback technique can enforce synchronization even during fast-forward or slow-motion playback. Synchronization can be enforced even during the playback of media strands recorded at different times (e.g. audio dubbing); in this case, the correspondence between media strands can either be provided explicitly, or they can be determined by playing back one of the media strands while simultaneously recording another. Although described in the context of multimedia storage at a central multimedia server, these techniques are extensible in a straight-forward manner for synchronization of media strands being retrieved from different multimedia servers to different display sites; the synchronization function itself may also be implemented in a distributed manner. If an external clock is used as the master, this technique can ensure absolute media synchronization as well.

3.3 Relationship to Clock Synchronization

Recent work on protocols for media synchronization has focussed on ensuring synchronous playback in the presence of synchronized clocks [2]. Since they assume that clocks are synchronized, such protocols do not consider the effect of mismatches in clock rates on synchronization during media playback. It can be shown that in environments in which network delays as well as fractional clock drifts are bounded, whereas the contribution of network jitter to the asynchrony is bounded, the contribution of clock drifts to the asynchrony may increase with duration of media playback[6]. The question that we address next is whether clock synchronization algorithms [1, 3] are necessary, sufficient, or even desirable for ensuring media synchronization?

In the discussion that follows, we argue that in order to synchronize the playback of pre-recorded media strands, (i) synchronization of clocks of the display sites alone may not be sufficient, and (ii) synchronization of clocks of each display site with the corresponding recording site, although sufficient for ensuring media synchronization, is neither necessary nor feasible in many environments.

In order to illustrate the first point, consider a pair of media strands S_1 and S_2, whose nominal period of recording is θ, but whose actual recording periods, θ_1^r and θ_2^r, respectively, differ owing to the relative drift between their clocks. If the clocks of the display sites used for playing back the media strands are synchronized, the playback periods, θ_1^p and θ_2^p will be equal. In this case, Equation 3 will not be satisfied, and playback *will not* be synchronous.

On the other hand, if media strands have been recorded at synchronous sites (i.e. $\theta_1^r = \theta_2^r$), then synchronization of the clocks of the display sites will be sufficient to ensure media synchronization (with the asynchrony being bounded by the network jitter), since clock synchronization ensures that $\theta_1^p = \theta_2^p$. Even when media strands are recorded at asynchronous sites, it is possible to ensure media synchronization if the clocks of each of the display sites is synchronized with that of the corresponding recording site. In this case, $\theta_1^r = \theta_1^p$ and $\theta_2^r = \theta_2^p$, and hence, Equation 3 is satisfied.

However, it may not always be possible to synchronize the clocks of the recording and display sites. This is because the recording and display sites may belong to different organizations, that may not want to synchronize their clocks across organizational domains. Furthermore, for stored media applications, it is possible that the media strands may have been recorded much earlier than the time of their playback. In such cases, it is possible that there may not be any commonality in times of existence of the recording and display sites. Based on this reasoning, we argue that media synchronization is not the same as clock synchronization in the case of playback of pre-recorded media strands, and hence, motivates specialized techniques; the feedback techniques that we have developed are a step in this direction.

4 Synchronization in Live Media Applications

Playback of live media strands presents an interesting contrast to that of stored media strands. In the case of stored media applications, since media strands are pre-recorded and stored at a multimedia server, the playback rates of the media strands may be much different from the corresponding recording rates (continuity during playback can be ensured by adjusting the the rates of retrieval of media strands from the multimedia server to match the playback rates). As a result of the independence between recording rate of a media strand and the playback rate of that strand, ensuring continuity of all media strands does not automatically guarantee synchronization among those strands. Mutual synchronization between playback of the media strands requires that the relative playback rates of the media strands match their relative recording rates (Equation 3).

In the case of live media applications, the playback rate of a media strand cannot be independent of its recording rate. Since recording and playback are in progress concurrently, a display site that consistently leads the corresponding recording site will starve, thereby effectively slowing down the playback rate at that site. Similarly, due to absence of an intermediate store, a display site that lags the corresponding recording site will experience buffer overruns due to excessive accumulation of media units at the display site, thereby effectively speeding up that site. Therefore, in both of the above cases, the effective playback rate must match the recording rate, i.e. $\theta_i^p = \theta_i^r, \forall i$. Since all media strands are being recorded simultaneously, matching of the playback rate of each strand with its recording rate ensures that the condition of Equation 3 is satisfied overall. However, the display sites may have drifted apart by an extent equal to the buffering capacity of the display sites before relative rate-matching due to buffer overruns comes into effect[4]. Hence, in the absence of any additional mechanisms, the display sites may be go out of synchrony by an extent equal to their buffering capacity very soon after the commencement of playback and may remain asynchronous for the remainder of their playback. Such a situation may be acceptable if the buffering capacity is less than the maximum tolerable asynchrony. If this is not the case, the buffer at the display sites has to be artificially resized to restrict asynchrony to within tolerable limits. Alternatively, the feedback technique described in Section 3 can be used to enforce synchronization even in the case of live media applications. Simulation studies indicate that the feedback-based synchronization technique entails lower average asynchrony than a bounded buffer-based synchronization technique. This is because, in the case of the bounded buffer synchronization technique, the asynchrony between display sites, once it reaches the maximum value, may remain very close to the maximum tolerable limit throughout the duration of playback (since resynchronization is effected only due to buffer overruns or starvation). On the other hand, the feedback technique precisely determines and corrects asynchrony, as and when it manifests itself, and therefore, entails a lower average asynchrony.

5 Conclusion

In this paper we have analyzed the problem of media synchronization and devised conditions that must be satisfied for ensuring media synchronization in stored as well as live media applications. We have presented feedback techniques for media synchronization in environments lacking globally synchronized clocks and have compared these techniques with techniques for clock synchronization. Initial performance studies are encouraging; they indicate that the feedback techniques are capable of enforcing media synchronization with little additional overhead.

References

1. F. Cristian. Probabilistic Clock Synchronization. *Distributed Computing*, 3:146–158, 1989.
2. J. Escobar, D. Deutsch, and C. Partridge. A Multi-Service Flow Synchronization Protocol. *BBN Systems and Technologies Division*, March 1991.
3. R. Gusella and S. Zatti. The Accuracy of the Clock Synchronization Achieved by TEMPO in Berkeley UNIX 4.3BSD. *IEEE Transactions on Software Engineering*, July 1989.
4. Srinivas Ramanathan and P. Venkat Rangan. Feedback Techniques for Intra-Media Continuity and Inter-Media Synchronization in Distributed Multimedia Systems. *To appear in The Computer Journal - Special Issue on Distributed Multimedia Systems*, December 1992.
5. P. Venkat Rangan, Srinivas Ramanathan, Harrick M. Vin, and Thomas Kaeppner. Media Synchronization in Distributed Multimedia File Systems. In *Proceedings of Multimedia'92 - 4th IEEE COMSOC International Workshop on Multimedia Communications, Monterey, California*, April 1-4, 1992.
6. P. Venkat Rangan, Harrick M. Vin, and Srinivas Ramanathan. Designing an On-Demand Multimedia Service. *IEEE Communications Magazine*, 30(7):56–65, July 1992.
7. P. Venkat Rangan, Harrick M. Vin, and Srinivas Ramanathan. Communication Architectures and Algorithms for Media Mixing in Multimedia Conferencing. *To appear in IEEE/ACM Transactions on Networking*, 1(1), February 1993.
8. D. C. Verma, H. Zhang, and D. Ferrari. Delay Jitter Control for Real-Time Communication in a Packet Switching Network. In *Proceedings of TriCom'91*, pages 35–43, 1991.

Short Paper Session II: Architectures and Environments

Chair: Duane Northcutt, Sun Microsystems

The two short papers in this session dealt with various broader and higher-level issues associated with distributed systems, of moderate to large scale, that support a significant amount of digital audio and video.

The first paper was presented by Mayer Schwartz of Tektronix Labs and described a research agenda which aims to explore the migration from the current state of broadcast-quality digital audio/video studios, to one which makes use of some of the latest advances in networking technology – in particular Asynchronous Transfer Mode (ATM) networks. In his presentation, Mayer described the digital studio application domain, and describe many of the technical challenges that must be addressed in order to create a practical studio system that makes use of ATM technology. The technical requirements of studio-quality digital video described in this paper are of particular interest as they represent a significant data point in the overall multimedia design space.

The second paper was presented by Eve Schooler of USC/ISI and addressed a subject whose importance is growing as distributed teleconferencing systems begin to spread – i.e., how can the problem of managing connections be done in a way that scales well across large numbers of participants, large numbers of conferences, and across a wide area. Eve's presentation described of the problem space by way of a conceptual model representing the various aspects of scale relevant to teleconferencing applications. She also discussed technical approaches to dealing with some of the key issues related to scale and teleconference connection management. It is clear that, as teleconferencing work progresses from small-scale local area networks, to larger-scale systems, the problems presented in this paper will become increasingly important in practical systems.

High Speed Networks and the Digital TV Studio

Guy Cherry, Jim Nussbaum and Mayer Schwartz

Tektronix, Inc., Beaverton, OR 97225 USA

Abstract. The high speed networks that will soon be available should have the bandwidth necessary to handle all of the networking needs of today's studio. With the advent of high definition television (HDTV) the necessary bandwidth will increase more than five-fold. We are interested in what is an appropriate networked-based architecture for the studio. Can we make do with just a single network for all the activities that are included in a studio? Assuming an affirmative answer to the preceding question, will an asynchronous transfer mode (ATM) network be up to the task? Will it meet the real-time needs of the studio? Will an ATM network have adequate bandwidth for the multiplicity of video signals, particularly for HDTV signals? Or, will other networking technology be necessary such as optical networks based on wavelength division multiplexing?

Although it is always easy to claim that great changes are coming to computer networking and telecommunications over the next few years it is somewhat more difficult to predict how these changes will impact specific application areas. An application area of interest to us is broadcast television production and distribution; our interest is due to Tektronix's ownership of the Grass Valley Group, a major supplier of broadcast television equipment.

As an application area, television production and distribution is sensitive to a variety of new technologies. Advances in workstation speeds and LANS should be able to improve the internal architecture of today's production studios. Advances in WANS and telecommunications have a potentially great impact on both the packaging and distribution of the resulting television signals as well as the acquisition of source material. Realizing that potential is as much a regulatory and political problem as it is a technical problem.

Our research focus is on the feasibility of using high speed, general purpose networks within the production studio to handle the distribution of video, audio and data traffic. A high speed network could replace the coaxial cable (unidirectional, one per audio/video signal), serial lines (used to control tape machines and other equipment) and intercom audio lines (used to facilitate communication between production personnel). We will also be looking into how the use of such a network might change the internal architecture of the studio itself.

Let us first describe the current studio architecture. The main component is a production switcher. A production switcher takes input signals from a number of different video sources, performs various types of mixes and cuts between the sources and outputs a number of different video signals. Representative video sources include: live in-studio camera feeds, live remote feeds (satellite, telecomm transmission), storage devices (various tape formats, disk recorders, and solid state storage) and character and effects generators. Representative mixes and cuts include: an instantaneous cut from one source to another, a user controlled gradual fade, a user controlled gradual cut using various cut patterns, chroma and luminance keying and many others. Output signals can go to a broadcast transmitter, a storage device and/or to monitors. A typical switcher might have several dozen possible inputs and maybe a dozen or so outputs. Audio is mixed and processed by separate equipment. Studio video may be distributed among the various pieces of equipment in analog, parallel digital or serial digital format. Video is distributed and processed in real-time.

In today's studio there is typically a switcher, separate networks for control, point to point wiring for the video signals (with a routing switcher to set up the connections), an intercom system, and a separate audio mixer. Our goal is to have a single physical network for all the functions of the studio. We envision an architecture where all the various components necessary in production are networked together over a single high-speed network. Components include the video sources and targets of output signals given above, as well as the mixers, keyers, character generators, and special effects generators. The components would be dynamically configured (and reconfigured) by means of control messages sent through the network. Components would communicate with each other through virtual channels over the network. For example, the input signals and output signal to and from a mixer would be sent over three separate virtual channels (one for each signal).

The bandwidth requirements of studio quality video are extreme. Studio quality NTSC 4:2:2 component 8-bit digital video (per CCIR 601) consumes 216 Mb/s per video signal [3]. HDTV will need approximately 1.5 Gb/s per video signal [2]. (To be more precise, the SPMTE-240M standard for 1125/60 HDTV with 10-bit samples consumes 1.485 Gb/s per video signal.) These are uncompressed signals. Studio users demand lossless transmission because of quality requirements for the final signal. Distortion and loss are generated by the repeated compression and decompression cycles that would be encountered in processing compressed video. (It is still a research topic whether there are practical schemes that allow multiple codec cycles without introducing progressive degradation of the signal.) Lossless compression can conceivably reduce the bandwidth requirements by a factor of 1.5. For monitoring purposes, compression might very well be acceptable. (Note that HDTV will be broadcast as a compressed digital image over a regular 6MHz TV channel; using 5-bit quadrature amplitude modulation (QAM) it will have an effective bandwidth of about 30 Mb/s. Work is also going on in codecs for HDTV transmission over a broadband ISDN network using ATM at an approximate 135 Mb/s rate [1].) For a HDTV-based studio, a realistic lower bound on total network bandwidth could be approximately 20 Gb/s; this assumes a dozen connections sending/receiving uncompressed HDTV signals plus a number of control paths and connections sending/receiving compressed signals; this set-up would be adequate for simultaneous use of 2 mixers (2 x 3 channels), a keyer (3 channels) and a special effects generator (2-3 channels) plus a number of monitors receiving compressed signals. An upper bound might be several times 20 Gb/s, say on the order of 50 Gb/s. Thus the television studio is an example of an application that requires what one might call "very" high-speed networks.

If we consider not HDTV but just NTSC quality, then the bandwidth requirements are much more modest since the bandwidth requirements are about one sixth those of HDTV which gives a minimum bound of 3+ Gb/s which is within reach of today's technologies.

In addition to the raw throughput requirements there is a very strict requirement that all processing be done in real time. A video signal being captured live (in a studio) and going through multiple processing elements must not appear to be delayed from the live action. This means that the video display of the end product generally must be delayed no more than 50-100 ms [3] from the capture of the original seed sources. The television production equipment industry is already dealing with concern over delay added by digitization and digital processing. Analog video processing and networking (using routing switches) adds only nanoseconds of delay. Digital processing (using the same network topology) introduces delays of microseconds. This low delay is achieved through sample based processing techniques which reduce latency and buffering requirements. Hardware pipelines are used to overlap steps in the processing algorithm. If a more general purpose and flexible network, not to mention a new processing architecture, is to be accepted it must not introduce perceptible delay. The possibilities for using compression are further reduced by this requirement due to the delay introduced by compression/decompression, especially if the codec needs complete frames or fields of information.

We will be looking at two possible networking technologies that offer hope for handling the studio bandwidth. The first is very high-speed cell switching networks, specifically Asynchronous Transfer Mode (ATM). The second is fully optical networks based on Wavelength Division Multiplexing (WDM). Our initial research will focus on the ATM technology in an attempt to decide to what extent it can handle all the networking needs of the studio – video, audio and data.

A frequently cited feature for ATM is it's ability to handle continuous media such as voice and video. Furthermore, the use of ATM as an internal bus or LAN for a TV studio should allow fairly effortless connection into the public telecommunications system for distribution of digital TV (regulatory and pricing issues aside). The teleconferencing and multimedia applications envisioned for ATM, however, rely heavily on compression techniques to drive the bandwidths down into the 1-10Mb/s per video channel range. In this range many channels of video can be delivered to the desktop through the 150Mb/s per port switches that are now beginning to appear commercially. Although we expect ATM switches that handle 600Mb/s ports to be commercially available within the next few years, this is still off by a factor of 2.5 for uncompressed HDTV. 600Mb/s is however adequate for two (and three with lossless compression) high quality digital signals for today's NTSC standard. One must also consider the total bandwidth of the ATM switch. The question that eventually must be answered is: Can an ATM-based TV studio architecture provide adequate bandwidth at an acceptable cost?

Variations on the "pure ATM" approach will also be studied. One interesting approach combines ATM with a multifrequency architecture based on Wavelength Division Multiplexing (WDM). In this model one or more ATM channels are used by all nodes in the network to manage studio control and low-bandwidth traffic. High-bandwidth traffic (such as uncompressed HDTV data) is moved "out of band" to other frequencies for high-speed transmission. Secondary channels might have other applications such as the maintenance of a global clock for channel synchronization.

References

1. Fleischer, P.E., Lau, R.C., Lukacs, M.E.: Digital Transport of HDTV on Optical Fiber. *IEEE Commun. Mag.*, **29**, 8, Aug., 1991, 36–41.
2. Lyles, J.B., Swinehart, D.C.: The Emerging Gigabit Environment and the Role of Local ATM. *IEEE Commun. Mag.*, **30**, 4, April, 1992, 52–58.
3. Pensinger, G., editor: *4:2:2 Digital Video — Background and Implementation*. Society of Motion Picture and Television Engineers, White Plains, NY, 1989.

The Impact of Scaling on a Multimedia Connection Architecture

Eve M. Schooler

USC/Information Sciences Institute

schooler@isi.edu

Abstract

As the last two meetings of the Internet Engineering Task Force (IETF) have shown, Internet teleconferencing has arrived. Packet audio and video have now been multicast to approximately 170 different hosts in 10 countries, and for the November 1992 meeting the number of remote participants is likely to be substantially larger. Yet the network infrastructure to support wide scale packet teleconferencing is not in place.

This paper discusses the impact of scaling on our efforts to define a multimedia teleconferencing architecture. Three scaling dimensions of particular interest include: (i) very large numbers of participants per conference, (ii) many simultaneous teleconferences, and (iii) a widely dispersed user population. Here we present a strawman architecture and describe how conference–specific information is captured, then conveyed among end–systems. We provide a comparison of connection models and outline the tradeoffs and requirements that change as we travel along each dimension of scale.

1 Overview of a Connection Management Architecture

We have proposed a multimedia connection architecture that has served as the basis for discussion within the IETF's Remote Conferencing Architecture BOF [3]. At the core of the architecture is the notion of a *connection manager* (CM), which resides at each end system to coordinate the orchestration, maintenance and interaction of multi–user sessions. It acts as a conduit for control information among peer CMs, and among other conference–related components as depicted in Figure 1. Media–specific details are relegated to underlying *media agents*, whereas functional commonality is distilled in the CM. Thus, the CM provides general mechanisms for session–related tasks (connect, invite, etc.) and brokers shared information across agents (participant lists, admission policies, etc.).

The CM's other principal responsibility is *configuration management* of end system heterogeneity. End system differences include asymmetries in available media, codec mismatches, variations in bandwidth capabilities, transport incompatibilities, etc. Accordingly, the CM's distributed control protocol negotiates a workable set of capabilities among group members [2].

The intent of the architecture is to facilitate interoperation among users' teleconferencing implementations across the Internet. Therefore, the connection manager is used to convey high–level configuration descriptions from users and peer CMs, into more detailed descriptions for media agents, which in turn translate them into real–time flow specifications [9]. In a simplified scenario, an application asks the CM for high quality audio and adequate quality video over moderate speed links for moderate cost. The configuration directory service is consulted by the CM and identifies media agents that both meet this specification and are available. This service might translate quality, speed and cost into agents that match encoding/data rate combinations. Once notified, the media agents reserve any devices upon which they rely (cameras, codecs, etc.). The initiator's CM then negotiates with the other participants' CMs. Finally, each CM instructs its local media agents to establish real–time transport sessions [8].

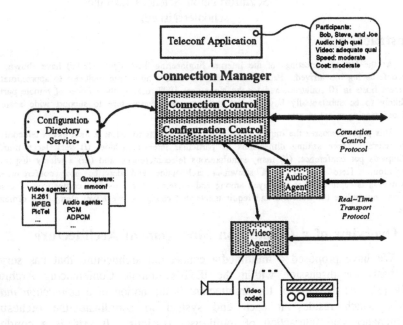

Figure 1. Flow of Control Information

2 The Problem of Scale

Most experimentation with teleconferencing systems has been conducted within LAN settings, with few users and with a modest degree of support for simultaneous conferences (Figure 2). Even projects that scale in one dimension, typically have architectural deficiencies in the other dimensions. To better understand the problem space, we analyze how conference requirements change as we travel along each axis of scale.

Scaling Up in Size. We briefly approximate three points along the horizaontal–axis that correlate to small, medium, and large conferences. A small number of participants (ones or a few tens of individuals) allows *impromptu sessions,* with full connectivity among all users in all media, flexible negotiation of conferencing parameters, authentication of participants, and the exchange of data encryption keys. As we approach medium sized sessions (hundreds or thousands of participants), we emulate *interactive seminars* that are too large for N–way sharing of either data or control. However, impromptu feedback channels are still needed, along with support for dynamic membership. At this size, authentication becomes impractical to provide. The IETF teleconferences were the first medium sized experiments in the Internet [1]. Large conferences (hundreds of thousands or millions of participants) are analogous to TV *broadcasts*. Information is disseminated in one direction, sessions are pre–arranged or even permanent, and descriptions of sessions remain static. All except the largest conferences should accommodate subconferencing.

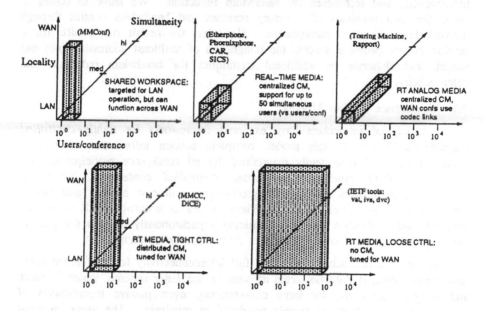

Figure 2. Axes of Scale: Current Teleconferencing Architectures

Extending Locality. Conferences within LANs often exploit the fixed community of user names, simplified authentication, and homogeneity among end system configurations. Farther along the axis, the inter–domain problem of obtaining unique user identifiers arises; with user mobility, the user–to–address mapping is further complicated. Furthermore, WAN conferencing brings greater likelihood of heterogeneity, less assurance of robustness, increased propagation delay, and movement away from centralized designs to ones that are replicated or hierarchical.

Simultaneity. This axis is not quite as straightforward, since the raw number of concurrent sessions is uninteresting. More intriguing is the competition among simultaneous sessions for limited resources both at end systems and inside the network. From the network's perspective, the main resource under contention is bandwidth, but group addresses, shared multimedia devices and users themselves also become commodities. From the user's perspective, resource discovery is needed to locate these shared commodities, and participation management is needed for call waiting, forwarding, suspension, merging, subconferencing, browsing, and filtering, among others.

3 Key Issues: Future Directions

Assessment of the problem space reveals a number of critical needs for a scalable teleconferencing architecture: a suite of directory services, a range of session control, multicast address management, detailed codification of heterogeneity and techniques for bandwidth reduction. We leave to others to solve the hard problems of directory services and bandwidth control through network–layer resource management. However, we revisit our choice for a session control protocol, discuss the integration of multicast addressing into our model, and elaborate on additional techniques for bandwidth reduction and heterogeneity.

3.1 Conference Session Models

A variety of researchers have explored frameworks for *tightly–controlled* conferences [3–7]. In this model, complete session information is actively shared among and consistently maintained by all conference participants. By comparison, IETF multicasts are *loosely–controlled* conferences, where an attendee simply "tunes in" to the agreed–upon multicast address and begins transmitting and/or receiving data. There is no coordination with other end systems, and conference state is constructed asynchronously through the passive (but regular) receipt of control messages from other group members.

Because the first scheme relies on full interconnectivity for conference setup and maintenance, it does not scale as well as the latter scheme, which is more lightweight. However, for large conferencing, even passive transmission of liveness messages leads to sizable overhead at receivers. The more apparent disadvantage to loose conferencing is that it lacks support for coordinated group interactions or consensus, *e.g.*, for authentication, floor control, invitations, or quality–of–service negotiations.

These two models represent but two points in a spectrum of services. They roughly are targeted at small and medium–sized conferences. Large conferences require yet a different session model that (in the extreme) has little (to no) setup, maintenance, or communication among participants. The characteristics which differentiate these models from one another (level of CM interconnectivity, flexibility in negotiations, reliability of communication, dynamics of session state, requirements for a consistent global views) need further scrutiny and organization, for they will ultimately influence the behavior of a more complete session protocol. Do we devise a protocol to adapt over the range of conference sizes and modes, or do we create a family of separate protocols for distinct circumstances? The trend for Internet standards is toward simplicity, which might suggest the development of a small number of simple protocols over one complex protocol.

3.2 Multicast Address Management

As teleconferences scale up in numbers of users, multicast addressing becomes essential for bandwidth reduction. As teleconferencing scale up along the other axes, management of these group addresses becomes more difficult. For initial IETF experiments, IP multicast addresses have been assigned manually and distributed out-of-band. The complication is that there are a fixed number of multicast addresses. Because most telecollaborations will be transient, address assignment and re-assignment will be highly dynamic. A global scheme is required to avoid unwanted address collisions and to promote reasonable address space sharing. A plan is presented in [8] to partition addresses among a hierarchy of multicast address servers; addresses are borrowed from other servers of equal stature in the hierarchy, and servers re-use addresses by exploiting locality. To offload dynamic addressing mechanisms, we can make use of fixed multicast addresses for static conferences and use unicast addressing in point-to-point calls.

We envision a local multicast address server being responsible for a single LAN. The request for a multicast address comes from an individual media agent, or comes from the connection manager if the address is being used to send control messages or to multiplex more than one media type. For conferences that are not publically registered, the CM distributes the multicast address(es) as part of the session configuration process.

3.3 Bandwidth Reduction and Heterogeneity

While multicasting reduces bandwidth usage by senders, mechanisms are also needed for reductions at receivers. A receiver may only want or be able to process M of N streams it is sent. To avoid wasting network bandwidth, we propose *application-level combination nodes*. These include software or hardware modules that embed functions for: mixing, as with audio streams; compositing, assembling the interesting pieces of several video flows into a single flow; selection, by a sender (chairperson) or receiver (individually tailored); translation, between encodings; reduction, when scalable coding is used; and combinations of these operations along the path from senders to receiver.

Combination nodes are likely to be separate from the end systems involved in the conference. As such, they must be incorporated into all aspects of session management, addressing and routing. They are likely to be described in terms of the function(s) they perform, to act as shared resources in the network and be located at branching points in the spanning trees of multicast routes. The drawbacks of using a combination node are control/routing complexity and increased transmission delay. Fortunately, necessity sometimes decides for us, as in the case of a slow link. A mixer upstream from the slow link would be located, then used to combine several streams into one to circumvent bandwidth limitations that would otherwise prohibit or restrict conference participation. The system behaves similarly when there are incompatibilities between end systems due to heterogeneity. For this case, a translator might be used to go between coding formats.

In either event, the component in the architecture that provides this capability is the *resource synthesizer*. It is intended to sit between the connection manager and the configuration directory service (see Figure 1). The quality of the service it provides is somewhat dependent on the configuration directory service, like the other directory services, belonging to a larger hierarchy of information that extends beyond the local domain.

A final comment about configurability: a *configuration language* for Internet teleconferencing must support highly stylized configuration descriptions, if a connection manager is to provide an abstraction beneath which we truly hide the details of end–system heterogeneity. Although we know that configuration translations occur en route from users to flow specification at local and remote end systems, we concede that much more work needs to be done to define useful configuration descriptions at each stage of the process.

4 Acknowledgment

These ideas stem from conversations with colleagues at ISI, discussions held within the IETF Audio/Video Transport working group and the RemConf Architecture BOFs, deployment and evaluation of prototype implementations, and comments on [2, 3] by H.Schulzrinne, Y.Chang and J.Whaley. IETF experimentation will continue to drive issues on scalability and will motivate more complete architecture and protocol solutions. This research was sponsored by DARPA under contract number DABT63–91–C–0001.

5 Bibliography

[1]   Casner, S., Deering, S., "First IETF Internet Audiocast", ACM SIGCOMM Computer Communications Review, Vol.22, No.3 (July 1992).

[2]   Schooler, E., "The Connection Control Protocol: Specification (v1.0)", Technical Report, USC/ISI, Marina del Rey, CA (Jan 1992).

[3]   Schooler, E., "An Architecture for Multimedia Connection Management", Proceedings 4th IEEE ComSoc International Workshop on Multimedia Communications (Apr 1992).

[4]   Arango, M., et al, "Touring Machine: A Software Platform for Distributed Multimedia Applications", Proceedings 1992 IFIP International Conference on Upper Layer Protocols, Architectures and Applications, Vancouver, Canada (May 1992).

[5]   Vin, H.M., Zellweger, P.T., Swinehart, D.C., Rangan, P.V., "Multimedia Conferencing in the Etherphone Environment", IEEE Computer, Vol.24, No.10 (Oct 1991).

[6]   Crowley, T., Milazzo, P., Baker, E., Forsdick, H., Tomlinson, R., "MMConf: An Infrastructure for Building Shared Multimedia Applications", Proceedings CSCW '90, LA, CA (Oct 1990).

[7]   Chang, Y., Whaley, J., "Remote Conferencing Architecture", IETF Teleconferencing Architecture BOF (July 1992).

[8]   Schulzrinne, H., "A Transport Protocol for Audio and Video Conferences and other Multiparticipant Real–Time Applications", Working Draft (Oct 1992).

[9]   Partridge, C., "A Proposed Flow Specification", RFC 1363 (Sept 1992).

Short Paper Session III: Networking and Protocol Support

Chair: Stephen Casner, Information Sciences Institute

An important requirement for network support for multimedia applications is real-time service. Network switches must schedule packets such that those belonging to real-time flows are delivered within the required delay bounds. Since the bounds cannot be met if the traffic presented is unlimited, an algorithm is required to control admission of traffic into the network.

This second aspect is the focus of the first paper, "An Admission Control Algorithm for Predictive Real-Time Service", by Sugih Jamin, Scott Shenker, Lixia Zhang, and David D. Clark. It was presented by the first author, who used imaginative viewgraphs depicting bandwidth allocation as freeway lanes to describe a packet scheduling algorithm with two classes of real-time service, guaranteed and predicted. Guaranteed service is the traditional form, providing an absolute delay bound at the expense of comparitively low bandwidth utilization in order to meet worst-case constraints. The premise of this paper is that many real-time applications can tolerate occasional violations of the delay bound, and that a predictive service, meeting the delay bounds for all but a very small percentage of the packets, can achieve much higher utilization.

The predictive service depends upon an admission control algorithm based on measurements of the current aggregate network load. The paper presents simulation results for a prototype admission control algorithm. The preliminary results claimed for the simulation are that the predictive bounds are reliable, and that the level of real-time utilization is significantly higher when using predictive service than when using only guaranteed service.

In the second presentation, by Daniel P. Ingold, we shifted to a higher level with the description of "MEGAPHONE: A Multimedia Application based on Object-Oriented Communication". The MEGAPHONE application combines enhanced video phone service with a "WhiteBoard" service providing a shared workspace for structured multimedia documents. Both services are combined in an integrated user interface on the workstation, displaying video in windows that can be manipulated like any others. Video communication is provided through an H.261 standard video codec that can be shared for live images and the playing of video clips from the multimedia documents in the shared workspace. The underlying network support is obtained from the MEGACOM network which provides dedicated channels for the various media, including isochronous service for the video and audio.

The final paper of the session, titled "System Support for Dynamic QOS Control of Continuous Media Communication," was presented by Hideyuki Tokuda. With co-author Stephen T.-C. Chou, he returns to the topic of providing service guarantees for continuous media applications. The paper describes the functions of the Capacity-Based Session Reservation Protocol (CBSRP) which covers not only network resources but also protocol processing cycles and packet buffers.

Strict delay bounds are met by considering worst-case resource requirements, corresponding to the more rigid guaranteed service of the first paper. However, a degree

348

of flexibility is retained to manage transient overloads through the use of dynamic control of quality of service (QOS) classes. Applications specify a range of QOS parameters over which they can operate; then, when faced with overload conditions or higher-priority tasks, the system can gracefully degrade the service toward the lower end of the parameter range, and return to higher quality service when resources are again available. CBSRP has been implemented on the ARTS operating system and tested in a LAN environment utilizing the synchronous mode of operation in FDDI to support time-critical traffic.

An Admission Control Algorithm for Predictive Real-Time Service (Extended Abstract)

Sugih Jamin[1], Scott Shenker[2], Lixia Zhang[2], and David D. Clark[3]

[1] Department of Computer Science, University of Southern California
[2] Palo Alto Research Center, Xerox Corporation
[3] Laboratory for Computer Science, Massachusetts Institute of Technology

1 Introduction

The technical and legal developments of the past decade have created the possibility of merging digital telephony, multimedia transport, and data communication services into a single Integrated Services Packet Network (ISPN). For an ISPN to be able to support this diverse set of applications, it must be capable of providing real-time service. Such real-time service will require network switches to implement special packet scheduling and flow admission control algorithms. In Reference [1] we proposed such a real-time scheduling algorithm (which, for convenience, we will call the CSZ scheme) for ISPN's, but did not fully define the required admission control algorithm. The purpose of this paper is to define an admission control algorithm for the CSZ scheme, and then analyze its performance through simulation. In the next section, we motivate the need for admission control and examine some of the issues involved in providing one. In Section 3 we describe a prototypical design, and, in Section 4, we report on some simulation results.

2 Real-Time Services and Admission Control

The ability of the network to meet its real-time service commitments is directly related to the criteria the network uses in deciding whether to accept another request for service. The admission control algorithm embodies these criteria and makes the decision to accept or reject a request for service.

The traditional form of real-time service provides a hard or absolute bound on the delay of every packet; in Reference [1], we labeled this *guaranteed* service. The hard bound in guaranteed service reflects the worst-case behavior of network traffic. Even though the algorithms themselves may be very complex, the underlying principle behind admission control algorithms for such guaranteed service is conceptually simple: if this request for service is granted, will the worst case behavior of the network violate any delay bound? An example of this form of admission control can be found in Reference [3]. In the CSZ scheme, guaranteed service is provided by the weighted fair queueing algorithm (WFQ) described in Reference [2], also called generalized processor sharing in Reference [6], which assigns a share of the bandwidth to each active flow; the admission control criterion is merely that the sum of the previously assigned bandwidths plus the bandwidth requested by the prospective flow does not exceed the link capacity.

In Reference [1] we made the observation that some real-time applications, such as many audio and video applications, can tolerate occasional violations of the delay bound. In some cases, this tolerance allows the application to adapt to moderate variations in the delivered delay, and thus take advantage of the fact that the actual delivered delay is often much less than the bound on delay. Such applications do not require an absolute delay bound, but merely desire a reliable bound; i.e., some very high fraction of the packets obey the bound. To meet the needs of these tolerant (and perhaps adaptive) clients, the CSZ scheme provides a type of service called predictive service which offers a fairly reliable but not absolute bound. When making an admission control decision, the validity of these bounds are assessed using the measured characteristics of the current traffic load, rather than the theoretical worst-case behavior. Thus, the admission control algorithms for predictive service must reliably predict the future behavior of the network based on its measured past behavior. The scheduling algorithm for this predictive class is described in Reference [1]; it attempts to minimize the maximal delays actually experienced, but does not guarantee an absolute maximum delay bound. Note that predictive service does not even provide a probabilistic bound. Probabilistic bounds, as discussed in Reference [3], are based on the statistical characterizations of the current and requesting traffic sources; these characterizations are given to the network by the flows when they request service. In predictive service, the validity of the delay bounds are based on actual network measurements of the current aggregate traffic load.

For guaranteed service, a flow can request any amount of bandwidth and thereby flexibly tune its resultant delay bound. In contrast, the delay bounds for predictive service are less flexible; each switch has a few predictive service classes which have pre-established target delay bounds. These delay bounds will typically be chosen to be roughly an order of magnitude apart, and prospective flows can choose which class of predictive service they desire. The key to predictive service is then having the ability to accept enough traffic to efficiently utilize the network, but yet not accept so much traffic that these target delay bounds are violated more than a tiny fraction of the time. Thus, the viability of predictive service depends on its admission control algorithm. We are not aware of any literature on admission control algorithms of exactly this type. The most similar work is found in Reference [5] where there is a small set of well-characterized traffic sources and the admission control is based on a feasibility graph which is derived from either simulations or approximate calculations. The key difference between that work and what we present here is that we are not limiting ourselves to a small and well-characterized set of traffic sources, and thus cannot rely on pre-existing measurements or calculations; in our situation, measurements of a traffic source can only be performed after the flow has started, and the characterization of the flow is limited to a perhaps rather loose worst-case description. Reference [4] also discusses related work on admission control. While the main emphasis is on analytical calculations for admission control, Reference [4] does raise the possibility of basing admission control decisions on measurements of the current aggregate network load.

The advantage of offering predictive service is that one can more fully utilize the network. The disadvantage is that the admission control algorithm, since it is based on measurement data and not on worst-case characterization of the traffic, is much more difficult to design and justify. In particular, in order to contend that predictive service is viable, one must demonstrate affirmative answers to the following two questions. First, can one provide reliable delay bounds for predictive service with an admission control algorithm that is based on measured history of the network? Second, if one can indeed achieve reliable delay bounds, does predictive service allow a higher level of network utilization as compared to the more traditional guaranteed service? The purpose of our study is to demonstrate that our approach of providing predictive service with measurement-based admission control can meet these two challenges. To this end, we have designed a prototypical admission control algorithm for the CSZ scheme, which is described in the next section.

3 Prototypical Admission Control Algorithm

For the network to make a service commitment to a particular client, it must know beforehand some characterization of the client's offered load. Based on the characterization of the offered load and the estimated current load of the network, the admission control algorithm decides whether to accept the new flow. Unfortunately, it is difficult to accurately characterize actual flows in advance. If admission decisions are based solely on clients' load characterization, then the characterizations either have to be very tight (which would severely constrain the burstiness of flows) or otherwise be very loose (which would lead to overly conservative reservations).

Our system is less dependent on the accuracy of a flow's claimed characterization because our admission decision is mainly based on the measured behavior of existing load. Only when processing a new request, before the system has any observed history of the new flow, are the client-specified parameters used in the decision process. Once a flow starts sending, however, the system will use the measured, rather than the given, values to characterize the current load in making future admission decisions.

We require that each real-time flow α be characterized by a token bucket filter with two parameters: the token generation rate r^α and bucket depth b^α. The load of a switch is represented by the measure of both bandwidth utilization $\hat{\nu}$ and maximal experienced queueing delay $\hat{d}$; following Reference [1], we use the convention that the hat symbol denotes measured quantities. Because the system provides different classes of service, each switch keeps separate utilization and delay measures for each class of service. For each predictive service class j, we keep the aggregate utilization $\hat{\nu}_j$, experienced queueing delay of the class $\hat{d}_j$, and a pre-defined delay bound for the class D_j. For guaranteed service, the switch keeps the sum R_G of the reserved rates of guaranteed flows and the aggregate utilization $\hat{\nu}_G$ of all guaranteed flows. Let μ denote the link bandwidth and n denote the total number of predictive classes. Our prototypical admission control algorithm is based on many heuristic and approximate criteria, which are detailed in the equations below. Due to space limitations, we cannot explain their rationale here; they will be more fully explained in a forthcoming paper. We hasten to note, however, that the validity of these heuristic and approximate criteria does not rest on the quality of the underlying rationale but on the ability of the algorithm to perform adequately in practice. With these caveats, our prototypical admission control algorithm is described below.

Consider some incoming flow α requesting real-time service from the network. The decision about accepting the flow depends, of course, on the quality of service requested by the flow. If the incoming flow α requests service in predictive class k, the admission control algorithm performs the following checks:

1. Determine if the bandwidth usage, after adding the new load r^α, will exceed the link capacity:

$$\mu > r^\alpha + \hat{\nu}_G + \sum_{i=1}^{n} \hat{\nu}_i \qquad (1)$$

2. Determine whether the worst possible behavior of the new flow (i.e., flushing the entire token bucket in one burst of length b^α) can cause violation of the delay bound of classes at lower or same priority levels:

$$D_j > \hat{d}_j + \frac{b^\alpha}{\mu - \hat{\nu}_G - \sum_{i=1}^{n} \hat{\nu}_1 - r^\alpha} \qquad k \leq j \leq n \qquad (2)$$

This specifies how we make the admission control decision when the incoming flow asks for predictive service.

If the incoming flow α requests guaranteed service, the admission control algorithm performs the following checks:

1. Determine that the total bandwidth usage is within capacity using Equation 1, then check that the reserved bandwidth of all guaranteed service flows will not exceed link capacity:

$$\mu > r^\alpha + R_G \qquad (3)$$

2. Determine that if all guaranteed flows use up their reserved bandwidth, the network will still be able to meet all the predictive service delay bounds assuming the current predictive load remains constant:

$$D_j > \frac{\hat{\nu}_j \hat{d}_j}{\mu - R_G - r^\alpha} \qquad 1 \leq j \leq n \qquad (4)$$

3. Determine that the the delay bounds of predictive service classes are still observed when the remaining bandwidth (the bandwidth not reserved for guaranteed flows) is decreased:

$$D_j > \hat{d}_j \frac{\mu - \hat{\nu}_G - \sum_{k=1}^{j} \hat{\nu}_k}{\mu - \hat{\nu}_G - \sum_{k=1}^{j} \hat{\nu}_k - r^\alpha} \qquad 1 \leq j \leq n \qquad (5)$$

The five equations above completely describe our admission control criteria. Their effectiveness requires that the heuristic criteria expressed in the above equations are appropriately conservative.

The above formulae are defined in terms of measured quantities; we now describe our method for measuring bandwidth utilization $\hat{\nu}$ and maximum queueing delay $\hat{d}$. The measurement process uses two tunable constants, T and a; T controls the length of the measurement period, which is expressed in terms of the number of packet departures, and a controls the rate of exponential averaging. When measuring utilization, we measure the average utilization A_j for each class for a period of T packet transmissions; after each measurement period we update the utilization measure $\hat{\nu}_j$ using exponential averaging as follows: $\hat{\nu}_j = (1-a)*\hat{\nu}_j + a*A_j$. To measure the maximum queueing delay for each class, we again consider a measurement period of T packet transmissions, although, as we clarify below, these periods are restarted after each flow acceptance. Define $d_j(i)$ to be the queueing delay experienced by packet i of class j. We update the delay estimate $\hat{d}_j$ upon every packet departure from class j. If the departing packet i is the last packet in the measurement period, we set $\hat{d}_j = MAX_{i'}[d_j(i')]$, where the maximum is taken over all departing packets i' in the present measurement period. If the departing packet i is not the last packet in the measurement period, then we set $\hat{d}_j$ according to the following formula: $\hat{d}_j = MAX[\hat{d}_j, d_j(i)]$. The measurement period length T controls how conservative the measurements of the delays are; increasing T will retain more history, making the $\hat{d}_j$ larger, and thus making the admission control more conservative and the delay bounds more reliable.

Immediately after a new flow is accepted, the measured utilization and queueing delay do not reflect the true load of the network because the measurements do not yet include the load of this new flow. To compensate for this, immediately following flow acceptance we artificially boost the measured values and then allow them to equilibrate. If the newly accepted flow α is in the predictive service class k, then we set $\hat{\nu}_k = \hat{\nu}_k + r^\alpha$ and increase additively the measured delay of predictive classes of priority equal to or lower than that of the new flow:

$$\hat{d}_j = \hat{d}_j + \frac{b^\alpha}{\mu - \hat{\nu}_G - \sum_{i=1}^j \hat{\nu}_i - r^\alpha} \qquad k \leq j \leq n \qquad (6)$$

If the new flow is a guaranteed service flow, then we set $\hat{\nu}_G = \hat{\nu}_G + r^\alpha$, and scale the measured delay of all predictive classes multiplicatively:

$$\hat{d}_j = \hat{d}_j \frac{\mu - \hat{\nu}_G - \sum_{i=1}^j \hat{\nu}_i}{\mu - \hat{\nu}_G - \sum_{i=1}^j \hat{\nu}_i - r^\alpha} \qquad 1 \leq j \leq n \qquad (7)$$

Since we replace the delay estimates $\hat{d}_j$ with the measured values at the end of a measurement period, we choose to restart the measurement period every time a flow is accepted. Each class has its own measurement period.

4 Simulation Results

Using the same simulator used in Reference [1], modified to generate dynamic arrival and departure of flows with random duration, we have simulated the admission control algorithm described above (with the CSZ scheme used in the switches). Our results can be summarized briefly as follows; in all of our simulations, which we performed on several different network topologies and load patterns, we find that (1) the predictive bounds are reliable (fewer than 0.1% of packets violate the bounds), and (2) if the delay bounds for the two classes are comparable, the level of real-time utilization is higher when using predictive service than when using only guaranteed service. Thus, our results, which are admittedly preliminary, support our conjecture that predictive service is viable.

To augment the above summary, we describe below the detailed results from two specific simulations. The simulations use a simple topology of two hosts connected through two switches. Flows traverse the network unidirectionally from the first host to the second host. The links connecting the hosts to the switches are infinitely fast. The bottleneck link between the two switches has a bandwidth of 10 Mbps and latency of 1 ms. All data packets in the simulation have a uniform size of 1Kbits. The simulations lasted for 30 minutes simulation time, with the measurements from the first 200 seconds discarded. We chose T to be 1000, and set $a = 2^{-5}$.

Datagram traffic is generated by a single TCP source with an unlimited amount of data and a TCP window size of 128 packets. Real-time flow requests are generated with exponential interarrival times, where the average interarrival time is 700 ms. Flow durations are uniformly distributed between 1 and 2 minutes. We use two kinds of real-time traffic sources, both constrained by a token bucket filter. A new flow (either guaranteed or predictive) will choose the first traffic model with probability 0.7, and the second model with probability 0.3.

The first real-time source model is a packet-train source modeled with two-state Markov processes. In each train, a geometrically distributed random number of packets are generated at some peak rate P. Let N denote the average train length and let I denote the exponentially distributed intertrain gap. The average packet generation rate G is given by: $G^{-1} = I/N + P^{-1}$.

The second real-time source model is similar to the first one, except that instead of just sending randomly spaced trains of packets, the source will occasionally send out much larger bursts. At the beginning of every train the source will, with probability p, remain quiescent for enough time to fully replenish its token bucket and then send the whole bucket of packets back-to-back. Although this second source model may not imitate the behavior of any real applications, it is used as a worst-case test of our admission control algorithm.

For our simulations, we let $N = 5$ packets, $G = 80$ packets/sec., $P = 160$ packets/sec., $I = 5/160$ sec., and $p = 0.3$. The token bucket used by all flow sources has a token generation rate r of 120 tokens/sec. with bucket depth b of 30 tokens. Sending each packet consumes one token; packets which do not pass the token bucket filter are discarded. Note that with these parameters for both the flow request generation process and the source model, the offered load is

Table 1. Average Link Bandwidth Utilization for Mixed Guaranteed and Predictive Real-Time Traffic.

Real-Time Traffic	Guaranteed	Class 1	Class 2	Datagram
89.0%	22.1%	24.9%	42%	10.7%

much higher than the link capacity. Thus, most flow requests must be denied. In the first experiment, we considered only a single predictive service class with a delay bound of 20 ms. When all incoming flows request predictive service, the link utilization due to real-time traffic is roughly 92.5% (with the datagram traffic using the leftover bandwidth). Furthermore, none of the 14.8 million packets violated the delay bound. For a guaranteed flow to obtain the same delay bound of 20 ms, it would have to request a reserved rate of 1500 kbits/sec. If all flows requested guaranteed service, the maximal network utilization would be 7.2%. Thus, predictive service in this case allows the network utilization to increase by an order of magnitude, while still providing reliable delay bounds.

The second experiment was designed to test whether the delay bounds are still reliable when faced with a heterogeneous set of service classes. Here, we used two classes of predictive service, with delay bounds of 20 ms and 80 ms, in addition to the guaranteed service. For a guaranteed flow, we further allow it to reserve either its token generation rate r (with probability 0.6) or its peak data generation rate P (with probability 0.4). When a new flow request is generated, there was a 0.2 probability of it being a guaranteed flow, a 0.32 probability of it being a class 1 predictive flow, and a 0.48 probability of it being a class 2 predictive flow. The utilization levels are displayed above in Table 1; there were no violations of the predictive service class delay bounds (out of 4 million class 1 and 6.7 million class 2 predictive service packets). We should also point out that the datagram traffic did not suffer starvation even though no bandwidth was preallocated for it.

Thus, our preliminary evidence indicates that our admission control algorithm allows predictive service to be sufficiently reliable while also increasing the network utilization. In the future we will continue to simulate this algorithm on a wider variety of network configurations. This is certainly not the definitive admission control algorithm. In particular, we hope that our additional simulations will lead to further refinements of the heuristic criteria embedded in the above equations.

References

1. Clark, D. D., Shenker, S., Zhang, L.: Supporting Real-Time Applications in an Integrated Services Packet Network: Architecture and Mechanism. Proc. ACM SIG-COMM '92, August, 1992.

2. Demers, A., Keshav, S., Shenker, S.: Analysis and Simulation of a Fair Queueing Algorithm. Internetworking: Research and Experience, 1 1990, 3-26.
3. Ferrari, D., Verma, D.C.: A Scheme for Real-Time Channel Establishment in Wide-Area Networks: IEEE JSAC, **SAC-8**:3 1990, 369-379.
4. Guérin, R., Ahmadi, H., Naghshineh, M.: Equivalent Capacity and Its Application to Bandwidth Allocation in High-Speed Networks: IEEE JSAC, **SAC-9**:9, 1991, 968-981.
5. Hyman, J. M., Lazar, A. A., Pacifici, G.: Joint Scheduling and Admission Control for ATS-based Switching Nodes: Proc. ACM SIGCOMM '92, August, 1992.
6. Parekh, A. K.: A Generalized Processor Sharing Approach to Flow Control in Integrated Services Networks: Tech. Report LIDS-TR-2089, Lab. for Information and Decision Systems, MIT, 1992.

MEGAPHONE:
A Multimedia Application based on Object-Oriented Communication

Daniel P. Ingold
ETH Zuerich,
Laboratory of Computer Engineering and Communication Networks
Gloriastr. 35, CH-8092 Zuerich,
Switzerland

Abstract

'MEGAPHONE' is a pilot multimedia application combining the functions of a video phone with the features of a workstation. It is built upon an object oriented framework that provides workstation-based personal multimedia communication. The video phone functions are fully integrated into the user interface of the workstation used. In addition to the face-to-face meeting mode known from video conference studios, the application offers a shared workspace for discussion, presentation and exchange of digital documents. The integration of telecommunication services with computer conference applications shows useful insights into the architecture of upcoming multimedia applications and their demands on the underlying operating system and network software.

Motivation

Within the scope of the project called 'ETHMICS', an advanced platform for multimedia communication is under development at our laboratory [1]. An important part of this effort is the development of adequate system software support for multimedia communication. The realization of pilot multimedia applications will help identifying bottlenecks and shortcomings in the system software of today's workstations. Another reason for the development of 'MEGAPHONE' was to investigate the future appearance of telecommunication services in public high-speed networks. As we did not want to change the existing window based desktop metaphor we integrated the telecommunication services homogeneously into the user interface of the workstation used.

The text continues with the following topics:
- MEGAPHONE basics
- CSCW aspects
- User interface for video and audio communication
- Network requirements for multimedia communication
- Key software components for multimedia applications
- Main 'MEGAPHONE' features
- Experiences

MEGAPHONE Basics

With MEGAPHONE a user should have access to a video phone facility and he should also benefit from the information processing capacities of his workstation. The best communication infrastructure for qualitatively good video conferencing in Switzerland is provided by MEGACOM. MEGACOM is a digital network run by the Swiss PTT. It supports individual access to up to 2 MBit/sec channels and offers also gateways to similar networks in other European countries. A possible alternative is the ISDN network (SWISSNET). However, the available capacities of only 2x64KBit/sec are not sufficient for the requirements of extensive interactive multimedia-applications. Thus, MEGACOM was chosen.

An application supporting multimedia communication implies the design and implementation of additional system-software support. In today's fast changing software environments this is best done with an object-oriented framework that provides the programmer with standard application objects like windows, menus, dialog boxes or buttons. The proper design of the framework guarantees a consistent application programmer interface (API) as well as a homogenous user interface. We strongly believe that such a consistency is very important for multimedia applications. Furthermore, the use of a multimedia-framework will speed up the development of new applications. For this reason, the presented application is based on a special multimedia application framework [2]. This framework implements class-hierarchies for real-time data streams, distributed 'CSCW'-features and multimedia documents. In particular, it provides an application with a 'channel' abstraction guaranteeing the synchronization of isochrounous source and destination processes.

Figure 1 shows a survey of the 'MEGAPHONE' architecture.

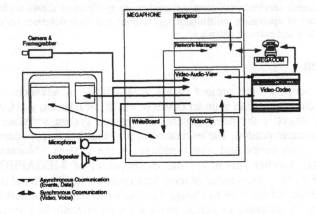

Figure 1. The 'MEGAPHONE'-Architecture

CSCW Aspects

Various models for computer-controlled video conferencing exist in the computer supported cooperative work (CSCW) literature [3,4,5]. The MEGACOM network offers enough bandwidth to carry combined audio, video and computer communication services. Therefore, we decided to implement an application providing an enhanced video phone service together with a shared workspace supporting electronic information exchange. This allows a user to 'videotalk' while publishing multimedia objects with relevance to an ongoing discussion.

The user interface of the shared workspace looks like the document model used in desktop publishing programs. The user can publish items in the shared workspace the same way he would assemble a publication. He places the items (chapters of styled text, drawings, pictures, movies) somewhere onto the document, which is then updated on all sites. The items can be freely arranged at all sites since an accidental overlapping of the items is of no harm. As a simple heuristic to employ locality, the item contents are internally maintained at the originating site. To guarantee consistent modifications, resource accesses are mutually exclusive per item.

Item modifications are mediated through the 'tools' palette which is usually found in desktop publishing programs. Each user can only select one tool at a time. If a user declares the intention to modify a text item (by selecting the corresponding tool), all text items currently under revision are grayed-out to mark them as unavailable.

User Interface for Video and Audio Communication

For an easy start with multimedia-communication, the application itself should be as simple to use as a common application of the workstation. A distinct improvement of the quality is a hands-free solution for the audio communication. Although the feature may disturb other collaborators in the same office, the gain predominates possible objections. The display of video views is handled similar to regular views of the workstation desktop. The user can scale, place, zoom, or hide video views. He also can have several video views within a window. The views can be placed in any order. They also may overlay one another. A user can zoom-in or zoom-out if desired, to inspect certain areas of a video view closer. Additional methods enable the manipulation of picture attributes like brightness, contrast and color. A video view also supports standard copy and paste methods for still pictures.

Network Requirements for Multimedia Communication

The main requirements for video and audio communication are bounded transmission delays. The delays should be less than 100 (150) msec and the transmission must maintain lip synchronization. Since the audio and video devices are connected through standard video conferencing protocols and interfaces (H.261, G.703 [6]) they adhere to our 'channel' abstraction [2] where each end-to-end chain of communication independently guarantees the synchronization of its isochrounous source and destination processes. From the application's point of view, a 'multimedia' channel is fully characterized by its maximal delay. The application has only to know the time when the multimedia object has arrived for sure. With this knowledge it can schedule its 'presentation'. All media specific signalling is done within the channel. The application deals only with the coarse grain orchestration of multimedia objects and is not concerned with the fine grain synchronization of a channel's source and destination process. This allows the channel architecture to take advantage of future improvements in network interface and workstation architecture. The event communication is realized by LAN 'tunneling'. LAN 'tunneling' means that the used LAN protocol for event communication, shared file access, etc., is routed over a dedicated channel of the underlying 'multimedia' network.

Key Software Components for Multimedia Applications

The application is composed of a suite of objects featuring the following functions:
- Basic primitives for network wide event communication
- Distributed resource management

- Distributed connection management
- Public view (shared workspace)
- Tool palette
- Audio video views
- Document handling for real-time audio and video recording (video clips)

The basic communication entity of the framework is a 'net-visible' object called 'PublicService'. This object offers methods needed for network wide event communication. An application wishing to cooperate with remote applications therefore has to publish a list of available 'PublicServices' [Fig.2]. A special object called 'Navigator' allows to search for matching services in the network. Two corresponding services are able to setup their real-time communication channels. The availability of 'PublicService' depends on the installed resources (channels, video codecs, frame grabbers, etc.) and their associated privileges.

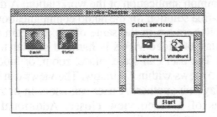

Figure 2. Dialog for selecting Partners and Services

The CSCW objects are also derived from the 'PublicService' class. They use the event oriented protocol of the 'PublicService' class to setup the shared workspace as well as the tool palette.

A video clip object consists of a sequence of frames together with two sound tracks and a textual descriptor. It is referenced when a view object has to store or play data in real-time. With the help of a special object named 'ClipEditor' it is possible to preview and build new clips [Fig.3]. The smallest possible clip is a single frame (a still picture).

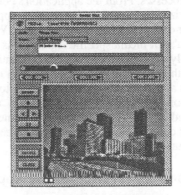

Figure 3. The Clip Editor

The Audio and video view objects are derived from our application framework [2].

Main MEGAPHONE Features

The 'MEGAPHONE' application itself consists of several services. After choosing one or more partners [Fig.2], the user can start a multimedia session with those services negotiated available at all involved sites. In the actual implementation a user is mainly aware of two services, namely the 'VideoPhone' and the 'WhiteBoard'. Both services may be used independently from each other. A 'VideoPhoneView' that is presented in a window, may thus be used as a simple video conference terminal when connect to any standard video conference site in the 'MEGACOM' network. The other available service, the 'WhiteBoard', offers a 'ToolBoxView' with buttons representing the above mentioned 'CSCW-functions' [Fig.4]. Dimmed buttons symbolize functions that are not available at the moment. A function is selected after the corresponding button was pressed. It negotiates the access privileges with corresponding 'WhiteBoard'-objects participating in the same session. The rest of the available window is used as shared workspace. A user can publish, delete or run documents (video-clips, multimedia documents) in previously reserved areas of the shared view. Often an application disposes only of a single video codec. Therefore, the 'WhiteBoard' and the 'VideoPhone' implement functions to switch the video channel from one service to the other.

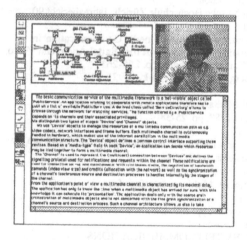

Figure 4. The 'WhiteBoard' during the replay of a multimedia document

Experiences

The approach described above allows a well-structured design of multimedia software. The hierarchical separation of the application, the underlying communication framework and the basic operating system core, turns out to be an efficient way to implement multimedia communication prototypes.

Although the communication framework meets all the requirements mentioned in the previous paragraphs, it is interesting to observe that some deficiencies finally appear at the user interface level. A closer look reveals that the event propagation and the scheduling mechanism of the operating system suffer from an unsuitable implementation for real time requirements.

Summary

This paper introduces the Megaphone application. The enhanced video phone-like service is shown to be operable over an existing public high speed network and its related infrastructure (H.261 video codecs, MEGACOM, etc.).

The object-oriented software design methodology has been successfully applied to the communication requirements of a multimedia application. It is reported to be an efficient way to implement a device independent multimedia communication abstraction.

Our experience shows that the operating system support is crucial for the event propagation mechanism in an object based distributed environment.

References

[1] A. Kündig
'Multimediakommunikation'
Eidgenössiche Technische Hochschule Zürich, Jahresbericht 90

[2] S. Frey, D. Ingold
'An Application Framework for Multimedia Communication' in
Network and Operating System Support for Digital Audio and Video
Lecture Notes in Computer Science, pp. 255-261
Springer, Heidelberg, Berlin, New York, 1992

[3] Robert W. Root
'Design of a Multi-Media Vehicle for Social Browsing'
CSCW'88, Portland 1988, pp. 25-38

[4] Kazuo Watabe et al.
'Distributed Multiparty Desktop Conferencing System: MERMAID'
CSCW'90, Los Angeles, pp. 27-38

[5] Hiroshi Ishii
'ClearFace: Translucent Multiuser Interface for
TeamWorkStation'
ECSCW'91, Amsterdam, pp. 163-174

[6] System 261 Video Codec
GPT GEC PLESSEY TELECOMMUNICATIONS

System Support for Dynamic QOS Control of Continuous Media Communication

Stephen T.-C. Chou and Hideyuki Tokuda*
School of Computer Science
Carnegie Mellon University
Pittsburgh, PA 15213

1 Introduction

Many modern workstations are equipped with specialized hardware capable of producing digital audio and displaying high resolution graphics. One of the current trends in multimedia computing is toward incorporating full motion digital video and audio into applications. We identify these multimedia data as continuous media. Each of data item such as a video frame of voice packet are presented in a continuous fashion. In a communication network, the continuous media data must be delivered in a timely fashion so that the system will not produce any observable jittering effect. We view such a continuous data stream as a sequence of data with its timing constraints attached to each data item.

Quality of service (QOS) in continuous media can be express in term of temporal and spatial resolution. The temporal resolution is the frame rate or periodicity of the data. The spatial resolution is associated with the data size or the compression ratio. The spatial and temporal characteristics of periodic and aperiodic data stream are mostly application dependent. Some generalizations can be drawn by observing continuous media applications. Large volume data are produced and sampled at high rate.

For the continuous media communication, the basic problem in achieving a bounded end-to-end delay are mainly due to unpredictable nature of the operating systems and the networks. Processing of the continuous media data can easily take up large amount of the system and network resources. Extensive system support is essential to insure timeliness of continuous media data delivery and manipulation. The system need to minimize the communication overhead by securing required processor and network resources base on the QOS requirements.

In order to overcome temporal and spatial constrains of the continuous media communication, several protocols such as ST II (Stream Protocol II)[10], SRP (Session Reservation Protocol)[2], Real-time Channel[4], and fast lightweight transport protocols such as VMTP (Versatile Message Transaction Protocol)[3] and XTP (Express Transport Protocol)[11] have been proposed. In general these protocol can be divided into two classes: reservation-based and non-reservation-based protocols. ST II, SRP, and Real-time Channel reserve necessary system resources and network bandwidth before transmitting data. On the other hand, VMTP and XTP transfer data in the best-effort basis.

We have developed, CBSRP, Capacity-Based Session Reservation Protocol in order to minimize end-to-end delay variance in a local area network environment[9]. When a

---

* {stc,hxt}@cs.cmu.edu

client want to open a session, CBSRP will attempt to reserve necessary system resources such as protocol processing cycles, packet buffers, and network bandwidth. CBSRP differs from ST II or Real-time Channel in its capacity of changing QOS parameters of a session dynamically. Each unidirectional communication entity for continuous media is called a session. CBSRP determines whether or not a session can be created base on user's requested QOS specifications. Quality of service is guaranteed unless a shortage of resources is caused by introduction of some urgent tasks. CBSRP has been implemented on the ARTS (Advanced Real-Time Systems)[6] testbed with a Fiber Distributed Data Interface (FDDI)[1] network.

2 Basic System Issues

In general, there are three system issues in continuous media communication: QOS specification and conversion to the system resource requirements, system capacity reservation and resource management, and transient overload management.

2.1 QOS Specification

Quality of service for continuous media data can be expressed with the spatial and temporal characteristics such as a number of frames per second (fps), sampling rate, data size, compression ratio, and so on. These parameters are as important as other characteristics which are common to all communications such as end-to-end delay, message ordering, reliable delivery of messages. Since the user demand can be very flexible, the specification of QOS should be able to provide a range of QOS values.

For instance, in digital video services, a user may choose a range from 1, 2, 3,5, 10, or 30 frames per second for its temporal resolution and range from the number of bits per pixel (bpp), such as 8, 16, or 24 bpp for the spatial resolution. By specifying the range, we will be able to degrade QOS of existing communication sessions to make a room for high priority tasks and restore to the original quality when the extra system capacity become available later.

The QOS parameters which the user should pass to the system are: minimum and maximum temporal resolutions, array of temporal resolution classes, minimum and maximum spatial resolutions, array of spatial resolution classes, priority/importance, maximum allowable end-to-end delay, and maximum packet loss rate.

These QOS requirements of applications has to be translated into a set of protocol attributes and system resource utilization. By quantifying temporal and spatial resolution into a finite number of QOS classes, we can list these classes in a monotonically increasing order based on their system resource requirement. This allows graceful gradual degradation of service when the system can no longer guarantee the resources for the high quality service. The dynamic control of these QOS classes is important to adjust a user's demand base on the availability of the system resources.

2.2 System Capacity Reservation and Resource Management

Successful resource reservation requires combination of admissions control and bandwidth enforcement. The admission control make sure that the requesting activities do not introduce overload situation to the system or do not affect the schedulability of other activities. For example, the admissions control applies schedulability test to the current task set with new task and prevent transient overload of processor resource. Bandwidth enforcement are mechanism to detect and prevent an activity which uses more than its own share of system resources. The processor bandwidth can be accomplish by discriminating real-time tasks from non real-time tasks by using preemptive prioritized scheduling and incorporating mechanism to detect task which exceeds its assigned processor cycles.

Priority inversion[7] is the case where high priority activity blocks for low priority activity to complete. Communication scheduling is especially vulnerable to a priority inversion problem because data delivery through multi-layered protocol modules and scheduling domains before reaching the receiver. Avoiding the priority inversion problems in real-time communication is essential for predictable management of resources.

2.3 Transient Overload Management

Transient overload is caused by over-utilized resource and results unpredictable delay to the system. In order to cope with the overload problems, the system should have a mechanism to detect the overload conditions and be able to produce a feedback to the admission control and user tasks. In the case of processor overload, an operating system must be able to regulate the total number of active tasks which are transferring continuous media data base on the availability of system resources. A static decision will return an error to a user task under such overload condition, while a dynamic decision policy may be able to adjust the allocation of the current resources accordingly based on the given set of QOS parameters. A user task also should be able to renegotiate its QOS requirement whenever the admissions control rejects its initial request. With a dynamic policy, low priority tasks may be reduced to a lower class of QOS so that higher priority task can be executed.

3 Capacity-Based Session Reservation Protocol

CBSRP is designed to minimize variance of delay for stable continuous media communication in a local area network. CBSRP provides guaranteed performance by reserving buffers, processor, and network bandwidth essential for bounded end-to-end communication session. In addition, resource management keeps track of system resource allocation so that no shortage of resources will occur. By preallocating system resources to the time critical activities, we can bound the worst case execution time by removing extra, often unbounded delay and overhead caused by resource contention. CBSRP also allows the dynamic mode change of a session to a higher or lower QOS class in response to the resource requirement of critical tasks. The quality of service which drops during shortage of resources, will be restored to a higher class based on the availability of the system resources.

A real-time session is a unidirectional communication path between a sender and a receiver with guaranteed performance. The sender uses an established real-time session and delivers data to a remote receiver. Each session has a unique session identifier and is registered at both sender and receiver.

Periodic real-time sessions can be distinguished by the periodic interarrival of data to the receiver. In general, messages are delivered through several scheduling domains before reaching the receiver: the sender's protocol processing domain, the network domain, and the receiver's protocol processing domain. In order to complete the message delivery within a bounded delay, processing and delivery at each domain must be met by individual deadlines and the sum of which must be less than or equal to the expected end-to-end delay. If processing within each domain is schedulable, the total delay of end-to-end communication is bounded. We use the deadline monotonic model based on the period and worst case execution time for the schedulability check.

4 Implementation

CBSRP is being implemented on ARTS, a distributed real-time operating system being developed at CMU. ARTS has an preemptable kernel which supports prioritized, preemptable protocol processing. By applying consistent priority mapping across protocol processing and network layers, priority inversion is minimized. This provides a predictable environment for resource reservation for the continuous media communication.

The session reservation service is implemented with the following servers: Session Manager (SM), System Resource Manager (SRM), and Network Manager. SM handles creation, termination, and reconfiguration requests from the users and negotiation with remote SM's. SRM and NRM reserve resources and apply admission control based on amount of the available resources.

4.1 Session Manager

The Session Manager handles creation, termination and reconfiguration requests from the users as well as accepting session establishment and deletion request from the remote session manager. Setting up a session between hosts requires cross domain coordination for negotiating the performance parameters. When a user application issues a session create request, SM converts the QOS specification into system resources requirements. SM keeps track of session parameters for all outstanding sessions of the host for negotiating parameters with the remote host. This allows quick dynamic adjustment of QOS class based on available system capacities.

The QOS parameters described above allows it to be mapped into a reasonable set of processor and processor and memory resource related information and lower-level protocol attributes in the system. These local system resources are checked and reserved using a greedy approach. For remote resources, SM forwards the request to the corresponding SM and NRM.

4.2 System Resource Manager

SRM receives processor cycle and protocol processing buffer requirements from the SM and applies the schedulability and buffer test. In order to prevent overloading of the system, SRM evaluates the schedulability of the task set before creating a new task. Processor bandwidth requirements are expressed in terms of period, execution time, and deadline. A new task can be accepted only if the cumulative computation time of same or higher priority tasks can be completed before the deadline.

4.3 Network Resource Manager

A centralized network resource manager is implemented for an FDDI network. The FDDI protocol supports two types of operation: synchronous and asynchronous mode. Since the transmission of synchronous frames takes precedence over asynchronous traffic, this allows separation of time critical communication from the non time-critical traffic. A station can reserve synchronous allocation (SA) to transmit frames in the synchronous mode. FDDI supports bounded end-to-end delivery of packets. With the synchronous service, the worst case access time can be bounded by twice of the target token rotation time (TTRT)[5]. It allows allocation of network bandwidth base on the expected usage of each node.

In order to create a session on an FDDI network, three criteria must be met. First, the sum of synchronous allocation must be less than TTRT minus the transmission overhead. Second, the minimum deadline of session must be longer than twice of TTRT. Third, the synchronous allocation of a node must be able to transmit the packets from all sessions at least twice of TTRT before the deadline. The NRM keeps track of all synchronous traffics from each node on the network and applies admission control based on the criteria above. The bandwidth is enforced by the synchronous allocation at each node.

4.4 Basic Evaluation

The basic cost of the session management and communication were measured using Sony NEWS 1720 workstation (25 MHz MC68030) which equips a FDDI board with an AMD Supernet chip set. The detailed analysis of the protocol processing overhead was measured by a timer board which is equipped with a timer/down counter of 1 μ second granularity.

The following table shows the cost of each the session manager's operation with no background traffics. Numbers of renegotiations shows how many established session are forced to reduce their classes by NRM. The cost of session creation depends on number of negotiation. For each additional renegotiation increases the session creation time by approximately 28 ms.

#renegotiations	0	1	2
Session_Create	90.2 ms	118.5 ms	146.5 ms
Session_Connect	74.6 ms	102.6 ms	131.0 ms
Session_Close	65.6 ms		
Session_Abort	42.2 ms		
Session_Reconfig	71.1 ms		
Session_Recalc	55.4 ms		

5 Conclusion and Future Work

This paper has presented a dynamic quality of service control for the continuous media communication. Unlike other reservation-based protocols which takes a static performance specification for opening a real-time stream and reserving system resources, CBSRP takes a range of QOS specification in spatial and temporal resolution and convert it into system resource requirements. CBSRP allows a range of QOS parameters which permit service quality to be degraded gracefully and later restored based on available system resources.

Cooperative system support in QOS specification conversion, system resources management, and transient overload management are essential for supporting CBSRP. We will continue to investigate dynamic control of QOS classes and effect of delay for the continuous media communication. We plan to evaluate CBSRP on our RT-Mach[8] platform using the XTP protocol.

References

1. American National Standard, FDDI Token Ring Media Access Control(MAC), ANSI X3.139-1987.
2. D. Anderson, S-Y Tzou, R. Wahbe, R. Govindan, and M. Andrews "Support For Continuous Media In The DASH System," Technical Report, University of California, Berkeley, Oct. 1989.
3. D. Cheriton, "VMTP: Versatile Message Transaction Protocol, Protocol Specification" Request for Comments:1045 Jan. 1987.
4. D. Ferrari and D. C. Verma, "A Scheme For Real-Time Channel Establishment in Wide-Area Networks," IEEE J.Select. Areas Comm.., vol.8, no.3, pp.368-379, April 1990.
5. K. Sevik and M. Johnson, "Cycle Time Properties of the FDDI Token Ring Protocol", IEEE Transaction on Software Engineering, Vol. 13, No. 3, 1987.
6. H. Tokuda and C. W. Mercer, "ARTS Kernel:A Distributed Real-Time Kernel," ACM Operating Systems Review, 23(3), July, 1989.
7. H. Tokuda, C. W. Mercer, Y. Ishikawa, and T. E. Marchok, "Priority Inversion in Real-Time Communication," Proc. of IEEE Real-Time Systems Symposium '89, December, 1989.
8. H. Tokuda, T. Nakajima, and P. Rao, "Real-Time Mach: Toward a Predictable Real-Time System," Proceedings of USENIX Mach Workshop, October, 1990.
9. H. Tokuda, Y. Tobe, S.T.C. Chou, and J.M.F. Moura., "Continuous Media Communication with Dynamic QOS Control Using ARTS with an FDDI Network," Proceedings of ACM SIGCOMM '92, September, 1992.
10. C. Topolcic et al., "Experimental Internet Stream Protocol, Version 2 (ST-II)," CIP Working Group, Request for Comments:1190 Oct.1990.
11. "XTP Protocol Definition, Revision 3.5", Protocol Engines, Inc., September, 1990.

Short Paper Session IV: Multimedia Retrieval

Chair: David Sincoskie, Bellcore

First paper: **NMFS: Network Multimedia File System Protocol** by *Sameer Patel, Ghaleb Abdulla, Marc Abrams, and Edward A. Fox*

Second paper: **A Continuous Media Player** by *Lawrence A. Rowe and Brian Smith*

Third paper: **Architecture of a Multimedia Information System for Content-Based Retrieval** by *Deborah Swanberg, Chiao Fe Shu, and Ramesh Jain*

NMFS: Network Multimedia File System Protocol

Sameer Patel, Ghaleb Abdulla, Marc Abrams, Edward A. Fox

Computer Science Department, Virginia Polytechnic Institute and State University
Blacksburg, VA 24061-0106 U.S.A.

Abstract. We describe an on-going project to develop a Network Multimedia File System (NMFS) protocol. The protocol allows "transparent access of shared files across networks" as Sun's NFS protocol does, but attempts to meet a real-time delivery schedule. NMFS is designed to provide ubiquitous service over networks both designed and not designed to carry multimedia traffic.

1 Introduction

This paper describes a protocol that allows a multimedia application running on a client machine to use multimedia files stored on a server reached through *any* type of network. For simplicity, we view the world as containing two categories of networks: *data networks*, which include existing local area networks (e.g., IEEE 802.3, 802.5, and FDDI) and the TCP/IP-based Internet, and *multimedia networks*, which are networks specifically designed to carry constant and variable bit rate traffic (e.g., ATM networks, and Synernetics and Starlight modified 10baseT Ethernet hubs). Therefore, our goal is remote access to multimedia files through both data and multimedia networks, whereas most research on distributed multimedia applications considers only multimedia networks.

Why consider delivery of multimedia documents over data networks? For a number of years into the future, the Internet will probably interconnect a combination of older data networks and newer multimedia networks. This must be the case, unless all networking infrastructure in the Internet is simultaneously upgraded to multimedia networks. A desirable goal is that of *universal access* by any host or client on the Internet to any multimedia server, even if the connection between client and host contains data network links that do not provide constant or variable bit rate traffic. The quality of presentation on the client workstation may suffer due to data network links, but this is a tradeoff caused by the economic decision not to replace all data networks by multimedia networks.

The paradigm that we use to allow multimedia applications to access media files on network-attached servers is that of Sun's Network File System (NFS) protocol. NFS "provides transparent access to shared files across networks"[1]. NFS provides asynchronous access to remote file systems with no guarantees on the latency or throughput of file access. This is adequate for many types of text and binary data and program files, but inadequate for multimedia.

This paper proposes the Network Multimedia File System (NMFS) protocol, which similarly "provides transparent access to shared files across networks," but also meets real-time requirements of delivery schedules with a high probability. To achieve synchronization, the protocol should deliver all data segments belonging to an interval within a certain real time delay as specified by the application. NMFS is suitable for multimedia files, containing audio, video, and other formats, in addition to text and binary files. The NMFS protocol, though similar in function to NFS, is not a modification of NFS but a new protocol. Like NFS, NMFS is an application layer protocol. NMFS can be ported to any client or server providing UDP and RPC. NMFS is intended to operate over any type or size of underlying network or internet, in conjunction with other traffic. In particular, NMFS can function with the minimum assumption that the underlying network offers datagram delivery with no bounds on latency and throughput.

2 File System Model

In our server file system model, a particular multimedia material is stored in a set of one or more files referred to as a *file group*, each of which can contain one or more *tracks*. Each track is divided into *blocks*, where a block is defined to be a contiguous portion of a track with constant quality of service (QOS) parameters. Blocks are divided into Application Data Units (ADU), as defined by Clark and Tennenhouse [4]. An ADU is the unit of error recovery. A set of ADU's in one or more tracks are logically grouped into a *frame*. Frames cannot cross block boundaries. The client specifies read calls in terms of frames. In addition, frames are used in specifying QOS parameters and in synchronizing presentation of tracks at the client. The relation of tracks, blocks, ADU's, and frames is illustrated in Fig. 1. *QOS* parameters specify the divergence vector (DV), inter-glitch spacing (IGS) and inter-frame pause (IFP), as proposed by Ravindran and Bansal [5].

Synchronization between tracks is specified by providing in an auxiliary file a *virtual time stamp* for each ADU in each track. Tracks are delivered to an application using a *virtual to real time stamp mapping or sequence of mappings*. The default mapping is 1:1, however an application can specify to the NMFS protocol that another mapping be used to play tracks at different or perhaps even variable rates relative to one another. Virtual time stamps are used as the fundamental synchronization mechanism because they can be used to synchronize different types of media.

For NMFS to function over data networks, with their inherent variances in latencies, NMFS tries to presend blocks that it anticipates are likely to be used by the application in the near future. Therefore, for each set of multimedia material, either the client or the server or both stores in an auxiliary file an *anticipated delivery schedule* (ADS). The ADS is an N by N matrix, where N is the total number of blocks in the multimedia material. Element i, j of the matrix contains several pieces of information, including an estimate of the probability

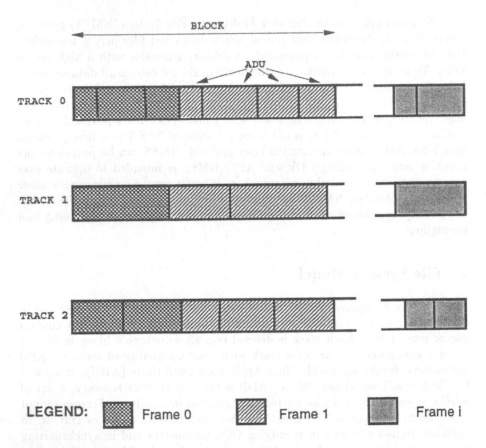

Fig. 1. The relation between tracks, blocks, ADU's and frames.

that the application will next request block j, given that it last requested block i. The ADS maybe stored on the client, in which case the probabilities are based on a per-client basis. A default ADS is always stored on the server and is used by clients that do not have their own ADS and the probabilities represent the aggregate behavior of all network clients. The probabilities in the ADS are derived in one of two ways: from an authoring system [2] or by measurement. In the first case the authoring system predicts client behaviour. In the second case a log file is written that records the sequence of blocks accessed by a client. When the client terminates the ADS is updated based on the log file.

3 NMFS Protocol Overview

3.1 Restrictions in Initial Version of NMFS

The initial version of NMFS contains two restrictions. First, all files are read-only; a client cannot write to the server. Second the protocol does not provide synchronization between traffic types coming from different servers. Therefore, all the tracks that have to be synchronized must be on the same server.

3.2 Service

Like the NFS protocol, the service interface for NMFS is a set of remote procedure calls. In fact, the NMFS calls are only slightly modified versions of the NFS calls. While an NFS file is byte addressable, an NMFS *file group* is frame addressible. In particular, in particular the NMFSPROC-READ call specifies an offset in unit of frames, and randomly accesses any frames in the *file group*.

Buffering: The protocol introduces the idea of buffering at the client's end a certain amount of data, that is likely to be referenced in the near future, in a *frame cache.* The cache size depends on the hardware resources of the client machine. Generally it is desired that the client has enough buffer space to buffer 1 or 2 seconds of presentation.

The frame cache is organized as a set of variable size frames. The frame cache is addressed with a frame number and an offset within a frame. The frame cache is used for two purposes. First, it is used to convert a network with variable latency into a network with constant latency. Second, it is used to hold frames which are anticipated to be used in the near future to reduce the network traffic. The frame cache is subject to a policy for replacing frames when the client reads a frame not in the cache; in this way the cache is conventional. However, the cache is unconventional in that the replacement policy will use the ADS to anticipate future frames that will be read, and notifies the server to presend these frames before the client reads them.

Flow Control and Error Recovery: Flow control is based on a model of the client buffer as a bucket with a spout at the bottom that delivers water to the client. This differs from a leaky bucket scheme [6] in that the buffer occupancy is used to set the server's transmission rate. The entry of water (e.g., ADU's) is sporadic. (For example, the delivery of ADU's may be slowed by a sudden burst in other network traffic, or interrupted entirely such as by a sequence of collisions in an Ethernet.) However, if the bucket never empties then the exit of water from the spout occurs at constant rate. In NMFS, the bucket corresponds to the set of frames within the cache that are currently being presented at the client. There are low and high water marks associated with the bucket. If the number of buffered ADU's reaches the low water mark, then the client side tells the server side to send ADU's at faster than real time. At the high water mark, the client tells the server to return to sending ADU's at real time (e.g., at the same rate at which the client presents them). If the client buffer falls below the low water mark, the NMFS client will modify the virtual-to-real time stamp mapping to slow down the presentation to avoid emptying the buffer.

For error recovery, NMFS allows an application to specify at the time of opening a file what user-defined algorithm should be used when reception of one or more ADU's occurs outside the QOS parameters. NMFS has default error recover for applications without error recovery algorithms. As proposed by Ravindran and Bansal [5], connections are broken and reestablished with tighter QOS parameters if continual errors and QOS violations occur.

3.3   Protocol Rules

Open Requests: A client begins by opening multimedia material on a server. In the open request, the NMFS client side requests a copy of the ADS from the server if the client does not have an ADS for the currently running application.

The open request also contains an estimate of the maximum and average QOS parameters that will be used in subsequent reads. The server will decide whether or not to accept the open request based on its estimate of wither the networks between the client and server can satisfy the estimated QOS. To do this, the server uses statistics collected in the past if available about the mean and variance of network load. For example, for a server on a LAN which has frequent interactions with LAN clients, the server can add the aggregate bandwidths of existing NMFS open multimedia material, add the mean background traffic, and use the variance observed in the recent past to estimate whether the LAN can sustain additional open multimedia material.

After sending the ADS if requested, the server responds by associating each track in the material with one or more *NMFS connections*. Each NMFS connection is mapped to an appropriate entity of the underlying network. For example, this is a UDP port if the underlying network is UDP/IP, or a virtual circuit in an ATM network. The reason that a track is delivered over possibly more than one connection is that portions of a track may have different quality of service parameters. For example, in a video track, reference frames are sent over one connection with the QOS parameter IGS set to a large number to prohibit losses, while remaining frames are sent over another connection with a small IGS value to allow a much higher loss rate.

Read Calls: The client then will perform NMFS read calls. The client side of NMFS will first search the client cache for the frame to satisfy the read. If the frame is not cached, the client uses its copy of the ADS to select m client caches to free, and then sends the NMFS server m read requests to fill the caches. If $m = 1$, the client is requesting only the frame specified in the NMFS read call. If $m > 1$, then the NMFS client is also requesting the server to presend anticipated future frame requests.

Each read call specifies a start frame number and the number of subsequent frames to read in either the forward or reverse direction. If the frame is in a different block than the preceeding frame requested from the server, then the client and server must negotiate the QOS parameters used for delivery in the new blocks. Afterwards the server begins transmitting the ADU's constituting the frame. Note that ADU's travel over 1 or more NMFS connections, each with its own QOS parameters.

4   Project Status

We are currently (October 1992) designing the protocol and expect to implement version 1 by early 1993. A model for providing real-time support of continuous media is being developed. Our environment consists of servers at the Computing

Center at Virginia Tech, clients that could be DVI machines, workstations or PC's located throughout the campus, and the interconnecting network which uses Ethernet and FDDI. In our final version of the protocol, we plan to relax constraints imposed by the initial version.

Acknowledgements: The authors wish to thank Scott Midkiff for his useful discussions. This work was sponsored in part by National Science Foundation grant NSF-CDA-9121999.

References

1. Norwicki,B.: NFS: Network File System Protocol Specification, RFC 1094.
2. Loeb, S.: Delivering Interactive Multimedia Documents Over Networks, IEEE Commun. Magazine **30** (1992) 52–59
3. Little T.D.C. and Ghafoor A.: Multimedia Synchronization Protocols for Broadband Integrated Services, IEEE J. on Sel. Areas in Commun. **9** (1991) 1368–1382
4. Clarck D.D. and Tennenhouse D.L.: Architectural Considerations for a New Generation of Protocols. Proc. SigComm '90 **20** (1990) 200–208
5. Ravindran K. and Bansal V.: Delay Compensation Protocols for Synchronization of Multimedia Data Streams, Technical Report, Dep. of Computing and Information Science, Kansas State University, Manhattan, KS
6. Bertsekas, D. and Gallager, R.: Data Networks, Prentice Hall (1992) 510–513

A Continuous Media Player[†]

Lawrence A. Rowe and Brian C. Smith
Computer Science Division-EECS, University of California
Berkeley, CA 94720, USA

Abstract. The design and implementation of a continuous media player for Unix workstations is described. The player can play synchronized digital video and audio read from a file server. The system architecture and results of preliminary performance experiments are presented.

1. Introduction

Our goal is to develop a portable user interface and continuous media support library that can be used to implement a variety of multimedia applications (e.g., hypermedia systems, video conferencing, multimedia presentation systems, etc.). A key component of these applications is a *continuous media* (CM) player that can play *scripts* composed of one or more synchronized data *streams*. Example data streams are: digitized video or audio, animation sequences, image sequences, and text.

The initial application we are implementing to test our abstractions is a video browser that allows a user to play high quality videos stored in a large database on a shared file server. Figure 1 shows a screen dump of the browser interface. The window on the left lists videos in the database, and the window on the right plays the video. The VCR controls below the video window allow the user to play the video forwards or backwards at several speeds or to access a particular position using the thumb in the slider.

The player runs on a Sun Sparcstation with a Parallax XVIDEO board which has a JPEG CODEC chip. The video stream is stored as a sequence of JPEG frames, and the audio stream is stored in a standard Sparc audio file.

The system has a flexible architecture that will allow other data representations and decompression technologies to be added. For example, we have implemented an MPEG video decoder in software that will be added to the system, and we are anxiously awaiting compression hardware for other Unix workstations.

A special-purpose datagram protocol was implemented to send CM packets from the file server to the client workstation. The current implementation runs on UDP, but it was designed to use the real-time IP protocol being developed by another research group at Berkeley [18]. We have run the player on a conventional ethernet and FDDI network.

The remainder of the paper describes the design and implementation of the player, the results of some initial performance experiments, and related work.

[†] This research was supported by the National Science Foundation (Grant MIP-9014940) and the Semiconductor Research Corporation with a matching grant from the State of California's MICRO program. Additional support was provided by Fujitsu America and Hewlett-Packard.

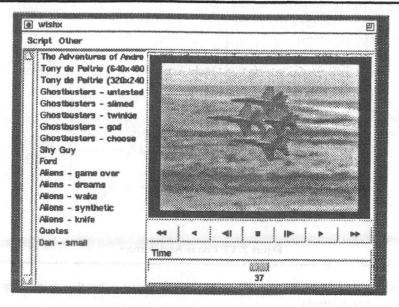

Figure 1. CM player user interface.

2. System Architecture

Figure 2 shows the architecture of the player. The playback application controls the user interface and the CM Server process. The application is responsible for creating windows, responding to input events, and sending commands to the CM Server.

The CM Server receives CM data from the CM Source and dispatches it to the appropriate output device (e.g., the DSP chip to play audio or the video window to play video). The CM Server has a time-ordered play queue to synchronize the playing of audio and video packets. It communicates with CM Source processes on the file server through interprocess communication channels, and it communicates with the X server through shared memory. The system clocks on the different systems are synchronized by the Network Time Protocol (NTP) [11] so that actions in the CM Server and Sources can be synchronized.

The CM Server will eventually be merged with the X server as in the ACME Server [1], but for now it is convenient to separate the functionality for several reasons. First, it makes the CM Server easy to change. Second, it reduces maintenance when a new X server is delivered since we do not have to retrofit our changes. Lastly, source code for the X server is not required which is important because we want to use commercial video boards. Commercial video boards usually include a modified X server for which source code is often difficult to obtain.

The CM Source processes read CM data and send it to the CM Server. CM data is sent in 8k packets on a UDP connection. We have implemented retransmission and adaptive flow control to improve reliability, throughput, and playback quality. Eventu-

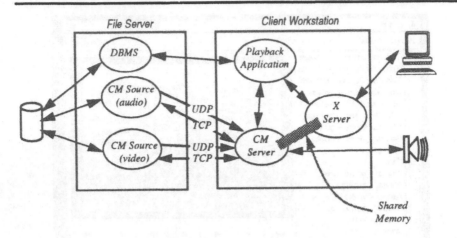

Figure 2. CP Player Architecture.

ally, we will implement a single CM Source server process rather than using a separate process for each stream.

Meta data about scripts is stored in a database. The raw CM data is stored in binary large object (*blob*) files. The meta data is separated from the raw CM data so that different scripts can include overlapping clips without having to make a copy of the CM data.

The remainder of this section describes the CM data model, the CM server abstractions, the CM network protocol, and their implementation.

2.1 CM Data Model

Figure 3 shows a logical picture of a script. Each stream is composed of a sequence of *clips* that represent a sequence of *frames*. A frame is a playable unit such as an image, a frame of video, or a block of audio samples. A clip is a contiguous sequence of frames stored in a blob file. The script has a *logical time system* (LTS) to which frames are synchronized.

CM data is stored in files and the meta data that represents the script is stored in a separate database. The database design for the meta data is shown in figure 4. The

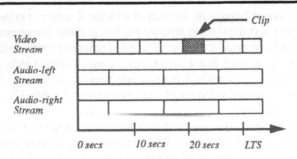

Figure 3. Script representation.

---

type cmtype: (AUDIO, VIDEO, ANIMATION, IMAGE, UNKNOWN)

type cmformat: (JPEGPLX, SPARCAUDIO, GIF, MPEGPARIS, ...)

class SCRIPT(scriptOID: *oid*, name: *string*, owner, *string*, source:*string*, duration: *ltstimeperiod*, micon: *image*, comment: *string*)

class STREAM(streamOID: *oid*, name: *string*, type: *cmtype*, script: *oid*)

class STREAMREP(format: *cmformat*, maxBufferSize: *integer*, maxFrameRate:*integer*, minFrameRate: *integer*, imageWidth: *integer*, imageHeight: *integer*, imageDepth: *integer*, audioSampleRate: *integer*, clipSeq: *oid*)

class CLIPSEQ(clipSeqOID: *oid*, seqNum: *integer*, blob: *oid*, startFrame: *integer*, endFrame: *integer*, duration: *ltstimeperiod*)

class BLOB(blobOID: *oid*, name: *string*, numFrames: *integer*, type: *cmtype*, format: *cmformat*)

class VIDEOBLOB(frameRate: *integer*, width: *integer*, height: *integer*, depth:*integer*, maxFrameSize: *integer*) **inherits from** (BLOB)

class JPEGPLXBLOB(qFactor: *integer*, timecode[]: *timecode*, startTime[]: *ltstime*, endTime[]: *ltstime*, frame[]: *byteoffset*) **inherits from** (VIDEOBLOB)

class MPEGPARISBLOB(startTime[]: *ltstime*, forwardEndTime[]: *ltstime*, reverseEndTime[]: *ltstime*) **inherits from** (VIDEOBLOB)

class AUDIOBLOB(sampleRate: *integer*) **inherits from** (BLOB)

class SPARCAUDIOBLOB() **inherits from** (AUDIOBLOB)

class GIFBLOB(width: *integer*, height: *integer*, depth: *integer*, colorTable[]:*color*) **inherits from** (BLOB)

Figure 4. Database design for script meta data.

---

schema is specified using an object-oriented data model with inheritance, object identifiers, and user defined attribute types including arrays. Several points should be noted about the design. First, a stream may have several representations (c.f. *STREAMREP*) so a script can be played on workstations with different compression hardware and output devices.

Second, time is represented either by a timecode (i.e., hours:mins:secs:frame) or an LTS time. Script and clip durations are stored so that applications can support a slider and operations to seek to a particular time. Video clips include a start and end time for each frame to support playing forwards and backwards using the mapping from logical time (i.e., LTS time) to system time described below. MPEG clips need both a forward and reverse end time to recover synchronization on dropped frames.

Third, an instance of a BLOB class such as *JPEGPLXBLOB*, *SPARCAUDIOBLOB*, and *GIFBLOB* represents a blob file that contains data in that format. We expect this meta data to be replicated in the blob file. The design accommodates the addition of new file types such as Apple's Quicktime files [14].

Lastly, we expect blob files to move between different levels of a storage hierarchy that will include local disks at a workstation, large video file servers, and near-line tertiary stores such as an optical disk or a robot tape jukebox. The mapping from the name of a blob file to a location on which it resides will be handled by a dynamic name server.

2.2 CM Server Abstractions

The CM Server is an event driven process that uses a time-ordered priority queue. Events come from many sources including system clock events, network events (e.g., receive packet or remote procedure calls), X events, and idle events (e.g., software decompression processing). The Server receives packets from the CM source, performs required local processing (e.g., assembles data from several packets into playable units, requests retransmission of missing packets, etc.), calculates the system time at which frames should be played, and queues the play request. Every queued play request has a time period during which it must be executed that is represented by an *earliest start time* and a *latest start time*.

At some later time, provided the Server was able to process the queued play request within the designated time period, the request is executed. Examples of play requests are "put image in video window" or "send packet to audio device." If the Server gets behind, the late request is dropped.

An important feature of the system is that audio frames will be played at the right time regardless of whether the synchronized video frames are played because audio play requests are given high priority. Consequently, audio plays smoothly even when video frames are being dropped.

The mapping from logical time to system time is

$$LTS = Speed \times (SystemClock - Start)$$

where *Speed* is the rate at which the script is being played and *Start* is the *SystemClock* time for *LTS* equal zero. The advantage of this abstraction is that conventional VCR controls can be implemented by setting the *Speed* and *Start* variables as follows:

Function	Implementation
stop/pause	Speed := 0
play forward	Speed := 1
play backward	Speed := -1
goto *lts*	Start := *now - lts*
step forward	Start := Start + 1 / *fps*
step backward	Start := Start - 1 / *fps*
fast forward	Speed := 2.5
fast reverse	Speed := -2.5

Speed represents the relationship between LTS time and *SystemClock* time. *Speed* equal one means LTS time advances at a real-time rate, *Speed* equals 2 implies that LTS time should advance at twice the rate of *SystemClock* time, and so forth. This definition is better than using frame rate as a metric for *speed* because frames per second (fps) can vary during a stream.

Notice that the fast forward and backward speed can be varied. This capability allows an application to implement a jog-shuttle control similar to the mechanical controls found on some video tape recorders.

Another feature implemented in the CM Server to produce high quality user interfaces is resampling audio data in real-time so that synchronized sound can be played when playing a script backwards or forwards at speeds other than normal. Taken together, prioritizing audio packets higher and audio resampling produce a perceptibly better user interface.

2.3 CM Network Protocol

The CM network protocol was implemented when we discovered that a normal TCP connection incurred too much overhead and was too slow.

CM Source processes send packets one second before they are needed. This delay is insignificant when the user begins to play the script, but it gives the Server a buffer against delayed packets due to network or file server load. Our experience has been that audio packets always are delivered on time, but that video packets are often delayed because the data volume and rate is beyond the capabilities of the system. This point is discussed in more detail in the next section. The CM Server periodically requests retransmission of lost packets.

The Server uses an adaptive feedback algorithm to match packet flow to the available resources.[1] The flow rate is based on the *fps* being played. Every 300 msec the CM Server calculates a *penalty* of 10 points if a frame is queued, but not played (i.e., missed) and 10 points if two consecutive frames are missed. For example, if two consecutive frames are missed, the *penalty* is 30 points. In addition, a 10 penalty is assessed if a frame was lost in the network. The maximum allowable penalty in a time period is 100 points. Thus, a *penalty* of 0 means every frame was played and a *penalty* of 100 means many frames were missed.

The *penalty* is sent to the CM Source. Each stream has a minimum and maximum frame rate specified either in the database or when play was initiated. The CM Source also maintains a current frame rate at which the stream is being played. The Source uses the *penalty* to adjust the current frame rate as follows

$$currentRate = currentRate \times (1 - \frac{penalty}{100}) + minimumRate \times (\frac{penalty}{100})$$

Thus if the penalty is 0, no adjustment is made. If the *penalty* is between 0 and 100, the current rate is reduced. If the *penalty* is 100, the current rate is set to the minimum rate.

---

[1] The real-time IP protocol will guarantee a bandwidth when the connection is established. However, this adaptive mechanism will still be required because the delivery rate that can be guaranteed may be below the rate required by the script.

At the same time the CM Source periodically increments the current rate until the maximum rate is achieved. The effect of this algorithm is to slow the rate quickly when the system is overloaded and to increase it incrementally to an achievable throughput. The system may be overloaded because of contention with other processes or because the video being played requires too much bandwidth. This point is discussed in more detail below.

We believe that reduced variability in the frame rate produces higher quality video playback than minimizing the number of dropped frames. Users are more sensitive to random frame drops than to regular drops. Consequently, the adaptive algorithm attempts to reduce the variation of *fps* played.

2.4 Implementation

All processes in the player are implemented with the Tool Command Language (Tcl) and the Tcl Toolkit (Tk) [12, 13]. Altogether, the player is approximately 20K lines of code of which 10% is written in Tcl. The application process uses both Tcl and Tk, includes 1.7K lines of code, and requires 1.5 MBytes at runtime. The CM Server and Source only use Tcl and a library of Tcl and C code developed for distributed applications (e.g., an RPC mechanism, client/server abstractions, CM abstractions, etc.). The library is 9K lines of C code. The CM Server and Source each have 500 lines of Tcl code and 4K lines of C code. The Server requires 1.8 MBytes at runtime and each Source is about 0.6 MBytes at runtime.

The majority of the communication between processes is accomplished by sending Tcl commands to a remote process to be executed. These commands are sent as strings, evaluated by the embedded Tcl interpreter in the remote process, and a string-valued result is returned. This mechanism is a simple remote procedure call (RPC). The application was very easy to develop because remote commands could just be defined and sent rather than requiring the definition of a shared header file that was compiled by a stub compiler as in other RPC mechanisms. Another advantage of Tcl/Tk is that it is very easy to prototype abstractions in Tcl, and when a time critical abstraction is discovered, it can be recoded in C. The entire application was written in under 10 person-months.

3. Performance

This section reports the results of some preliminary performance experiments. Two video scripts were used: "The Adventures of Andre and Wally B" and "Tony De Peltrie" [4]. Both are 24 *fps* computer graphics generated videos. Wally was digitized at 320 by 240 pixels and Tony was digitized at 640 by 480 pixels. Both streams were JPEG compressed using the XVIDEO board. The following table shows the static size of the data.

Video	Total Number Frames	Minimum KB/Frame	Maximum KB/Frame	Average KB/Frame	StdDev KB/Frame
Wally	1806	7.7	12.7	11.3	1.3
Tony	2530	12.8	24.9	20.9	1.6

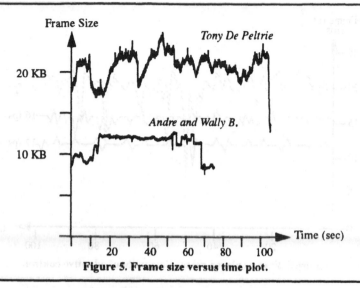

Figure 5. Frame size versus time plot.

Figure 5 plots frame size versus time. Notice that Wally remains somewhat constant at 12KB/frame whereas Tony varies between 13KB/frame and 25KB/frame.

When we play these videos, all audio packets are received and played at the right time, but depending on which video is being played and the system load at the time, video frames are dropped. Wally plays correctly most of the time, but Tony always drops many frames. The problem is the maximum throughput required to play the videos. For example, the maximum throughput required by Wally is 305KB/sec (2.4 Mbit/sec) whereas Tony requires 598KB/sec (4.8 Mbit/sec). The problem is either in the network or the file server, because essentially all frames received by the CM Server are played (e.g., one video frame is received but not played about every 2 seconds). We ran experiments on both ethernet and FDDI networks and the same problems were observed, so it is not the throughput on the network. Since essentially all packets sent by the server are successfully received, we conclude that the bottleneck is in the file server.

Figure 6 shows the effect of varying the requested play rate of the Tony video without the adaptive frame rate control algorithm. The figure plots the number of frames played per second at requested playback rates of 12, 16, and 24 *fps* versus time. As you can see, at 12 and 16 *fps* most frames are played but that many frames are dropped at 24 *fps*. Further investigation suggests that the problem is the number of packets per second that must be sent by the CM Source. The limiting factor appears to be the overhead of sending a packet.

Figure 7 plots the requested frame rate, shown by the thick line, and the actual frame rate, shown by the thin line, with the adaptive frame rate control algorithm. This plot shows the requested frames and the played frames when playing Tony with bounds of 11 to 16 *fps*. You can see that the play rate closely follows the requested rate. Although it does not show in these graphs, the quality of the video is perceptibly better when rate control is used. However, playback could still be improved if predictive data on the

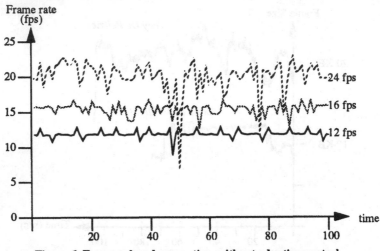

Figure 6. Frames played versus time without adaptive control.

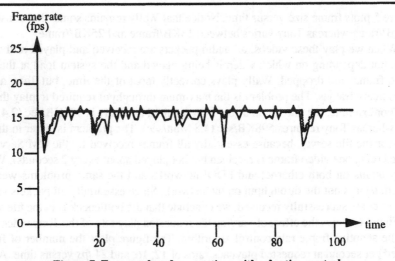

Figure 7. Frames played versus time with adaptive control

bandwidth that will be required during the next few seconds was available so the feedback algorithm could begin reducing the frame rate in anticipation of a change in the required bandwidth.

4. Related Work

Many groups are working on multimedia applications that include playing continuous media [1, 2, 3, 6, 7, 8, 14, 15, 19]. None of these systems report an application-level adaptive control algorithm to vary frame rate dynamically. More research is needed to validate the claims made here and to explore algorithms that look ahead at future resource requirements as suggested by Little and Ghafoor [9]. In addition, we believe that this system is the first one to use NTP and globally synchronized clocks to synchronize transmission of CM data between processes on different machines. The LTS abstraction we used is similar to an analogous abstraction used in Apple's Quicktime.

The synchronization model is similar to the time-based models described by others. We eventually plan to add hierarchical synchronization similar to the model suggested by Steinmetz [16].

Lastly, the data model we developed uses ideas from many sources including [5, 10, 14, 17]. One important point in our design is the replication of meta data in a database that will allow us to manage a tertiary store where CM data can be archived.

5. Conclusions

The design and implementation of a CM player was described. The player uses a globally synchronized clock to synchronize the process that plays the CM data on a client with the file server processes that delivery the data. This design along with synchronizing streams on a logical time system simplified the implementation and made it possible to play audio packets correctly even when video packets were being dropped. Lastly, preliminary performance experiments were reported that illustrate the need for adaptive control of the rate at which video is played and the effect of a simple feedback control algorithm.

Acknowledgments

Several other people contributed to the CM Player software: Steven Yen implemented the playback application, Dan Wallach implemented the audio resampling code, and Greg Paschall worked on an earlier prototype system.

References

[1]   Anderson, D.P. and G. Homsy, A continuous media I/O server and its synchronization mechanism, *IEEE Computer* (October 1991).

[2]   Blakowski, G., et.al., Tools for specifying and executing synchronized multimedia presentations, *Proc. 2nd Int. Wkshp on Network and OS Support for Digital Audio and Video*, Heidelberg (November 1991).

[3]   Coulson, G., et.al., Protocol support for distributed multimedia applications, *Proc. 2nd Int. Wkshp on Network and OS Support for Digital Audio and Video*, Heidelberg (November 1991).

[4]   Dream machine the visual computer - an anthology of computer graphics, Laserdisc, The Voyager Company (1986).

[5]   Gibbs, S., Composite multimedia and active objects, Proc. OOPSLA '91, Phoenix (October 1991).

[6]   Gusella, R. and M. Maresca, Design considerations for a multimedia network distribution center, *Proc. 2nd Int. Wkshp on Network and OS Support for Digital Audio and Video*, Heidelberg (November 1991).

[7]   Hodges, M.E., et.al., Athena Muse: a construction set for multi-media applications, *IEEE Software* (January 1989).

[8]   Lamparter, B, et.al., X-MOVIE: transmission and presentation of digital movies under X, *Proc. 2nd Int. Wkshp on Network and OS Support for Digital Audio and Video*, Heidelberg (November 1991).

[9]   Little, T.D.C. and A. Ghafoor, Scheduling of bandwidth-constrained multimedia traffic, *Proc. 2nd Int. Wkshp on Network and OS Support for Digital Audio and Video*, Heidelberg (November 1991).

[10]   Little, T.D.C. and A. Ghafoor, Synchronization and storage models for multimedia objects, *IEEE Journal on Selected Areas in Comm.*, 8 (3) (April 1990).

[11]   Mills, D., Measured performance of the network time protocol in the internet system, Network Working Group, RFC 1128 (October 1988).

[12]   Ousterhout, J., Tcl an embedded command language, *Proc. 1990 Winter Usenix Conference* (1990).

[13]   Ousterhout, J., An X11 toolkit based on the tcl language, *Proc. 1991 Winter Usenix Conference* (1991).

[14]   Quicktime, software product, Apple Computer Inc. (1991).

[15]   Rosenberg, J., et.al., Presenting Multimedia documents over a digital network, *Proc. 2nd Int. Wkshp on Network and OS Support for Digital Audio and Video*, Heidelberg (November 1991).

[16]   Steinmetz, R., Synchronization properties in multimedia systems, *IEEE Journal on Selected Areas in Comm.*, 8 (3) (April 1990).

[17]   Stevens, S.M., Embedding knowledge in continuous time media, *Proc. 2nd Int. Wkshp on Network and OS Support for Digital Audio and Video*, Heidelberg (November 1991).

[18]   Verma, D., and H. Zhang, Design documents for RTIP/RMTP, unpublished manuscript (1991).

[19]   XMedia Toolkit, software product, Digital Equipment Corp. (1992).

Architecture of a Multimedia Information System for Content-Based Retrieval

Deborah Swanberg, Chiao Fe Shu, Ramesh Jain

University of Michigan, Ann Arbor Michigan 48109, USA

Abstract. This paper describes a multimedia information system (MIS) that bridges the gap between the low level multimedia signal and high level semantic information which is useful in retrieving the data. We first describe the functionality expected of information systems and the types of data the system must manipulate. We then describe a four tiered data model for mapping between the signal and the semantics, and an architecture which integrates databases, knowledge bases, and vision systems to support the data model. Finally, we discuss some of the challenges to be addressed in underlying architecture to support the development of multimedia information systems.

1 Introduction

Current research addressing the integration of video and audio into computer systems focuses on giving us access and control of video and audio data. Once given access, we will need tools to help locate desired data. Due to the massive amounts of data we will face in the future, manual indexing will not be possible. Consequently, we must develop tools to analyze the data and describe its content.

The InfoScope project is researching techniques and systems for bridging the gap between the raw data and the user's descriptions. One project is developing a structured video[1] parsing system, which segments structured video into clips and episodes, and stores the information in a database for future retrieval.

The primary goal in developing the system is to explore issues inherent in systems that support content-based analysis such as knowledge modeling, image segmentation, and database models. In addition to the analysis process, there are several research issues inherent to the supporting system architecture which should be addressed. Like most multimedia systems, the information system requires data storage and compression, synchronized deliver and play, and tools for manipulation. However, unlike current authoring applications, the data analysis component imposes new demands on the system.

This paper disucsses the new demands required by the analysis of video data placed on the underlying operating system and network. The first section discuss the structure of the information system. This is followed by a discussion of the system requirements.

---

[1] Structured video has a well specified structure, such as CNN News, which can be used to identify individual video clips and episodes (sequences of video clips).

2 Structured Video Information Systems

Our video information systems, still in initial development, is intended to segment structured video streams into clips and episodes[2], and store the segmentation information in a database for future retrieval and analysis. Due to the structured nature of the video stream, it is possible to build models of the video. For example, in CNN, the "Headline News" icon appears in every anchorperson clip. These icons can be used to construct models of the structured video.

The models can be used during the segmentation process to help identify the video clips and episodes. For CNN, it could segment and store data according to clips of "anchorpersons" and "newsreels", and episodes of "Headline News", and "Dollars and Sense". There is some information the system cannot parse. For example, it is not possible to identify clips of the President speaking about the economy. The information which can be derived by the parser is limited to structural content of the video images, and cannot parse the meanings of the images and video streams.

The video models can also be used in the retrieval process to help the user specify queries. It could retrieve video data according to date and time; video clips by the type of the clip such as a commercial; and episodes by title such as "Dollars and Sense".

3 Datamodel: From Bytes to Semantics

A system which supports content-based analysis and retrieval must be able to map from the raw data such as the video, to the high level semantics such as episodes of "Headline News". To describe this process, we have defined a four layered data model. Each layer consists of a set of objects upon which new objects and relationships in the higher layer can be based. This data model is described in more detail in ([2, 1]). A brief description of each of the layers follows:

The Image Representation level stores and manages the results of segmentation routines applied to the data. For example, it stores spline curves and chain link encodings.

The Image Object layer provides a generalization to the representation layer. In terms of datamodeling on the higher level, the low level representation may not be relevant. This layer hides the detail of the representation from the higher layer, making it possible to extend the segmentation system with better representations as they become available without changing the upper level modeling.

Two Domain layers represent and manage the data according to the users semantic understanding. Domain Objects are the objects of the users knowledge, Domain Events are the events that happen in the domain.

---

[2] A video clip is a segment of the video stream, such as a commercial. Here, it also refers to a segment without a scene change. An episode is a set of video clips, such as "Headline Sports"

4 Architecture

Multimedia information systems must provide three types of basic functions over the data: segmentation of the information, navigation of the datamodel, and storage of the information. Technology offers three different solutions to these problems. We use image processing techniques of vision systems to segment the data into features such as points, lines, and regions. We use knowledge modeling to guide the traversal of the data model, where models of the domain are matched to features of the data. We use database technology to store the information from the segmentation routines.

A description of the insertion process will best explain the interaction of the different components. The video parser uses vision techniques to segment the video stream into generic clips. Once clips are generated, the knowledge module feeds components of the video model to the video parser, where the parser compares the models to the clips. When the parser identifies a clip or episode, it communicates with the knowledge module, which specifies which features to calculate for the database. After the parser has calculated the features, it delivers the information to the knowledge module, which stores it in the database.

The individual components are briefly discussed below. A thorough discussion of the system is beyond the scope of this paper. For more information, see ([3]) for more details.

4.1 Video Parsing Library

The video parsing library contains routines and processes which segment the data. Specifically, it contains segmentation routines to identify objects, as well as segmentation routines to derive features of identified objects.

In terms of video, the vision library contains routines that segment the video stream into clips. Generally speaking, this is done with a frame comparison routine that compares the histogram of nine different quadrants of two video frames. The calculation of a similarity measure below a specific threshold determines that a scene change has occurred between these two frames. This allows generic clips to be identified in a video stream. These clips are free of any semantic or domain information.

This library also houses routines for calculating the similarity of a video clip model to a generic video clip, or for parsing a video episode model from a set of video clips. Currently, histogram comparison techniques are being investigated, for comparison of the models, but other methods may be used in the future.

4.2 Knowledge Module

This module maintains three types of knowledge:

- Information to aid the segmentation and insertion process
- Information about what objects and features to store in the database
- Information about query optimization techniques

It provides the connection between the vision system and the database. It describes the transition between the layered data model objects and their relationships, both within and between the layers. For the user, it translates the user's semantic language to the underlying system, guiding the user on insertion and querying. For the system, it describes methods of deriving objects from the underlying raw data, methods of representing and storing those objects in a database, and methods of retrieving those objects from the database. These methods use a combination of object-oriented techniques such as semantic labeling, complex object, generalization and subclassing. It also use knowledge base techniques which address issues of similarity measurements, fuzzy logic to manage different definitions of similarity, and weighted object modeling.

4.3 Database

The database stores and manages the information across all layers of the data model: from Domain Objects such as "Headline News" episodes, including the general Image Objects such as regions and curves, and the specific Image Representation objects such as splines and histogram averages. It also maintains the indexes across single and multi-dimensional features to speed object retrieval. Each stored feature maintains a reference to the data from which it was derived. Similarly, every image in the database references the features which it contains.

5 System Challenges

There are many system level challenges presented by video and audio data. Network delivery, operating system synchronization, and authoring tools are being addressed by researchers today. However, when a system has to analyze the data in addition to retrieving and displaying it, the basic authoring and access capabilities may be met in better ways that that provided by generic systems. Listed below are areas of research that should be continued to support video and audio analysis.

5.1 Models of video for segmentation and editing

Segmentation should allow the representation of relationships in video data as complex as those used to edit and create the video tape. Ideally, this should address both the temporal and spatial nature of modern video construction. For example, news broadcasts such as CNN use many different temporal transitions from one scene to the next, such as clean cuts, fades, swipes, etc. A system that models video information for either analysis or editing should allow the representation of these different temporal transitions.

Spatial relationships should also be represented. For example, frequently, broadcasts show pictures in pictures, or multiple speakers in different quadrants of the screen. In addition, static images may be projected for a period of time. All of these structures should be representable within the system that segments/edits video.

5.2 Persistent Objects vs. Databases

Our applications are working with objects retrieved from the database, but unlike common databases, our data is not typically alpha-numeric data, rather it is multimedia/scientific data. Databases, historically, have interfaces defined only by text strings. This harkens to the days when terminals were only alphanumeric. But, today, standard worstations are capable of working with many types of media. Databases should evolve away from the text-based query interfaces, especially if they are to begin working with non-alpha numeric types of data.

Persist objects provide a better paradigm in the computing and operating system world, but here again, for multimedia information systems, the paradigm is not complete. Systems that manipulate persistent objects allow single objects to be stored and retrieved; they do not allow the construction of queries where multiple objects can be retrieved.

5.3 Distributed processing vs. Shared memory

It is a vast understatement to claim that video and audio are costly to load and access. Consequently, distributed architectures may be it may be necessary to ease unnecessary communication burdens. Because it is unreasonable to routinely ship 100Meg files around the network to different computers, which may not be able to store the data, it makes more sense to do the processing at the location of the data. This processing could be controlled from a remote sight, through a message passing paradigm. If distributed processing is desirable, few computers can spare a random 100Meg, so the networked file system should support caching of the relevant bytes of data. We found that the AFS policy of shipping the whole file to the local disk was untenable for operations with large data, so we used NFS to support our distributed work.

Currently, the vision system and the knowledge base are designed as different process which communicate via object names and codes. Specifically, once an object has been identified, the segmentation system queries the knowledge base to determine what features should be calculated for storing the object in the database. There may be other methods for achieving this objective, such as shared memory.

5.4 Storage

There are several different issues involved in working with massive amounts of data. The ubiquitious compression merits discussion in that analysis routine currently operate only over decompressed data. The may be ways of finding analysis routines that operate over compressed data. This will work, only if the desired features are not lost in the compression algorithm, and the data is parsable from the compressed representation.

Other issues regarding storage merit discussion. The mapping of the data to the hard disk should be considered, as some mappings may result in more efficient operations than others. Such patterns may consider operations like as zoom and pan over the data.

Finally, there may be ways of representing signals of data is entirely different forms than the current digitized data. For example, Sandy Pentland's work models objects as shape deviations from standard physical shapes. Faces can be described with these differences, For image analysis applications, different data representations have been developed, such as the quad tree and R* trees. These representations can speed data analysis of some operations. Similar research should be pursued with video information.

6 Conclusion

References

[1] A. Gupta, T. Weymouth, and R. Jain, "Semantic queries with pictures: the VIM-SYS model," in *Proceedings of the 17th International Conference on Very Large Data Bases*, September 1991.

[2] D. Swanberg, T. Weymouth, and R. Jain, "Domain information model: an extended data model for insertions and query," in *Proceedings of the Multimedia Information Systems*, pp. 39–51. Intelligent Information Systems Laboratory, Arizona State University, February 1992.

[3] D. Swanberg, C. Shu, and R. Jain, "Knowledge guided parsing and retrieval in video databases," in *Proceedings of the conference Storage and Retrieval for Image and Video Databases*, EI 1993.

Short Paper Session V: Object Oriented Systems and Toolkits

Chair: Jean-Bernard Stefani, CNET

First paper: **Application Construction and Component Design in an Object-Oriented Multimedia Framework** by *Simon Gibbs*

Second paper: **Audio and Video Extensions to Graphical User Interface Toolkits** by *Rei Hamakawa, Hidekazu Sakagami, and Jun Rekimoto*

Third paper: **An Introduction to HeiMAT: The Heidelberg Multimedia Application Toolkit** by *Thomas Kaeppner, Dietmar Hehmann, and Ralf Steinmetz*

Application Construction and Component Design in an Object-Oriented Multimedia Framework

Simon Gibbs

Centre Universitaire d'Informatique, University of Geneva
24 rue du General-Dufour, Geneva 1211, Switzerland
simon@cui.unige.ch

Abstract. An approach to constructing multimedia applications by connecting groups of high-level components is presented. The components, and their connections, are software abstractions provided by an object-oriented multimedia *framework* – a set of related classes that provide basic functionality and composition mechanisms. Several examples of components and a working application constructed using these components are described. We also consider design issues when components are used under a wide range of conditions.

1   Multimedia Framework

The multimedia framework is a set of C++ classes that provide the building blocks for multimedia applications (an early version of the framework is described in [1]). The framework contains two class hierarchies: MediaValue and MediaObject. Specializations of MediaValue correspond to different media types – such as image, audio and video – instances of these classes are particular media values. Specializations of MediaObject correspond to devices and processes that produce, consume or transform media values. Because the object-oriented approach separates interface from implementation, the user of a media object need not be aware of how the object is implemented – whether it corresponds to a piece of hardware, a software process, or some combination of the two.

In addition to the two core classes, MediaValue and MediaObject, the framework contains classes that support connecting media objects and composing media values. For example, an audio/video recording system might be built by connecting various media objects, some responsible for capturing audio values, others for video, and still others for saving this data. The recorded data, containing temporally correlated audio and video information, provides an example of composing media values – this particular form of composition is known as "temporal composition", other forms include "data flow composition" and "event composition" (composition is discussed in more detail in [1]).

To summarize, the framework provides the "glue" allowing more complex media values and more complex media objects to be built up by composition and connection – we call these "multimedia values" and "multimedia objects" respectively. (Since multimedia objects, like the recording system above, can be viewed as applications, we often refer to media objects simply as "components".)

In order to use the framework for application development on a particular platform, specializations of the framework classes, accounting for the functionality of the platform, must be defined and implemented. For example, if the platform contains some audio capture board, then a MediaObject class for the board and a MediaValue class for the audio data format used by the board, would be needed. We call such an extension to the framework a "multimedia environment".

2 Example Components

At the University of Geneva we have constructed a small multimedia lab containing workstations and audio/video equipment. Since the framework encapsulates the implementation of components and their connection, it is possible to build applications where components run on different processors and use different forms of connection. For instance, we have configured applications with components running on a mix of Silicon Graphics, HP, Sun, NeXT and Amiga hardware. Connections correspond to such communication mechanisms as sockets, communication via a device "file descriptor", or connection by physical cable, e.g., video cables. (In the latter case, it is still possible to connect components from software – all our video components are routed through a 16x16 switch which is software controllable.)

Generally these components have "ports" allowing media values to enter and leave the component. Components are connected by connecting their ports, this is possible provided one port is an output and the other an input port, and the same type of data flows through the ports.

Table 1 lists many of our components. Some are hardware based while others are software based, however in either case the component can be represented by a C++ class which inherits from the MediaObject class. We also indicate the MediaValue subclasses associated with component ports.

Table 1. Components and port descriptions.

component	input port(s)	output port(s)
6D input device		GeoEvent
navigator	GeoEvent	AnimationEvent
animator		AnimationEvent
modeller	AnimationEvent	AnimationEvent [SoundEvent]
model tee	AnimationEvent	AnimationEvent AnimationEvent
model recorder	AnimationEvent	
model player		AnimationEvent
2D and 3D renderers	AnimationEvent [NTSCVideo]	Image
framebuffer	Image	RGBVideo
workstation monitor	RGBVideo	
video monitor	NTSCVideo	

Table 1. Components and port descriptions.

component	input port(s)	output port(s)
read-only video disc		NTSCVideo
write-once video disc	NTSCVideo	NTSCVideo
video overlay	RGBVideo NTSCVideo	RGBVideo
scan converter	RGBVideo	NTSCVideo
video mixer and effects processor	NTSCVideo [NTSCVideo]	NTSCVideo
audio renderer	SoundEvent	PCMAudio
audio DAC	PCMAudio	
audio ADC		PCMAudio
audio file reader		PCMAudio
audio file writer	PCMAudio	

3   Example Application

We have developed a test application as a means of evaluating the framework and allowing us to experiment with different forms of composition and connection. We call the application the "virtual museum" (see [3]). It is an example of a virtual world – a 3D world model which is rendered at interactive rates. In our case the world is a "museum" building (the Barcelona Pavilion from the 1929 World Exhibition, the model of this building was developed by Silicon Graphics). We have populated the museum with "artifacts," these include: stored raster images, images grabbed from video sources, video disc clips, depth images (raster images with a depth channel), geometric objects, and digital sounds. A user, or users, can explore the world in two ways. First, by manipulating a 6-degree of freedom trackball-like device to navigate though 3D scenes, and second, by using a mouse to navigate through a 2D floor plan.

In developing the virtual museum we have gone through a series of configurations, starting with a simple configuration using only a few components and running up to the final configuration using most of the equipment we have available. At each stage, adding new components adds functionality to the application, furthermore the application can evolve from a single-user centralized configuration to a multi-user distributed configuration. Some possible configurations are shown in Fig. 1.

An application configuration can be visually represented by a graph in which nodes correspond to media objects (components) and arcs their connections. At the moment we are developing an editor for these graphs. In addition to handling the display and editing of the graph itself, the editor will instantiate, connect and activate actual components, thus allowing new configurations to be created and tested.

(a)

A	Animator	R_2	2D Render
D	Display	R_3	3D Render
F	Framebuffer	SC	Scan Converter
G	6D Input Device	TS	Video Effects
M	Modeller	V_D	Video Disc
M_R	Model Record	V_R	Video Record
M_T	Model Tee	V_T	Video Tee
O	Video Overlay	N	Navigator

(b)

(c)

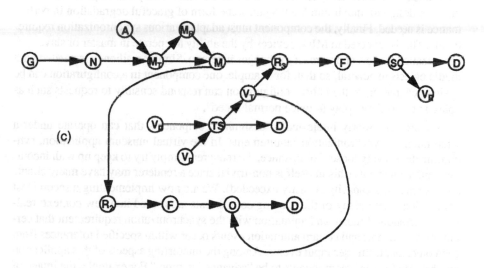

Fig. 1. Two configurations for the virtual museum. Abbreviations for the components
are listed in (a). The first configuration (b) consists of an input device followed
by rendering and presentation components. The second configuration (c) incorporates two renderers, various video components, and more extensive
modelling capabilities.

4 Component Design Issues

The key to constructing multimedia applications by connecting components is to develop large collections of components that can be reused in different contexts. A compo-
nent's context is not just its neighboring components, but the entire ensemble of components in which it finds itself embedded. Since the context is not known when the com-
ponent is designed, there must be a high degree of robustness in component operation.
Some of the problems facing the designers of components include:

- the resources available to the component may not be known,
- the load to be placed on the component may not be known,
- synchronization requirements may not be known.

For example, consider a component that synthesizes audio from some other representation, for example a MIDI synthesizer. Suppose we want to implement this component in software. First consider resources, those needed by the component would include buffers and processor cycles. If sufficient resources cannot be obtained the component should not simply terminate, but rather attempt to continue at some reduced performance level. Now consider load on the component, in the case of MIDI there is a maximum on the number of MIDI messages which can be received per second, so we can estimate the maximum load on the component. This is not true in general though. For example, one could envisage an "n-track" MIDI synthesizer or one where message rates are related to network characteristics. In either situation it may not be practical to estimate, or design to, maximum load; again some form of graceful degradation in performance is needed. Finally, the component must adapt to various synchronization requirements. This is reflected in MIDI devices by the ability to operate in master or slave synchronization modes, and to react to changes in tempo. Similar resiliency is needed with media objects in general, so that, for example, one component in a configuration can be designated master, or the entire configuration can respond sensibly to requests such as "play in reverse" or "play at twice normal speed".

We are particularly interested in designing components that can operate under a wide range of synchronization requirements. In the virtual museum application, synchronization is very loose. For instance, the renderers simply try to keep up with incoming animation events (this in itself is non-trivial since a renderer may have many clients and its rendering capacity is easily exceeded). We are now implementing a second test application where many of the existing components are reused in a new context: real-time generation of music and animation with the synchronization requirement that certain musical events and certain animation events occur within specified tolerances from each other *and* from user input events. Among the interesting aspects of this application are the need for some components to be "advanced in time." For example, the image of a drum being hit must be ready for display when the user actually hits the beat. The component designer cannot simply assume that the rendering period is less than the synchronization tolerance. Should this condition not hold, the only way in which synchronization can be accomplished is by "advancing" the renderer in time and have it anticipate animation events.

References

1   de Mey, V., Breitenoder, C., Dami, L., Gibbs, S., Tsichritzis, D.: Visual composition and multimedia. *Proc. EUROGRAPHICS'92*, 9-22
2. Gibbs, S.: Composite multimedia and active objects. *Proc. OOPSLA'91*, 97-112
3. Tsichritzis, D., Gibbs, S.: Virtual museums and virtual realities. *Proc. Intl. Conf. on Interactivity and Hypermedia in Museums*, (Pittsburgh, Oct. 1991), 17-25

Audio and Video Extensions to Graphical User Interface Toolkits

Rei Hamakawa[1], Hidekazu Sakagami[1] and Jun Rekimoto[2]

[1] C&C Systems Research Laboratories, NEC Corporation
1-1, Miyazaki 4-chome, Miyamae-ku, Kawasaki, Kanagawa, Japan 216
[2] Software Engineering Laboratories, NEC Corporation
11-5, Shibaura 2-chome, Minato-ku, Tokyo, Japan 108

Abstract. This paper describes audio and video extensions to graphical user interface (GUI) toolkits for multimedia systems. These extensions are based on a new object-oriented model for handling multimedia data. The introduction of "temporal glue", and a mechanism for constructing composite multimedia data hierarchically in the model, make it quite easy to edit and reuse composite multimedia data. Programs created on the basis of this model are versatile in the range of their potential application and highly adaptable to changes in hardware. The model can help answer the need for toolkits that will ease the development of multimedia systems and the editing of multimedia data.

1 Introduction

Recent advances in input/output hardware have made possible significant progress in the development of application systems for handling multimedia data. However, with multimedia systems becoming complicated and large, several problems have also become apparent. One of them, described here, is the lack of toolkits for developing multimedia systems and editing audio and video data within these systems. This means it takes programmers a long time to build multimedia systems, and it is difficult for end users to handle multimedia data.

In response to this problem, we have propose an object-oriented model consisting of two sub-models: an object composition model and a playback model. The introduction of "temporal glue", an extension of TeX's glue, and a mechanism for constructing composite multimedia data hierarchically in the model, make it quite easy to edit and reuse composite multimedia data. Programs created on the basis of this model are versatile in the range of their potential application and highly adaptable to changes in hardware. The model can help answer the need for toolkits that will ease the development of multimedia systems and the editing of multimedia data.

After briefly describing our model, we present here an implementation program (a set of C++ multimedia class libraries) based on a model which provides audio and video extensions to graphical user interface (GUI) toolkits.

2 Multimedia Object Model

Models for multimedia toolkits need to be considered from both the programmer's point of view and from that of the user.

For the programmer, application programs created with toolkits based on a model should be highly versatile (i.e. those created for one application should ideally be reusable, as is, for other applications), and adaptable (i.e. it should not be necessary to change a program each time a piece of hardware is changed). Users, on the other hand, want ease of editing (i.e. the ability to edit data with as few operations as possible).

Our proposed model consists of two sub models: an object composition model and a playback model [1]. The *object composition model* is used in the construction of multimedia objects. One important feature is that, unlike previous approaches [1][2], the precise time-line location for each object is not determined until playback; only the relative locations of time and space among objects are defined. Another feature is the introduction of a new object called *temporal glue*. While it is similar to the glue in T$_E$X[5], it is dissimilar in that it can stretch or shrink in temporal space, but not in two dimensional positional space. These features allow users to edit composite objects very easily while at the same time maintaining necessary constraints (*e.g.* synchronization, etc.) in the relationship between objects. For example, Fig.1 illustrates a case in which a user wishes to maintain the constraint that a new audio sequence will begin each time a new video sequence begins (*e.g.* synchronization.) By placing temporal glue objects after audio objects 1 and 2, he is able to accomplish this effect easily.

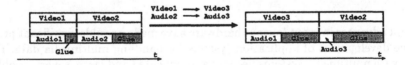

Fig. 1. Composite Object Example

The playback model is used in playing back multimedia objects, which requires treating static information regarding those objects (*i.e.* that which is unaffected by time or physical position on a display) separately from dynamic information (*i.e.* that which is dependent time and/or position). To achieve this, the playback model is equipped, first of all, with media class management objects (for managing such static playback data as object duration time, composite objects structure etc.) and with context class management objects (for managing such dynamic playback data as current frame number of a video object, the

---

[1] We adopt an *object-oriented* approach, so an elementary object can be regard as multimedia data, such as audio and video.

position of objects displayed in a window etc.). Additionally, a viewer class of
management objects is provided for use in displaying objects on a display screen.
Viewer class management objects are capable of such functions as stop, pause,
and play, etc. (see Fig.2) [2] . Context object is generated whenever a multimedia
object is played back. This structure of classes clearly separates the multimedia
static data from its temporal state. Normally, one media has its corresponding
context object, but it is possible that two or more context objects share one
media. This means that we can have several different temporal contexts on the
media; we can playback different portions of the media simultaneously through
different windows.

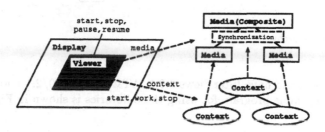

Fig. 2. Playback Model

3 Audio and Video Extensions to GUI

To realize the above model on a computer, we constructed audio and video extensions for the above model, using InterViews [3] [3]. Written in C++, InterViews
provides a set of C++ class libraries with high-level abstractions for implementing interactive programs The main reason why InterViews was selected as our
base is that it also adopts a Glue model, as well as its implementation being
purely object-oriented. Therefore, it was easy to transfer our model to programs
as a natural extension to the InterViews class library. Figure 3 shows our extension to the class hierarchy for audio and video. Classes beginning with Hs are
extension classes, while others are original InterViews classes. The extended library consists of about 40 classes[4], and is about 4,500 lines of C++ code. To use

---

[2] One easy way to understand these relationships is by considering a music performance. Media is to context what a music score is to a player. Media is a layout of
a composite object which should be played in a temporal dimension, like a music
score. It does not contain any information depending on current situation. Otherwise, context can be considered as people performing music while reading a music
score. Also, viewer corresponds to a stage for performance.
[3] InterViews was developed by Dr.M.Linton's group, Stanford University, and is a
graphical user interface (GUI) system for building and using interactive software.
[4] Original InterViews has about 70 classes

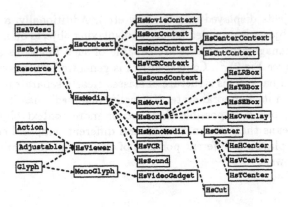

Fig. 3. Class Hierarchy

it, it is necessary to have InterViews 3.0.1, AT&T C++ 2.0/2.1 and X window R4/R5. An example program for using these libraries is shown in Fig.4. This is

```
HsMedia* m = HsAVdesc::read(''paris.av'');   // Create multimedia object
HsViewer* v = new HsViewer(m);               // Create media, and set media
Glyph* g = new Background(                    // Set display information
  new TBBox(v, new HsVideoGadget(v, style)),
  background_color
);
Window* w = new ApplicationWindow(g);        // Create window
session->run_window(w);                       // Event processing loop
```

Fig. 4. Program Example

the program code for constructing the multimedia application in Fig.5. Current available media types are: Glue, (non-compressed) digital video, still objects (pictures, characters..), audio for SUN Sparc station. **HsViewer** is a subclass of **Glyph**[5], so it is able to use audio objects and video objects with everything constructed by **Glyph**. For example, it is easy to make a movie icon or a movie button. Also a class **HsAVdesc** was provided, which reads an AV file (Fig.6), a text formatted description of a multimedia objects structure, and generates multimedia objects. AV file contains information about the hierarchy of composite objects, but not the objects themselves.

Without special OS support such as real-time scheduling, our library can

---

[5] **Glyph** is an object that draws, and the basic building block for the presentation side of a user interface in InterViews.

Fig. 5. Application Example

achieve accurate playback rate for sampled data such as digital video. This is implemented simply by calling context's playback task method very frequently (about 10 times as actual frame rate). On each playback task method invocation, the context object checks current real time by `gettimeofday` system call and decides which frame in the video should be displayed. This mechanism is simple but quite robust. Even when other process steals CPU time or other anomaly occurs, context object skips obsolete frames and can keep ideal playback rate. A user can still operate an application while it is playbacking video data, because the task method is invoked only when there is no user input event.

Portability and extensibility of the library is very high because we clearly separate device dependent classes and independent classes. When new hardware (MPEG etc.) is introduced, which is necessary to incorporate such hardware is only adding new context and media classes for it. Based on our experience, this is about 100 line programming in C++. Such modification does not affect on existing applications because class interfaces for applications (such as `HsViewer`) are totally device independent. Furthermore, since our extension fits into InterViews' design policy so naturally, only less effort is needed to add multimedia capability to an existing application developed with InterViews. For example, only about 40 lines of C++ had to be added for the audio and video extension to `Doc`[6], a WYSIWYG document editor based on InterViews [6] . We also developed a multimedia editor system, *mbuild*, using these classes. *Mbuild* is developed based on our composition model, and helps users create and edit composite objects visually. An editing result can be saved as AV file, and reused. Figure 6 shows the *mbuild* user interface.

---

[6] The total amount of program for doc is more than 10,000 lines.

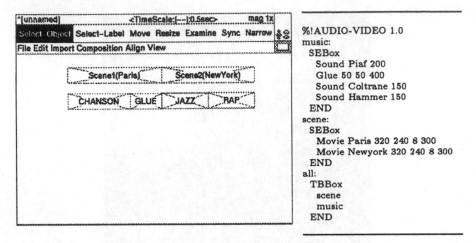

Fig. 6. Mbuild User Interface and AV File Description Example

4   Conclusion

We proposed an object-oriented model for handling multimedia data and showed experimental C++ multimedia class libraries based on these models. These libraries allow end-users to handle multimedia easily, especially audio and video data. Currently, we are developing various multimedia application systems, such as a multimedia presentation system, and a multimedia electronic mailing system. Finally, we are planning to distribute this class of extension libraries as free software until the end of this year [7]. It is hoped that this will be a standard toolkit for constructing multimedia applications.

References

1. Apple Computer Inc. *QuickTime*, 1991.
2. MacroMind Inc. *Director User's Manual*, 1990.
3. M. Linton, J. Vlissides, and P. Calder. *Composing user interfaces with InterViews*, Computer, Feb. 1989.
4. Jun Rekimoto, Hidekazu Sakagami, and Rei Hamakawa. *Design and Implementation of the Multimedia Object Model*, the proceedings of JSSST 92 (Japan Society for Software Science and Technology), 1992.
5. Knuth, D.E. *THE TEXbook*, Addison-Wesley Publishing Company, 1986
6. P. Calder. *The Object-Oriented Implementation of a Document Editor*, the proceedings of OOPSLA 92 (Conference on Object-Oriented Programming Systems, Languages, and Applications), 1992.

---

[7] If you are interested in this software, please mail to rei@tsl.cl.nec.co.jp

An Introduction to HeiMAT:
The Heidelberg Multimedia Application Toolkit

Thomas Käppner
Dietmar Hehmann
Ralf Steinmetz

IBM European Networking Center
P.O.Box 10 30 68
D-6900 Heidelberg 1
Germany
{kaeppner, hehmann, steinmet} @ dhdibm1.bitnet
Phone: +49-6221-404-403 / -214 / -280
Fax: +49-6221-404-450

Abstract: HeiMAT is a toolkit that supports the development of distributed multimedia applications. It provides adaptable communication services, distribution transparency, and cross platform interconnectivity (between various operating and window systems). HeiMAT makes extensive use of product and prototype level available multimedia communications and operating system environments. This paper motivates the major design decisions and presents the system structure of HeiMAT.

1 Introduction

1.1 Motivation

At the IBM European Networking Center (ENC) in Heidelberg, Germany, the HeiProjects have been established to develop prototypes that support distributed multimedia applications on RISC System/6000s under AIX as well as on PS/2s under OS/2 [Herrtwich, 1992] [Luttenberger and Steinmetz, 1992]. Within this framework three related areas have been worked on:

- HeiCoRe: The Heidelberg Continuous-Media Realm is concerned with providing local support services for multimedia on the workstations mentioned above. Those services consist of buffer management, an operating system shield, resource management, and a real-time environment for stream handlers [Herrtwich and Wolf, 1992] [Herrtwich, 1992a] that handle multimedia data. In the initial phase this system support was designed and implemented for AIX as well as OS/2. However, with the introduction of appropriate multimedia products, some of HeiCoRe's earlier prototype code will be replaced. This has already happened with IBM's Multimedia Presentation Manager/2 (MMPM/2) [IBM, 1992], i.e., on OS/2, HeiCoRe stands for MMPM/2 with an additional resource management and an operating system shield.

- HeiTS: The Heidelberg Transport System [Hehmann et al., 1991] [Herrtwich and Delgrossi, 1992] is designed to transfer continuous media data between systems over today's networks such as Token Ring or FDDI in real time. The kind of media and its properties are specified by the transport service user employing quality of service (QoS) parameters, which are then negotiated between the different HeiTS stacks. HeiTS runs as a stream handler in the HeiCoRe environment.

- HeiMAT: The Heidelberg Multimedia Application Toolkit interfaces HeiCoRe providing a uniform distribution mechanism on both platforms. It allows for abstractions of multimedia data, giving access to functions that are commonly needed in multimedia applications like synchronization and mixing of streams. On top of these generic abstractions, modules are provided that are ready to use for development of applications that belong to specific classes. It is aimed to supply AIX as well as OS/2 applications with a homogeneous interface. This paper focuses on this third project area.

Today's multimedia application products are usually programmed in conventional languages (such as C), augmented with hardware-specific multimedia libraries. Replacing any underlying continuous-media device, even with a functionally-equivalent component from another vendor, often requires re-implementing a substantial part of the application programs. If, on the other hand, applications were built using high level abstractions, they would not be affected by these replacements. These applications could even be ported to other platforms where the abstractions are provided without any effort.

1.2 Related Work

We believe that various levels of abstractions for multimedia are required starting with the low-level operating system extensions like [Herrtwich, 1990] and [Leung, 1988]. Newly available multimedia products, such as IBM's MMPM/2, Microsoft's Multimedia Extensions [Microsoft, 1991] and Apple's QuickTime [Apple, 1991] provide the next layer of abstraction for programming of applications. They are however, unlike HeiMAT, not aimed at the distributed environment, and HeiMAT goes beyond their interfaces in providing top-layer abstractions such as a video conferencing module that serves as a complete building block for multimedia applications.

Several experimental systems provide subsets of HeiMAT's functionality. Among them are ACME [Anderson et al., 1990], VOX [Arons et al., 1989][Arons et al., 1989a], [Angebranndt et al., 1991], and [Sventek, 1987]. Those systems cover the distribution aspect of HeiMAT, in that they provide access to an I/O-server within a distributed environment. HeiMAT does not only provide network transparency, but also serves as a toolkit to allow for application development that is independent of multimedia encoding and devices. Those existing toolkits that yield similar support [Anderson and Chan, 1991] are solely aimed at a homogeneous environment in terms of windowing and operating system which usually is UNIX. HeiMAT covers two different environments.

In the subsequent section we introduce our major design principles. These design decisions lead to the system structure of HeiMAT as discussed in Section 3. Section 4 describes the generic functional blocks and Section 5 introduces the concept of support for special classes of applications in HeiMAT.

2 Design Principles

In [Steinmetz and Fritzsche, 1992], the authors note that multimedia programming today is typically based on low-level constructs with strong hardware dependencies. The available multimedia operating system extensions relieve this problem to a certain degree for local applications. Application programmers who want to build, e.g., a multimedia tutoring system have to bridge a wide gap between the functions available to them and the service they want to provide. The complexity they have to deal with increases when the multimedia application is distributed and network access has to be programmed as well. We experienced this with our first OS/2 based integrated multimedia communication system [Cramer et al., 1992] [Steinmetz and Meyer, 1992]. Relying on this experience and the HeiProjects framework we designed the following set of principles.

2.1 Flexibility of Services

At the highest level it should be very easy to develop applications. The toolkit requires only the essential knowledge from the application in order to supply the demanded services. All the details involved with establishing data streams using devices in the distributed system are hidden from the application.

Multimedia applications that are to be supported by HeiMAT form a very diverse set. Even applications, which belong to one class at the first sight, can range from very simple to full featured professional systems. Imagine e.g. a simple multimedia editor for mailing purposes compared to a professional system used for editing commercials. Their requirements with regard to timeliness, quality of edited data streams, and exactness of presentation differ substantially. For some developers fast development is important, whereas others demand the usage of special compressed data types. Thus, services offered by HeiMAT should not only be easy to use, but must also be flexible enough to be useful for every application. This flexibility imposes a set of subsequently discussed transparency requirements which also belong to the design principles.

2.2 Device Transparency

The independence from multimedia devices requires to hide the characteristics of physical devices, thus, allowing for the development of portable applications and making the development much more worthwhile. It also includes the provision of a source/sink paradigm as part of the HeiMAT upper layer interface.

Generally, using audio/video data in applications causes the establishment of streams on the lower level of the system [Herrtwich and Wolf, 1992]. Since streaming of data is very illustrative, it serves also well as a programming abstraction [Steinmetz and Meyer, 1992]. Using this abstraction, an application specifies the endpoints of streams they want to establish. However, details of the stream, which the application is not interested in, can be hidden by HeiMAT.

Information about endpoints can be specified using the level of detail demanded by the application. Imagine e.g. a voice mail system establishing a data stream from the central server to the workstation. It can issue a request to HeiMAT naming the involved devices and a quality for the data stream of *high quality voice*. HeiMAT will map this specification to lower level data formats which not only correspond to the expressed general quality, but are also appropriate for the devices. In contrast, a more powerful retrieval system can utilize special audio formats and compression schemes by explicitly requesting them.

Independently of whether the information about endpoints and filters is specified at

a high or low level, we can distinguish three different origins:
- The application is not interested in that specification: HeiMAT will select an appropriate value.
- The application wants to specify it by itself.
- The application lets the user make the decision.

Programmers of applications can decide which one of the alternatives is used. If parameters are simply omitted, HeiMAT will assign a value. Using the corresponding interface functions of HeiMAT, the application can specify any value. In order to let the user decide about the information, HeiMAT provides elements that an application can construct a user interface of. These elements, applied by an application, will let the user decide on a specification transparently for the application.

2.3 Presentation Transparency

Today's coding technology and standards are still evolving, compete again de-facto standards, and provide diverse algorithms for the same coding problems. The ISO JPEG standard defines compression and coding of single images like many available de-facto standards. ISO MPEG video specifies compression and competes with, e.g., CCITT H.261 and the DVI de-facto compression standard. MPEG-2 is still to be defined for video compression with high er quality. MPEG audio covers audio compression similar to CCITT (e.g. G.721 and 722). However, it also allows for a compressed data rate of more than 64 kbit per second.

Applications need to be developed without constraining to certain coding standards. They should be allowed to specify quality parameters in a manner, which is independent of the real encoding of data. Requesting a new data stream, the application can connect devices with different data types. If a conversion needs to be applied, HeiMAT will insert a filter in the data stream converting from source to sink type. This filter is fully transparent to the application. Thus, HeiMAT allows applications to use lower levels of specification and even explicit usage of different encoding schemes along a data stream will be handled gracefully by the toolkit.

Explicit specification of filters (e.g. mixer, multiplexer) can either be made by type or by name of the device to be used. If only the type of the device is specified, the application is not interested in location and implementation of that specific unit. On the other hand using device names, an application can choose from several different implementations of the same type of a device (e.g. choose distributed or centralized mixing for their respective advantages [Vin et al., 1991][Anderson and Chan, 1991]).

2.4 Distribution Transparency

Applications do not need not know about the actual location of devices. For the application there should not exist any difference between usage of local and remote devices. Since development of applications is more complex being aware of their distributed nature, HeiMAT must provide mechanisms to hide the distribution from the application where desired. Hiding the distribution means
- Dispatching requests of an application to the corresponding remote servers
- Managing necessary transport connections for multimedia data.

Imagine a distributed application, which establishes a data stream between two or more hosts. Such an application can be realized by one application entity residing on any host. The application would have to contact the respective servers, which are responsible for devices on the involved hosts. However, HeiMAT supports device names consisting of a host and a device name local to that host. Any request with regard to a specific device is routed to the responsible server by HeiMAT, thus leaving the application unaware of the need to communicate with possibly several servers.

During establishment of data streams HeiMAT will set up a transport connection, wherever a data stream crosses host boundaries, thus completing distribution transparency.

Hence, applications can be developed totally unaware of their distributed nature. Changing device names is sufficient to switch between local and remote operation. The application can run on any host accessible in the network without any changes. However, applications are free to use their knowledge about device distribution if they decide to do so. For example, an application could insert special compression and decompression filters in the data stream before and after the network devices, respectively.

Note, HeiMAT does not preclude any architecture for an application. Distributed applications based on the client/server approach can also be build easily on top of HoiMAT. The advantage of this approach is that local and remote execution of functions can be triggered independently. For example, a *stop stream* operation could be performed locally, immediately freezing a video image on the screen, whereas in the distribution-transparent case, the request would always be sent to the application, wherever it resides. This will cause the video not to stop before one round-trip time.

3 The System Structure

This set of design principles of HeiMAT together with the availability of HeiMAT on two different platforms imposes hard design requirements on it's system structure. HeiMAT must be a seamless integrator of the available paradigms on both platforms.

The OS/2 platform provides the MMPM/2 as local multimedia extension to the operating system. MMPM/2 allows the definition of sources and sinks of streams to be used by the application. HeiCoRe adds resource management and an operating system shield. The resource management provides the reservation and scheduling of reserved resources in a distributed system [Vogt et al., 1992]. The implementation of the full HeiCoRe environment as a port from AIX with adaptation to the MMPM/2 is in progress. MMPM/2 closely interacts with the Presentation Manager, the OS/2 native window system. This window system was designed for fast response with a large set of functions in a local environment.

For the input and output of discrete media in a UNIX environment, the X window system is the most prominent and widely used system. It provides abstractions from both the hardware dependencies of text and graphics I/O and network transport. It introduces various layers of abstraction from the basic Xlib over the X toolkit to a widget set such as OSF/Motif. Most application programmers base their code on the highest abstractions; however, they can access the lower layers if they need to do so. We envisage multimedia support in a distributed environment to be architected similar to X. In addition to the traditional X server, handling discrete media, there is another server in that architecture, which we call the AV server. It deals with continuous media and communicates with X for presentation on the common display for video output.

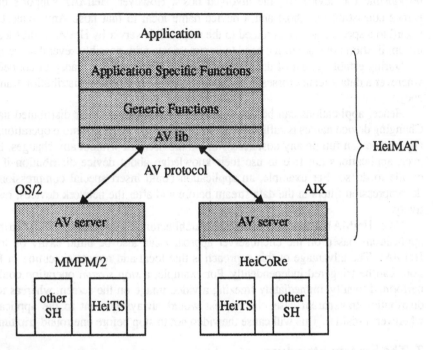

Fig. 1: HeiMAT and it's enviroment on different platforms.

The AV server encapsulates the functionality of HeiCoRe providing access to all types of stream-oriented multimedia devices/filters through a consistent interface (see Fig. 1). As in X, communication between server and application program (client) is supported by an AV protocol which itself is hidden by an AVlib. A typical function set provided by the AVlib includes operations for creation, modification, connection, control, and destruction of the logical multimedia devices, which run in a dedicated real-time environment.

The base layer of HeiMAT comprises the top layer of the AV server that handles the AV protocol, thus, HeiMAT being the first level that allows for remote access to HeiCoRe functionality. On top of this base layer HeiMAT provides services that belong to two distinguished layers:

- Generic functions are common to most multimedia applications (see Section 4). They are provided in a network and device transparent fashion and form a general basis for the development of applications.
- Application specific modules are built on top of generic functions (see Section 5). They provide support for specific classes of applications. The classes include conferencing, collaboration, editing, mailing, and retrieval of multimedia data.

However, HeiMAT will also support the development of multimedia interfaces through user interfacle elements, which partly are already provided in MMPM/2. For this purpose in AIX, HeiMAT itself uses the X toolkit stack. Thus, HeiMAT is residing on both the AV server and the X toolkit stack to provide it's services.

In OS/2 the base layer of HeiMAT is being built around the native Media Device Interface which is a component of the MMPM/2 capabilities in HeiCoRe. It distributes calls to the local multimedia devices. In the first release of the distributed OS/2 system, HeiCam (Heidelberg Remote Camera Control) uses this transparent distributed access to remote multimedia devices.

4 Generic Functions of HeiMAT

Generic functions are used in all types of applications. They are provided according to the design principles and form the basis for application development.

4.1 Abstract Devices

Audio and video devices differ in their capabilities and interfaces. To allow applications to work device independently, logical devices are introduced, which hide the detailed characteristics of physical devices. Classes of devices are defined, in which they are grouped by type (e.g. audio input, mixer, multiplexer, weaver). Every member of a class provides a similar interface to the application. However, devices can implement subsets of the class' functionality and can generally differ in their implementation. It turned out to be a rather difficult task to define a common interface across several similar audio or video devices [Steinmetz and Fritzsche, 1992].

In order to give the user control over a device, HeiMAT provides user interface elements, readily applicable by the application. Thus, the user can control e.g.

- start and stop of a device by a button,
- volume and balance of a stereo device by circular sliders and mixing ratios of several channels by slider bars,

all in a device control window, communicating with HeiMAT transparently for the application.

4.2 Data-Type-Related Functions

During establishment of a data stream HeiMAT has to arbitrate the capabilities of the devices/filters involved. If data types that the devices can work with do not match, HeiMAT tries to apply a conversion. Data type conversion comes in different flavors:

- Degradation of quality: This has to be applied, if in a data stream the quality of the data at the source is higher than the quality which can be displayed at the sink (e.g. the source supplies an stereo audio signal and the sink can play mono only). Generally, downgrading the quality should be done at the source, since degraded streams use less resources in the workstations as well as in the network [Anderson and Chan, 1991]. Note however, the JPEG hierarchical coding mode comprises the joint compression with several resolutions. Since the compression makes use of the redundancy at the different resolutions, the choice of the picture to be displayed can be done at the sink only.
- Multiplexer/Demultiplexer: A data stream contains either one medium only or con-

sists of several interleaved media (e.g. PCM audio and RTV video in an DVI AVSS stream). The conversion between these kinds of formats is referred to as weaving and unraveling.

- Mixer: When several audio streams arrive at a single sink, they generally must be mixed digitally, since only one device is used to apply the conversion to an analog signal.

- Compression/Decompression: In today's multimedia systems, multimedia data streams are normally compressed, due to storage and data rate constraints. Compression and decompression have to be negotiated and applied by HeiMAT. This can include conversion between different compressed data formats.

4.3 QoS Management

During establishment of a data stream, HeiMAT derives the QoS parameters that are valid for the lower layer services. These are passed to HeiCoRe, where they are used to evaluate whether and how the necessary resources can be provided by the real-time environment.

QoS parameters do not only depend on the kind of media used in a data stream or on the classes of service the application provides (as defined in CCITT Study Group XVIII, Draft Recommendation I.211, which include conversational, distribution, retrieval).

The great diversity of CCITT services and applications within individual CCITT service classes yield a great variability of communication requirements. Hence, it remains difficult to classify applications such as distributed tutoring or joint editing based on the CCITT scheme alone.

Other groups of performance criteria have been proposed by [Wright and To, 1990]: Delay sensitive, loss sensitive, and delay and loss sensitive. However, this classification is very coarse.

We suggest that a service must be treated within its context; coding and compression must be taken into account. The QoS depends on the kind of media as well as the service class, the type of sink and source (live -nonpersistent- or stored -persistent-), as well as the configuration of the application [Steinmetz and Meyer, 1992].

4.4 Synchronization

Most applications need synchronization for the correct display of several separated data streams. HeiMAT provides abstractions to group data streams in order to (1) express time relations between them and (2) allow for concurrent execution of operations on all streams in a group.

Synchronization can be expressed (1) implicitly, by combining audio and video (which easily solves many synchronization issues), or (2) explicitly using separate connections for different streams.

Utilizing HeiTS for transport connections, synchronization can be guaranteed by imposing the same end-to-end delay on related streams (choosing an absolute end-to-end delay and limiting the jitter of the data information units to about 0 msec). In practice, it is neither possible nor necessary to guarantee service with such tight tolerances. Audio can be played ahead of video for about 120 msec, and video can be displayed ahead of audio for about 240 msec [Murphy, 1990]. Both temporal skews will sometimes be noticed, but can easily be tolerated without any inconvenience by the user. Note, this asymmetry is very plausible: In a conversation in which two people are located 20 m apart, the visual impression will always be about 60 msec ahead of the acoustics due to the fast light propagation compared to the acoustic wave propagation.

In the next version a more sophisticated implementation of synchronization allows to guarantee timing relations between various sources: A logical time system (LTS) is being introduced as suggested by [Anderson et al., 1990]. Presentation of the data is performed based on a comparison of the LTS with the real-time clock of the destination workstation.

5 Application Specific Functions

The upper layer modules are described in the following by looking at modules designed to provide parts of the functionality needed for conferencing.

An application-specific module can be understood as a complete application itself. However, the module can as well be used as one building block in other applications. Imagine e.g. the development of a banking application in which the customer at an automatic teller machine communicates with specialists located at the remote support center. By utilizing this conferencing module, the development of the banking application can be drastically reduced towards adaptation and enhancements of the available conferencing module.

On the other hand, a module usually comprises several independent application specific functions. Thus, an application can reuse some of the functions provided and implement others on top of it. That yields full flexibility and support for the application.

For conferencing we envisage the following functions:

Conference Environment

Conference directories are utilized to store information about potential participants in conferences. Group information in the directory indicates a special relationship between participants. It contains the information who may assume which role in a conference instantiated on that group. Groups allow users to conveniently invoke conferences with people they need to meet more often.

Conference Scheduling

Scheduling of conferences allows to make appointments for and to plan conferences. These mechanisms are not only needed to announce future conferences to many participants, but can also be used to reserve scarce resources in advance.

Conference Management

There are two different approaches to conference management: Centralized and distributed architectures. Independently of the approach many aspects are involved:

- Invitation protocols must be provided, which can handle concurrent requests by several participants.
- Negotiation is needed about media types to be used, roles participants can assume in a conference, whether mixing of media streams is applied, which floor mechanisms are used etc.

- Streams are established, using the lower layer functions of HeiMAT.
- Floor passing comprises mechanisms and strategies to determine which participant in a conference can provide input at a given time.

All of above functions must be accessible via natural graphical interfaces, e.g. a virtual conference room (represented by a window), which can be left by participants simply by dragging out their picture.

6 Concluding Remarks

HeiMAT, the Heidelberg Application Toolkit, will provide a set of functions that greatly simplifies development of multimedia applications. In analogy to the X Window toolkit stack it will allow application programmers to use a high level of abstraction by which details of distribution and multimedia devices are handled transparently. However, applications can be build refining this high level specification and thus utilizing advanced features of the underlying system. The system structure allows for the integration of two different system environments, AIX and OS/2. Note, the described work is still in progress.

We would like to thank Ralf Guido Herrtwich for his many useful comments on an earlier version of this paper.

7 References

[Arons et al., 1989]
Arons, Barry; Binding, Carl; Lantz, Keith; Schmandt, Chris: "A Voice and Audio Server for Multimedia Workstations", Proceedings of Speech Tech '89, May 1989.

[Arons et al., 1989a]
Arons, Barry; Binding, Carl; Lantz, Keith; Schmandt, Chris: "The VOX Audio Server", 2nd IEEE COMSOC International Multimedia Communications Workshop, Montebello, Quebec, Canada, Apr. 1989

[Anderson et al., 1990]
Anderson, David; Govindan, Govindan; Homsy, George: "Abstractions for Continuous Media in a Network Window System", Technical Report UCB/CSD 90/596, UC Berkeley, Sep. 1990.

[Anderson and Chan, 1991]
Anderson, David P.; Chan, Pamela: "Toolkit Support for Multiuser Audio/Video Applications", in: Herrtwich, R.G. (Ed.): Proc. Second International Workshop on "Network and Operating Systems Support for Digital Audio and Video", Heidelberg, Germany, Nov. 1991, 230-241.

[Angebranndt et al., 1991]
Angebranndt, Susan; Hyde, Richard L.; Luong, Daphne Huetu; Siravara, Nagendra; Schmandt, Chris: Integrating Audio and Telephony in a Distributed Workstation Environment", Proceedings 1991 Summer Usenix Conference, 59-74.

[Apple, 1991]

Apple: "QuickTime Developer's Kit Version 1.0", Apple Document Number 030-1899.

[Cramer et al., 1992]

Cramer, Andreas; Farber, Manny; Hehmann, Dietmar; Jungius, Christiane; Luttenberger, Norbert; Markgraf, Frank; McKellar, Brian; Mengler, Stefan; Meyer, Thomas; Reinhardt, Kurt; Sander, Peter; Sandvoss, Jochen; Schütt, Thomas; Schulz, Werner; Steinbeck, Werner; Steinmetz, Ralf; Stüttgen, Heiner; Vogt, Carsten: "The Heidelberg Multimedia Communication System: Multicast, Rate Enforcement and Performance on Single User Workstations", IBM ENC Technical Report no.43.9209, July 1992.

[Herrtwich, 1990]

Herrtwich, Ralf Guido: "Time Capsules: An Abstraction for Access to Continuous-Media Data", IEEE Real-Time Systems Symposium, Orlando, December 5-7, 1990, pp.11-20.

[Hehmann et al., 1991]

Hehmann, Dietmar, Herrtwich, Ralf Guido, Schulz, Werner, Schütt, Thomas, Steinmetz, Ralf: "Implementing HeiTS: Architecture and Implementation Strategy of the Heidelberg High-Speed Transport System", in: Herrtwich, R.G. (Ed.): Proc. Second International Workshop on "Network and Operating Systems Support for Digital Audio and Video", Heidelberg, Germany, Nov. 1991, 33-44.

[Herrtwich, 1992]

Herrtwich, Ralf Guido: "The HeiProjects: Support for Distributed Multimedia Applications", IBM ENC Technical Report No. 43.9206, 1992.

[Herrtwich, 1992a]

Herrtwich, Ralf Guido: "An Architecture for Multimedia Data Stream Handling and Its Implication for Multimedia Transport Service Interfaces", 3rd IEEE Workshop on Future Trends of Distributed Computing Systems, Taipei, April, 1992.

[Herrtwich and Delgrossi, 1992]

Herrtwich, Ralf Guido, Delgrossi, Luca: "Beyond ST-II: Fulfilling the Requirements of Multimedia Communication", 3rd Intl. Workshop on Network and Operating System Support for Digital Audio and Video, San Diego, Nov. 1993.

[Herrtwich and Wolf 1992]

Herrtwich, Ralf Guido; Wolf, Lars: "A System Software Structure for Distributed Multimedia Systems", 5th ACM SIGOPS European Workshop, Le Mont Saint-Michel, France, September 1992.

[IBM 1992]

IBM: "Multimedia Presentation Manager/2: Programming Reference" IBM Document Number 41G2920.

416

[Leung et al., 1988]
Leung, W. H.; Luderer, G. W. R.; Morgan, M. J.; Roberts, P. R.; Tu, S.-C.: "A Set of Operating System Mechanisms to Support Multi-Media Applications", Proc. Intern. Seminar on Digital Comm., Zurich, Mar. 1988, 71-76.

[Luttenberger and Steinmetz, 1992]
Luttenberger, Norbert; Steinmetz, Ralf: "Videocommunication over the IBM Token Ring", leaflet for CeBIT '92 and Didacticum no. 14, July 1992, 20-23.

[Microsoft 1991]
Microsoft Corporation: "Microsoft Windows: Multimedia Programmer's Reference", Microsoft Press, 1991.

[Murphy, 1990]
Murphy, Alan: "Lip Synchronization" Personal Communication on a Set of Experiments.

[Steinmetz and Fritzsche, 1992]
Steinmetz, Ralf; Fritzsche, J., Christian: "Abstractions for Continuous-Media Programming", Computer Communications, vol. 15, no. 4, July/August 1992.

[Steinmetz and Meyer, 1992]
Steinmetz, Ralf; Meyer, Thomas: "Modelling Distributed Multimedia Applications", Intl. Workshop on Adv. Comm. and Appl. for High Speed Networks, Munich, Germany, March, 1992.

[Sventek 1987]
Sventek, J. S.: "An Architecture for Supporting Multimedia Integration", IEEE Computer Society Office Automation Symposium, Apr. 1987, pp.46-56.

[Vin et al., 1991]
Vin, Harrick M; Rangan, P. Venkat; Ramanathan, Srinivas: "Hierarchical Conferencing Architectures for Inter-Group Multimedia Collaboration", Proceedings of Conference on Organizational Computing Systems (COCS'91), November 1991.

[Vogt et al., 1992]
Vogt, Carsten; Herrtwich, Ralf Guido; Nagarajan, Ramesh: "HeiRAT: The Heidelberg Resource Administration Technique, Design Philosophy and Goals", IBM ENC Technical Report no.43.9101, March 1991.

[Wright and To 1990]
Wright, David J.; To, Michael: "Telecommunication Applications of the 1990s and their Transport Requirements", IEEE Network Magazine, vol.4, no.2, March 1990, pp.34-40.

Springer-Verlag
and the Environment

$\mathbf{W}$e at Springer-Verlag firmly believe that an international science publisher has a special obligation to the environment, and our corporate policies consistently reflect this conviction.

$\mathbf{W}$e also expect our business partners – paper mills, printers, packaging manufacturers, etc. – to commit themselves to using environmentally friendly materials and production processes.

$\mathbf{T}$he paper in this book is made from low- or no-chlorine pulp and is acid free, in conformance with international standards for paper permanency.

Lecture Notes in Computer Science

For information about Vols. 1–629
please contact your bookseller or Springer-Verlag

Vol. 669: R. S. Bird, C. C. Morgan, J. C. P. Woodcock (Eds.), Mathematics of Program Construction. Proceedings, 1992. VIII, 378 pages. 1993.

Vol. 670: J. C. P. Woodcock, P. G. Larsen (Eds.), FME '93: Industrial-Strength Formal Methods. Proceedings, 1993. XI, 689 pages. 1993.

Vol. 671: H. J. Ohlbach (Ed.), GWAI-92: Advances in Artificial Intelligence. Proceedings, 1992. XI, 397 pages. 1993. (Subseries LNAI).

Vol. 672: A. Barak, S. Guday, R. G. Wheeler, The MOSIX Distributed Operating System. X, 221 pages. 1993.

Vol. 673: G. Cohen, T. Mora, O. Moreno (Eds.), Applied Algebra, Algebraic Algorithms and Error-Correcting Codes. Proceedings, 1993. X, 355 pages 1993.

Vol. 674: G. Rozenberg (Ed.), Advances in Petri Nets 1993. VII, 457 pages. 1993.

Vol. 675: A. Mulkers, Live Data Structures in Logic Programs. VIII, 220 pages. 1993.

Vol. 676: Th. H. Reiss, Recognizing Planar Objects Using Invariant Image Features. X, 180 pages. 1993.

Vol. 677: H. Abdulrab, J.-P. Pécuchet (Eds.), Word Equations and Related Topics. Proceedings, 1991. VII, 214 pages. 1993.

Vol. 678: F. Meyer auf der Heide, B. Monien, A. L. Rosenberg (Eds.), Parallel Architectures and Their Efficient Use. Proceedings, 1992. XII, 227 pages. 1993.

Vol. 679: C. Fermüller, A. Leitsch, T. Tammet, N. Zamov, Resolution Methods for the Decision Problem. VIII, 205 pages. 1993. (Subseries LNAI).

Vol. 680: B. Hoffmann, B. Krieg-Brückner (Eds.), Program Development by Specification and Transformation. XV, 623 pages. 1993.

Vol. 681: H. Wansing, The Logic of Information Structures. IX, 163 pages. 1993. (Subseries LNAI).

Vol. 682: B. Bouchon-Meunier, L. Valverde, R. R. Yager (Eds.), IPMU '92 – Advanced Methods in Artificial Intelligence. Proceedings, 1992. IX, 367 pages. 1993.

Vol. 683: G.J. Milne, L. Pierre (Eds.), Correct Hardware Design and Verification Methods. Proceedings, 1993. VIII, 270 Pages. 1993.

Vol. 684: A. Apostolico, M. Crochemore, Z. Galil, U. Manber (Eds.), Combinatorial Pattern Matching. Proceedings, 1993. VIII, 265 pages. 1993.

Vol. 685: C. Rolland, F. Bodart, C. Cauvet (Eds.), Advanced Information Systems Engineering. Proceedings, 1993. XI, 650 pages. 1993.

Vol. 686: J. Mira, J. Cabestany, A. Prieto (Eds.), New Trends in Neural Computation. Proceedings, 1993. XVII, 746 pages. 1993.

Vol. 687: H. H. Barrett, A. F. Gmitro (Eds.), Information Processing in Medical Imaging. Proceedings, 1993. XVI, 567 pages. 1993.

Vol. 688: M. Gauthier (Ed.), Ada-Europe '93. Proceedings, 1993. VIII, 353 pages. 1993.

Vol. 689: J. Komorowski, Z. W. Ras (Eds.), Methodologies for Intelligent Systems. Proceedings, 1993. XI, 653 pages. 1993. (Subseries LNAI).

Vol. 690: C. Kirchner (Ed.), Rewriting Techniques and Applications. Proceedings, 1993. XI, 488 pages. 1993.

Vol. 691: M. Ajmone Marsan (Ed.), Application and Theory of Petri Nets 1993. Proceedings, 1993. IX, 591 pages. 1993.

Vol. 692: D. Abel, B.C. Ooi (Eds.), Advances in Spatial Databases. Proceedings, 1993. XIII, 529 pages. 1993.

Vol. 693: P. E. Lauer (Ed.), Functional Programming, Concurrency, Simulation and Automated Reasoning. Proceedings, 1991/1992. XI, 398 pages. 1993.

Vol. 694: A. Bode, M. Reeve, G. Wolf (Eds.), PARLE '93. Parallel Architectures and Languages Europe. Proceedings, 1993. XVII, 770 pages. 1993.

Vol. 695: E. P. Klement, W. Slany (Eds.), Fuzzy Logic in Artificial Intelligence. Proceedings, 1993. VIII, 192 pages. 1993. (Subseries LNAI).

Vol. 696: M. Worboys, A. F. Grundy (Eds.), Advances in Databases. Proceedings, 1993. X, 276 pages. 1993.

Vol. 697: C. Courcoubetis (Ed.), Computer Aided Verification. Proceedings, 1993. IX, 504 pages. 1993.

Vol. 698: A. Voronkov (Ed.), Logic Programming and Automated Reasoning. Proceedings, 1993. XIII, 386 pages. 1993. (Subseries LNAI).

Vol. 699: G. W. Mineau, B. Moulin, J. F. Sowa (Eds.), Conceptual Graphs for Knowledge Representation. Proceedings, 1993. IX, 451 pages. 1993. (Subseries LNAI).

Vol. 700: A. Lingas, R. Karlsson, S. Carlsson (Eds.), Automata, Languages and Programming. Proceedings, 1993. XII, 697 pages. 1993.

Vol. 701: P. Atzeni (Ed.), LOGIDATA+: Deductive Databases with Complex Objects. VIII, 273 pages. 1993.

Vol. 702: E. Börger, G. Jäger, H. Kleine Büning, S. Martini, M. M. Richter (Eds.), Computer Science Logic. Proceedings, 1992. VIII, 439 pages. 1993.

Vol. 703: M. de Berg, Ray Shooting, Depth Orders and Hidden Surface Removal. X, 201 pages. 1993.

Vol. 704: F. N. Paulisch, The Design of an Extendible Graph Editor. XV, 184 pages. 1993.

Vol. 705: H. Grünbacher, R. W. Hartenstein (Eds.), Field-Programmable Gate Arrays. Proceedings, 1992. VIII, 218 pages. 1993.

Vol. 706: H. D. Rombach, V. R. Basili, R. W. Selby (Eds.), Experimental Software Engineering Issues. Proceedings, 1992. XVIII, 261 pages. 1993.

Vol. 707: O. M. Nierstrasz (Ed.), ECOOP '93 – Object-Oriented Programming. Proceedings, 1993. XI, 531 pages. 1993.

Vol. 708: C. Laugier (Ed.), Geometric Reasoning for Perception and Action. Proceedings, 1991. VIII, 281 pages. 1993.

Vol. 709: F. Dehne, J.-R. Sack, N. Santoro, S. Whitesides (Eds.), Algorithms and Data Structures. Proceedings, 1993. XII, 634 pages. 1993.

Vol. 710: Z. Ésik (Ed.), Fundamentals of Computation Theory. Proceedings, 1993. IX, 471 pages. 1993.

Vol. 712: P. V. Rangan (Ed.), Network and Operating System Support for Digital Audio and Video. Proceedings, 1992. X, 416 pages. 1993.